FISHES OF THE STRAIT OF GEORGIA

FISHES OF THE STRAIT OF GEORGIA

More than 240 Life Histories

Dick Beamish
and **Jeff Marliave**

HARBOUR
PUBLISHING

1 2 3 4 5 — 29 28 27 26 25

Harbour Publishing Co. Ltd.
P.O. Box 219, Madeira Park, BC, V0N 2H0
www.harbourpublishing.com

Dust jacket illustrations: From *Pacific Fishes of Canada* by J.L. Hart. Chinook Salmon; Northern Anchovy; Northern Smoothtongue; Longfin Smelt; Gray Starsnout; Pacific Sardine; California Headlightfish; Rockhead.
Interior illustrations: Unless otherwise credited, all illustrations are from *Pacific Fishes of Canada* by J.L. Hart.
Map by Roger Handling, Terra Firma Digital Arts
Edited by Emma Skagen
Indexed by Colleen Bidner
Dust jacket design by John Montgomery
Text design by Libris Simas Ferraz / Onça Publishing
Printed and bound in South Korea

Canada | Canada Council for the Arts / Conseil des arts du Canada | British Columbia Arts Council | British Columbia

Harbour Publishing acknowledges the support of the Canada Council for the Arts, the Government of Canada, and the Province of British Columbia through the BC Arts Council.

Thanks also to the Pacific Salmon Foundation, the Raincoast Conservation Foundation, the Pacific Biological Station and Chrys Neville from Fisheries and Oceans Canada for their support.

Library and Archives Canada Cataloguing in Publication
Title: Fishes of the Strait of Georgia : more than 240 life histories / Dick Beamish & Jeff Marliave.
Names: Beamish, R. J. (Richard James), 1942- author | Marliave, Jeffrey B., 1949- author.
Description: Includes bibliographical references and index.
Identifiers: Canadiana (print) 20240525833 | Canadiana (ebook) 20240525841 | ISBN 9781990776830 (hardcover) | ISBN 9781990776847 (EPUB)
Subjects: LCSH: Fishes—Georgia, Strait of (B.C. and Wash.) | LCSH: Fishes—Ecology—Georgia, Strait of (B.C. and Wash.) | LCSH: Fishes—Development—Georgia, Strait of (B.C. and Wash.)
Classification: LCC QL626.5.B7 B43 2025 | DDC 597.09164/33—dc23

ABOUT 70 PERCENT OF BRITISH COLUMBIANS LIVE IN CLOSE PROXIMITY TO THE Strait of Georgia, one of the world's most productive inland seas, which lies between Vancouver Island and the mainland. This is a book about the fishes that live there. In it we provide a general appreciation of each species with the hope that British Columbians will see that we all play a role in ensuring there is a natural ecology in the complex relationships among the fish species and the other animals and plants in the strait.

The more we know about this amazing group of fishes, the more all of us will prioritize our responsibility to keep the populations healthy. There is no question that ocean ecosystems are changing rapidly, which will affect fish communities in a way not seen in the past. There is some public awareness about iconic species such as salmon, herring, a few species of rockfish and perhaps lingcod. But most of the community of fishes is out of sight. Our book is intended to make many of the species in the strait better known.

We want British Columbians to know that these mostly unseen fish species have survived for thousands of years and are now enduring unprecedented habitat changes as well as human interference in their normal lives. Evidence of these changes is seen in more familiar species such as chum salmon. Juveniles of this species are the most abundant juvenile Pacific salmon that rear in the Strait of Georgia, although in years ending in an even number chum salmon share their abundance with pink salmon. Catches and spawning abundances of chum have collapsed virtually everywhere around the Strait of Georgia. Alarmingly, this collapse extends throughout British Columbian chum salmon populations as well as many other chum salmon populations around the rim of the northern Pacific Ocean. The issue is that there is indisputable evidence that oceans are warming. Any internet search of "ocean heat content" immediately provides a shocking graph of a warming ocean. We believe that British Columbians who read this book will understand the complexity and urgency of the responsibility we all have to be stewards of a community of fishes that we literally live around.

We provide an up-to-date account about the life of each species, along with a description and a life history. Our information about the presence and location of the species comes from research surveys that have fished from the bottom to the top of the water column with a variety of gear. The federal government's groundfish and shellfish database at the Pacific Biological Station in Nanaimo has excellent records of catches and locations. Scuba observations were extremely useful, as were faunal surveys of major estuaries. We were selective in our approach to published identifications and accepted only reports from those known to be experienced at identifying fishes and reports of species with features that would make misidentification unlikely. Our information also comes from a variety of books and scientific papers. At the back of this book we list further readings that provide additional technical information. Much of the life history information and descriptions come from these references. We have avoided the use of citations in the main text to make the book more enjoyable reading.

Our focus is on the life of a fish species, but defining the word "species" is not as straightforward as one might think. Traditionally, a competent observer would use measurements or counts of stable biological parts of a specimen to convince colleagues that the specimen was new to science. Recently it has also been possible to use genetic structure to identify species. For differences to develop into a new species there needs to be some form of isolation in distribution that is seen in the life history. Importantly, this need for isolation shows that all species of fish in the Strait of Georgia have some unique ways of using the strait: they share an ecosystem but use the ocean differently for portions of their lives. We are writing about how each species uses the Strait of Georgia ecosystem.

The species in the Strait of Georgia occur in other areas, and there are publications listed at the back of our book that provide detailed taxonomic descriptions and keys for their identification. The recently published three-volume *Fishes of the Salish Sea* provides detailed taxonomic descriptions and illustrations that collectively may be the best ever produced. We identify the key features that distinguish species, but we do not provide taxonomic keys. One reason for this is that correctly identifying fishes requires a specialized vocabulary and training. Another reason is that it is only a matter of time before it will be possible to scan a specimen with a cellphone and get a possible identification.

We list 243 species that range from the primitive lampreys to the more evolved flounders. We estimate if a fish is abundant, common, uncommon, rare or a visitor. These are educated guesses based on hundreds to thousands of research surveys that have fished throughout the Strait of Georgia, as well as scuba observations, commercial catches and published reports. Abundant species such as Pacific herring, Pacific hake and northern smoothtongue were easy to categorize, as they are everywhere compared to visitors such as the salmon shark, which are extremely rare. Fish in the common category are distributed in a number of locations but not abundant. Fish in the uncommon category were reported from only a few locations. Rare species were seldom seen, with only a few reports, but could be reproducing in or near the Strait of Georgia. Visitors are those fishes that were only occasionally reported straying into the strait. There are also species at the margins of these categories, indicating that abundance estimates can be approximate.

Our book was written to be enjoyable reading within anyone's grasp. Our focus was to produce a biography for each species, which in many cases is available for the first time. We encourage readers to refer to the glossary at the back of the book if they encounter any unfamiliar terminology in our descriptions. We included longer and more detailed accounts for some of the better-known fishes, but in general we have kept the accounts short. There will be species not listed in this book that arrive in the Strait of Georgia, show up on beaches or are seen by scuba divers sometime in the future. Scuba divers are particularly important observers, as their records of new species can contribute to improved understanding, especially with the use of high-resolution images suitable for taxonomic identification. Even beach walkers can capture good images with nothing more than a modern cellphone.

Fishes are a link in a chain of a healthy ecosystem and not a separate entity. Our book provides a new understanding of how fishes live in the Strait of Georgia. These fishes depend on a healthy ecosystem in ways that we need to understand. There is no question that ocean ecosystems are changing, and British Columbians can use information in this book to help ensure there is a commitment to protecting this amazing group of species. British Columbians are the true stewards of the Strait of Georgia, and their fascination with fish is already evidenced by the popularity of the Vancouver Aquarium, koi ponds, tropical fish stores and Siamese fighting fish in bowls. We hope this book extends that interest to the community of fishes that is virtually on our doorstep. Readers of this book will increase their ability to make good personal and political decisions about the future health of the fishes in the Strait of Georgia.

BC AND THE STRAIT OF GEORGIA

Legend

Active Pass 65
Annacis Island..............53
Baker Passage 3
Baynes Sound 15
Big Qualicum River.....24
Boat Harbour................56
Boundary Pass 72
Burrard Inlet 22
Cape Lazo.......................7
Capilano River.............32
Cheakamus River.......... 6
Chehalis River 35
Chrome Island 18
Clover Point 76
Crescent Beach............61
Crofton 66
Departure Bay..............46
English Bay....................41
Englishman River........42
Entrance Island............47
Esquimalt Harbour.....74
False Creek 51
First Narrows 33
Flora Islet 16
Foley Head 5
Gabriola Island 57
Halibut Bank.................29
Hammond Bay.............38
Hornby Island..............17
Icarus Point 36
Inch Creek 62
Indian Arm 23
Jericho Beach...............50
Ladysmith Harbour....59
Lantzville 44
Little Qualicum
River 25
Mayne Island................68
McCall Bank..................30
Mitlenatch Island......... 4
Montagu Channel 14
Nanaimo River 55
Neck Point.....................37
Northwest Bay 43
Pacific Biological
Station 45
Pender Harbour...........11
Pender Island 70
Pipers Lagoon 28
Pitt Lake 34
Point Atkinson 31
Porlier Pass...................64
Porpoise Bay.................13
Portage Inlet.................75
Puntledge River............ 9
Qualicum Bay...............26
Quinsam River 1
Round Island 58
Sabine Channel 10
Sakinaw Lake 8
Saturna Island 71
Secret Cove...................12
Shelter Point.................. 2
Silva Bay 48
Spanish Bank................40
Stanley Park 52
Starboat Cove 21
Stave River.....................54
Stuart Channel.............63
Sunshine Coast............19
Swanson Channel.......69
Taylor Bay 39
Village Bay 67
Whytecliffe Park 20
Willis Point.....................73
Winchelsea Islands.....27
Wreck Beach.................49
Yellow Point..................60

A fish-collecting trip in the Strait of Georgia on the research ship *W.E. Ricker*. Left to right: John Fraser, Jeff Marliave, Dick Beamish, Rusty Sweeting.
Marja DeJong Westman

A BRIEF HISTORY OF THE FISHES IN THE STRAIT OF GEORGIA

THE MAXIMUM OF THE LAST GLACIATION OVER BRITISH COLUMBIA WAS THE Cordilleran Ice Sheet about 15,000 years ago. From about 14,000 to 10,000 years ago, glaciers over the Strait of Georgia melted and there were dramatic changes in sea levels. It is generally accepted that 14,000 years ago there were no fish in the area now known as the Strait of Georgia; certainly, there is no evidence of fish remains in the fossil record prior to that time. An impressive fossil record of a Steller sea lion pup was discovered near Courtenay in 1995. Found with the well-preserved skeleton were the fish remains of Pacific cod, walleye pollock and Pacific salmon. The fossils were all dated to be about 12,500 years old, which is evidence that a fish fauna was well established by this date. The fact that Steller sea lions existed in the area of what is now Courtenay shows that there was sufficient prey available by this time.

A number of researchers have looked at the information from archaeological sites around British Columbia and reported on the timing of fish appearance as well as the species in the diets of the first human residents. Most of the coastal archaeological sites, dating from about 11,000 years ago, are within 120 km of the present shore.

In a study of the sediments in Saanich Inlet, marine biologist Verena Tunnicliffe and colleagues did not find evidence of fish before 14,000 years ago but did find evidence of Pacific herring from 12,000 to 13,000 years ago. Their 2001 report of the first fish in the Strait of Georgia is consistent with the fish remains found with the juvenile Steller sea lion fossil. Thus, it appears that between 12,000 and 13,000 years ago there were Pacific herring, Pacific cod, walleye pollock and even Pacific salmon in the area now known as the Strait of Georgia. The Saanich Inlet study identified a greater diversity of fishes after 9,000 years ago, with a maximum diversity occurring between 6,000 and 7,500 years ago.

The archaeological records show that the exploitation of marine fishes by Indigenous Peoples increased during the period from 7,500 to 6,000 years ago. In Saanich Inlet, the sediment records showed that Pacific herring and Pacific hake appeared 12,000 to 13,000 years ago, followed by rockfishes and sardines about 10,000 years ago. Pacific spiny dogfish and skates appeared about 8,000 years ago, followed by sand lances about 1,000 years later. Lingcod arrived 2,000 to 3,000 years ago. The arrival of Pacific salmon is always of interest. It is known that Pacific salmon were in the Columbia River between 8,000 and 9,000 years ago. When most of the coast was ice-free and rivers were more stable, about 11,000 years ago, Pacific salmon could have first appeared in rivers in the Strait of Georgia area.

The species of fishes in the diets of Indigenous Peoples around the Strait of Georgia are not well known before about 6,500 years ago, when there were well-established First Nations populations along the British Columbia coast. At a site near Namu 5,000 to 6,500 years ago, 80 percent of the middens contained Pacific salmon. Other species included Pacific herring, Pacific cod, Pacific spiny dogfish, flounders, rockfishes, sablefish, greenlings, sculpins and ratfish. At Bear Cove near Port Hardy at about the same time, in addition to Pacific salmon, there were bones of rockfishes, Pacific cod, walleye pollock, greenlings, sculpins, dogfish and ratfish. In a site in the Fraser River delta about 5,000 years ago, there were remains of Pacific salmon, flatfishes and eulachon. More recent research summarizes the fishes in the diets of Indigenous Peoples in British Columbia after the last glaciation, identifying about 90 different species.

The first European commercial fishing in British Columbia began in 1830 when the Hudson's Bay Company exported barrels of salted or cured salmon to Asia and Hawaii. The first cannery was created in 1866, and by the late 1800s there was concern about the declining abundance of salmon. In these early years, the commercial fishery for salmon was virtually unregulated. By the 1920s there were over 7,000 vessels in the salmon fishery. What is seldom recognized is that lower salmon abundance reduces the number of juveniles entering the Strait of Georgia, and this changes the ecology of the strait. There could have been four to five billion young salmon of all species entering the Strait of Georgia in the spring prior to the beginning of the commercial fishery. These juvenile salmon were prey for lingcod, rockfish, dogfish and a variety of other species. Declines in the abundance of adults resulting from fishing would have a major effect on the population dynamics of the fish community, and the commercial salmon fishery was only one of the new fisheries started by the European settlers.

Beginning in the mid-1800s, fishing started for species that we now lump into a category called groundfish. Pacific halibut was one of the first species to be exploited. It appears that they were considered abundant in English Bay, Burrard Inlet and what is now called Halibut Bank. In 1892, the inspector of fisheries for British Columbia wrote that the fishery for halibut was capable of being increased to an almost unlimited extent. In 1898, 850 t were landed in Vancouver. In 1908 halibut catches were declining, and by 1923, when the International Pacific Halibut Commission (IPHC) was founded, halibut had virtually disappeared from the strait. There are some halibut today, but nothing like there used to be. In the past, halibut would have been one of the largest fish predators in the strait. They also possibly reproduced in the Strait of Georgia, which would represent a southern population that no longer exists, according to the IPHC.

Lingcod is another species with a story that was close to being tragic. Lingcod were readily caught with a hook and line, and later with bottom trawls. Initially, lingcod were kept alive in tanks and marketed in Vancouver. In the early 1900s, the commercial catch was about 300 t, and in 1936 the peak landing was 3,000 t, which represented the fourth-largest commercial fishery in British Columbia for that year. Conservation measures were finally established in the early 1990s, when it was estimated that the abundances were about 5 percent of the historical levels. Today, the population may still be only about 15 percent of those levels. It is possible that prior to the commercial fishery, lingcod were the major fish predator in the Strait of Georgia. Removing large abundances from the strait would be the equivalent of removing the wolves and foxes on land and letting the rabbits take over an ecosystem.

The dogfish, which is now called the Pacific spiny dogfish, is another species in the Strait of Georgia that is generally unloved but a major part of the ecosystem. Readers can refer to the summary of dogfish biology in this book. Pacific spiny dogfish could be considered to be the equivalent of crows or gulls that remove dead and dying organisms from an ecosystem to help keep it healthy. In 1877, a commercial fishery started to fish dogfish to produce oil. Markets collapsed and then increased again in the late 1920s for vitamin A from dogfish livers. At one point in the late 1940s, the dogfish fishery in all areas of the province was the third most valuable fishery in Canada. The fishery collapsed again when synthetic vitamin A was produced in 1950. In an effort to maintain a fishery for Pacific spiny dogfish, and without any understanding of the role of dogfish in the ecosystem, the federal government embarked on a program to try to eradicate dogfish from the Strait of Georgia. In fact, there were 13 attempts up to 1973 to eradicate dogfish or at least reduce their numbers so that they were no longer a nuisance to humans. The federal government paid cash and gave herring licences to people to kill dogfish. This is another example of reducing the abundance of a major species in the Strait of Georgia with the certainty that the ecosystem will change but with uncertainty on *how* it will change.

Other groundfish fisheries for the English sole, other flounders, Pacific cod and rockfish started in the early 1900s. By the 1930s, there were new technologies that allowed bottom trawling at deeper depths, which permitted fishing on the entire populations. The abundances of these species declined with time, and generally they are no longer fished with trawls. The story of the decline of rockfish is a little different as it was not until the mid-1980s that it was discovered that most species of rockfish were substantially older than previously thought. Instead of being 7 to 15 years old, they were found to be 20, 40, 60 years old, with the oldest being a 206-year-old rougheye rockfish caught off the west coast. These fish evolved to live for a long time because their reproduction is not as regularly successful as that of short-lived species such as Pacific herring. This means that relatively few long-lived fish could be fished from a population in any one year. There now are conservation areas in the Strait of Georgia where fishing for rockfish is not permitted, but it will be decades before populations of some species can rebuild.

In 1979, it was discovered that Pacific hake and walleye pollock were abundant in the strait. A fishery for Pacific hake peaked at 11,860 t in 1995. Both species are still abundant, but the interest in fishing them has diminished. It is speculation, but the fact that the large abundances were discovered late in 1979 probably means that one of the changes in the ecosystem as major predators were fished to low levels was that hake and pollock became abundant. The dramatic increase in harbour seal abundance beginning in the 1970s probably resulted from the increase in hake abundance, as hake became the major food for the seals.

Pacific herring may be one of the most difficult species to understand. They are a major food for a large number of other species and they are abundant. However, most herring live in the Strait of Georgia for only their first year, and they leave to rear off the west coast of Vancouver Island until they mature at age three. They return in the late fall to spawn the following spring and then leave again. There are some resident populations that remain in the Strait of Georgia all year, but these populations are small. There may have been more resident populations in the past that have since been lost, as the fisheries for herring have been extensive. Commercial fishing started for Pacific herring in 1877 with the export of salted and dried product to Asia. Fishing continued with exceptionally large catches in the late 1950s and early 1960s. The fishery was closed in 1967 when it was realized that abundances had collapsed, which years later was found to have resulted from too much fishing at a time of poor survival of the larval herring. Today it is believed that abundances are at high levels, but only a few years ago they were declining to levels just higher than what occurred in the early 1970s. It is clear that the factors regulating the abundances are poorly understood. The fishery is managed according to accepted international strategies, but more needs to be understood about the role of herring in the Strait of Georgia ecosystem.

The future of the lives of the fishes in the Strait of Georgia is in the hands of British Columbians. It is one more concern for all of us, but future generations will decide the success of our stewardship.

THE NAMES OF THE FISHES

ALL LIVING ORGANISMS ARE NAMED AND ORGANIZED INTO A TAXONOMIC HIERarchy. The taxonomy of fishes provides a framework for humans to understand relationships among species. Changes in these names and relationships occur regularly, but most are of little significance to the general public. Species in this book are reported in taxonomic order according to a traditional format such as that used in the book *Fishes of Alaska*, published by the American Fisheries Society. We use the scientific and common names from *Fishes of Alaska* as well as from other references, including scientific papers. We provide brief descriptions of the general features that distinguish between orders and families to introduce the species descriptions. The common names are the consensus of a committee, and some may not be familiar to British Columbians. It is the convention in fisheries science to retain the singular form of the common name of a fish when more than one individual is reported. A plural is used only when there is more than one species involved. For example, "two salmon" means two individuals of the same species, while "two salmons" means there are two different species. This can appear to be "unnatural" for a general reader, and we will occasionally make a species' common name plural when there is more than one individual.

THE FISHES

JAWLESS FISHES

JAWLESS FISHES INCLUDE THE CLASS MYXINI, ORDER MYXINIFORMES AND FAMILY Myxinidae, which are hagfishes, and the class Petromyzonti, order Petromyzontiformes and family Petromyzontidae, which are lampreys. These are the oldest living groups of vertebrates on the planet. Their taxonomy continues to be revised. There is no doubt that they have existed for at least 360 million years, but some estimates go as high as 500 million years. Clearly, they have a remarkable ability to survive major changes in the planet's environment.

Hagfishes

Hagfishes are in the family Myxinidae and have a worldwide marine distribution. There are about 76 species of hagfish, and they are mostly found on soft sandy or muddy bottoms in cool water at depths to 1,800 m. Hagfishes lack jaws and have long eel-like bodies with no fins, vertebrae or scales. They were once known as the only vertebrates on the planet without vertebrae, but recent studies identified rudimentary vertebrae. They have an eversible tongue with rasp-like teeth that tear tissue from dead animals. However, they are best known for their production of slime when captured. Most are about 50 to 60 cm long, but a maximum length of 1.27 m has been reported. Hagfish have well-developed senses of smell and touch. They are extremely successful species, having survived for about 430 to 480 million years. We have only a rare report of hagfish in the Strait of Georgia.

BLACK HAGFISH

Eptatretus deani RARE ★

IN THE STRAIT OF GEORGIA

The only record of hagfish in the Strait of Georgia is a recent report in the groundfish database at the Pacific Biological Station. The report is for the black hagfish, which is less common overall than the Pacific hagfish (*Eptatretus stoutii*). The record in the Strait of Georgia is for two specimens that were feeding on a Pacific spiny dogfish caught in the commercial longline fishery in November 2006 near Nanaimo, just off Neck Point at 149 m. Since hagfish have no eyes, jaws or fins, there is no mistaking them, but the species of hagfish may be misidentified. There is no record of the reasons behind the identifications, but it may have been the colour.

DESCRIPTION

Black hagfish are immediately distinguished by their dark-purple or black skin as opposed to the tan or light-brown skin of the Pacific hagfish. The darker skin of the black hagfish is particularly evident in the contrast around the lighter areas of the eye-spots. Black hagfish have 4 to 10 slime pores posterior to the gill openings, while Pacific hagfish have 10 to 16. Black hagfish often

Adobe Stock

have light spots, and both species have loosely attached skin. Both species are easily distinguished from all other fishes by their elongate, cylindrical eel-like shape, with no pelvic or pectoral fins, no eyes, no scales and a mouth with no jaws. The skeleton is cartilaginous. There is a single large nostril situated dorsally on the snout with a pair of tentacles, and a mouth that is surrounded by four fleshy tentacles. The eversible tongue has paired rasp-like teeth that remove flesh from prey. These fish have no eyes, but they have eye-spots that are sensitive to light. The dorsal and ventral fin folds extend from well behind the snout onto the caudal fin. Perhaps the most distinguishing feature of hagfish is the production of copious amounts of slime when handled, resulting in the alternative common name "slime eel."

LIFE HISTORY

The black hagfish and the Pacific hagfish occur from southern Alaska to Baja California, though they are rare off Alaska. They can be very abundant in some areas. Hagfishes are bottom-dwelling species that mostly occur in burrows in soft sediments but are also in sand, in gravel and even among boulders. Maximum depths have been reported as 2,743 m for black hagfish and 960 m for Pacific hagfish, but most occur from about 100 m to 1,100 m. A maximum length of 82 cm has been reported for Pacific hagfish and 63.5 cm for black hagfish. Off Vancouver Island, the largest Pacific hagfish was 75.4 cm, with average lengths of 42.8 cm, and a 43.2 cm average for black hagfish. Very little is understood about the biology of either species as it has not been possible to determine age, except in one Japanese study that used tagging results. Based on this study, maturity is believed to occur after 8 to 12 years, with 50 percent of males mature at about 35 cm and females at about 40 cm. Spawning is thought to occur throughout the year in batches, perhaps with a peak in late summer. Pacific hagfish can release 5 to 82 eggs, hooked together in a strand, but most releases are fewer than 30 eggs. Black hagfish generally release fewer eggs, ranging from 8 to 30. The young resemble the adults when hatched and may grow about 4 to 5 cm a year. If these estimates are approximately correct, they indicate that hagfishes are relatively slow-growing, long-lived species.

Hagfish are mostly scavengers feeding at night on small fishes, whale carcasses, cods and spiny dogfish, but will also feed on shrimp, squid and polychaetes. They are eaten by cods, Pacific spiny dogfish, marine mammals and octopuses. When they produce large amounts of slime, they tie their bodies in knots to scrape off the slime. In 1988, fisheries were started for hagfish from California to British Columbia to supply a market in South Korea.

Lampreys

Lampreys, which are in the family Petromyzontidae, are very common in the Strait of Georgia. All lampreys begin life in fresh water. Some species remain in fresh water for their entire life, while some are anadromous and migrate into the ocean. There are currently about 41 species of lamprey distributed throughout most temperate regions worldwide. There are nine species that migrate into the ocean to feed. Some of these feeding juvenile adults travel long distances, and some, such as the very large Pacific lamprey in the Skeena River, may home to annual spawning areas hundreds of kilometres up the river. All lampreys are eel-like but are not true eels, and none are strictly marine. All lampreys begin life in fresh water as ammocoetes, which are worm-like larval forms without eyes or teeth. Ammocoetes feed on freshwater plankton-like organisms for about three to seven years. Beginning in the early summer, they metamorphose into the adult form. Adults do not have jaws, paired fins or scales. They do have one or two dorsal fins, seven pairs of gill pores, and teeth that are a major taxonomic character. There are three major life-history types: freshwater and nonparasitic, freshwater and parasitic, and anadromous parasitic. There are two species of lampreys in the Strait of Georgia.

PACIFIC LAMPREY

Entosphenus tridentatus COMMON ●

IN THE STRAIT OF GEORGIA

Pacific lamprey commonly occur in the Strait of Georgia, but it is likely they would be observed only when they are attached to a fish caught in the sport fishery as they are usually in deeper water. The Pacific lamprey begins life as a freshwater larva (ammocoete). Pacific lamprey occur in almost every major river flowing into the Strait of Georgia. A good rule of thumb is that if a river supports Chinook salmon, it will also have Pacific lamprey ammocoetes. The Fraser River is the major source of Pacific lamprey in the Strait of Georgia. Pacific lamprey are also resident in the strait throughout the year. The Squamish People have stories about not eating salmon with sores. It is possible that the sores were wounds from Pacific lampreys.

DESCRIPTION

The Pacific lamprey is readily distinguished from the western river lamprey, the other species of lamprey in the Strait of Georgia, by its colour and its teeth. Pacific lamprey are uniformly dark grey or brown, while western river lamprey are silver dorsally and white ventrally. Teeth in the mouth are the major character used to identify species. In the central area of the mouth, there is a dorsal bar with teeth called the supraoral lamina. Pacific lamprey have three teeth on the supraoral lamina, compared to two teeth for western river lamprey. The middle tooth is much smaller than the other two. The Pacific lamprey has four groups of lateral teeth. There are also teeth on a ventral bar called the infraoral lamina. The Pacific lamprey has five teeth on the infraoral lamina, compared to seven for the western river lamprey.

There are two other species of lamprey ammocoetes in rivers in British Columbia that flow into the Strait of Georgia, including a species that does not

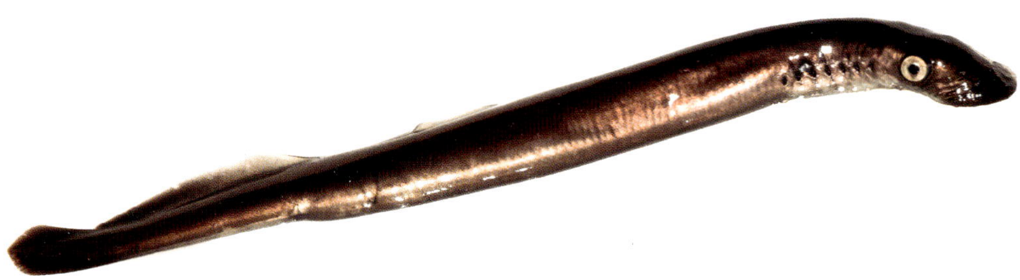

Bill Pennell

enter the ocean. Pacific lamprey ammocoetes are distinguished from these two species by their lightly pigmented area in the middle of the tail.

LIFE HISTORY

This species occurs in the ocean off Hokkaido, Japan, and Kamchatka, Russia, but there are no records of the Pacific lamprey in fresh water in these areas. They occur in the Bering Sea, the Chukchi Sea and south to northern Baja California. There is a record of a Pacific lamprey tagged off the coast of Russia and detected returning to the Columbia River. Once in the ocean, Pacific lamprey immediately begin feeding. Pacific lamprey are parasitic and use suction and their oral disc to grip onto fish while the teeth on the tongue rasp away flesh, allowing the lamprey to feed on tissue and body fluids. Initially they feed on herring and salmon in the Strait of Georgia, but as they grow they move into deeper water, where they feed mainly on walleye pollock, Pacific hake and Chinook salmon. Pacific lamprey are routinely found in catches of walleye pollock in the Strait of Georgia. In one research trawl net catch in July, at a depth of 45 m, there were 464 walleye pollock and 132 Pacific lampreys. The Pacific lamprey also commonly feeds on sockeye salmon returning to the Fraser River. In September 2018, a trawl net fishing at the mouth of the Fraser caught 429 adult sockeye salmon, of which about 14 percent had lamprey wounds. There

A Pacific lamprey in its first ocean year is attached to the tail of this humpback whale in the Strait of Georgia, October 2021. Andrea Scheffler

were 85 Pacific lamprey captured in the same catch, with an average length of 20.7 cm from a range of 15.3 to 30.5 cm.

Mouth of a Pacific lamprey.
Bill Pennell

Pacific lamprey in the Strait of Georgia grow to an average size of about 28 to 30 cm; the largest recorded in the Strait of Georgia was 42 cm. Pacific lamprey in other areas, such as the Skeena River in northern British Columbia, can reach lengths of 75 cm. It is estimated that the larval stage lasts between four and seven years. It is not known how long juvenile Pacific lamprey are in the ocean, but in the Strait of Georgia it is probably between one and two years. Thus, a maximum age could be six to nine years. Migration back to the Fraser River has been observed from late June to early September, with most migration being in the first two weeks of July. Lamprey that return to fresh water in the fall may spawn in the spring of the next year, while those that return in the early spring might not spawn until the following spring. All lampreys die after spawning. Eggs hatch within a few weeks, and the larval lampreys prefer a sandy, muddy substrate that has a gentle current. After about four to seven years and beginning in early July, the ammocoete changes into the adult in a process called metamorphosis. During metamorphosis, the fish develops eyes, teeth, and a number of new organs. Metamorphosis is complete around the end of September. During metamorphosis, the larval lamprey will move from the sandy, backwater habitat to a small gravel substrate that is usually in a more open area of the river. After metamorphosis, some lamprey migrate into the ocean in the fall, but most do so in the spring of the next year. The average size at ocean entry is about 12 to 13 cm.

WESTERN RIVER LAMPREY

Lampetra ayresii COMMON ●

IN THE STRAIT OF GEORGIA

Western river lamprey are commonly found throughout the surface waters of the Strait of Georgia from June through to September. However, small abundances have been found as late as November. There are no observations of the western river lamprey in the Strait of Georgia in the winter. They can be abundant in the southern strait and in the Fraser River plume.

DESCRIPTION

These fish have countershading that is silver dorsally and white ventrally. They are easily distinguished from Pacific lamprey by their colour as well as by their dentition, which has two supraoral teeth, three groups of lateral teeth and seven teeth on the infraoral lamina. The larval ammocoete form of the western river lamprey lives in fresh water. It has a very distinctive pattern of a pinkish colour with weak pigmentation on the body and tail.

Bill Pennell

LIFE HISTORY

The western river lamprey occurs from San Francisco Bay and the San Joaquin and Sacramento Rivers to southeast Alaska north of Juneau. Western river lamprey begin feeding immediately on juvenile herring and small Pacific salmon when they enter the Strait of Georgia. They are voracious eaters, attaching themselves to the top and side of their host. Their sharp teeth quickly remove flesh, in much the same way we bite corn off the cob. They are not strong swimmers but are very efficient at catching fish. In early July, an observer with a sharp eye can see these lamprey in schools of Pacific herring swimming around

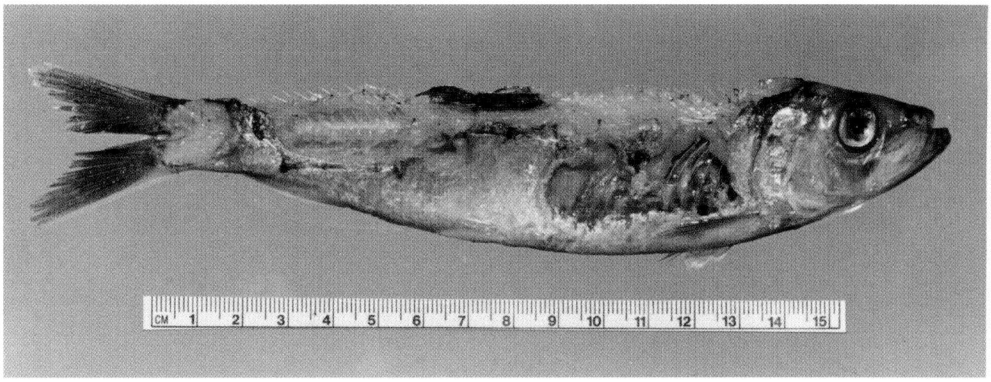

Pacific herring held in captivity and eaten by western river lamprey also held in captivity. Dick Beamish

docks or close to shore. In one study in July near the mouth of the Fraser River, a small school of Pacific herring was captured with a seine net and found to contain 40 western river lamprey ranging from 10 to 15 cm in length. Western river lamprey grow quickly, reaching an average size of about 25 cm and 34 g by the end of July to mid-August. The largest western river lamprey found in the Strait of Georgia was captured in the northern strait in September 2021 and was 35 cm long.

Western river lamprey kill millions of fish because the wounds they inflict are so severe. Studies indicate that their diet is approximately 20 percent juvenile salmon and 80 percent other fishes, which are almost all Pacific herring. The number of fish killed obviously depends on the abundance of lampreys; however, each lamprey probably kills about 65 prey during the period it is feeding and growing in the Strait of Georgia.

Western river lamprey probably begin to return to the Fraser River in August or September. Only one upstream-migrating western river lamprey has ever been captured in the Fraser River. Catches of western river lamprey in the Strait of Georgia diminish in September, suggesting that most return to fresh water in August and September after three to four months in the ocean, but do not spawn until the following spring. Spawning probably occurs in June, but the exact times remain to be determined. At present, all estimates of the age of western river lamprey are guesses based on growth, and it is estimated that the larval stage lasts between five and seven years. Therefore, spawning adults would have a maximum age of seven to nine years, two years older than larvae at metamorphosis.

CARTILAGINOUS FISHES

CARTILAGINOUS FISHES INCLUDE CHIMAERAS, RATTAILS, SHARKS, SKATES AND rays. These fishes have skeletons consisting of cartilage and not true bone. They have teeth that are frequently replaced. They also have paired spiracles, or nostrils, that are equipped with a valve to allow water to flow in and out. Males have external claspers that allow for internal fertilization of females. Resulting embryos are in a tough, leather-like egg case. There is no swim bladder, but livers are large and rich in oil to help maintain buoyancy. The skin is tough, with tooth-like scales called denticles. It is estimated that there are about 1,100 species of cartilaginous fishes, and we list 17 as occurring in the Strait of Georgia.

Chimaeras

Chimaeras are in the class Holocephali, order Chimaeriformes and family Chimaeridae which has about 45 species worldwide in temperate marine waters. Lengths range from about 60 cm to 2 m. There is one species of chimaera in the Strait of Georgia, the spotted ratfish, and it is very common. Chimaeras are believed to be the most primitive of the cartilaginous fishes. They have four pairs of gills that have a single pair of openings to the exterior. The skin is smooth with no scales. Males have both pelvic claspers and a club-shaped clasper on the head. The large pectoral fins are used in swimming almost like the motion produced by the wings of a large bird. The swimming motion has been described as "sluggish," but "persistent" might be a better description. The gradually tapering body that ends in a filamentous tail is made up of joined dorsal and anal fins. There is a large spine in front of the dorsal fin with a poison gland at the base that produces venom. The big head and prominent protruding teeth that are fused into three pairs of plates give chimaeras a rabbit-like appearance. Most chimaeras are deep-sea fishes, found below 200 m and as deep as 2,600 m, but the spotted ratfish is an exception and can be distributed throughout the water column.

Spotted ratfish, *Hydrolagus colliei*. Danny Kent

SPOTTED RATFISH

Hydrolagus colliei ABUNDANT ○

IN THE STRAIT OF GEORGIA

The spotted ratfish is one of the most abundant and attractive fishes in the Strait of Georgia. There are no abundance estimates for the strait, but there are an estimated 200 million in Puget Sound, and it is likely there are just as many, if not more, in the Strait of Georgia. They are mostly in the midwater and down to near the bottom in the strait.

DESCRIPTION

Taxonomically, spotted ratfish are classified as cartilaginous fishes, but they also have similarities to bony fishes. The only species of chimaera in British Columbia waters, spotted ratfish have smooth, scaleless skin, large heads with a prominent snout, and rodent-like protruding teeth that are joined to form a plate. They have large, paddle-like pectoral fins, and the prominent first dorsal fin has a sharp, strong anterior spine with fine teeth on the posterior edge. The spine can inflict a cut that may also receive venom that is produced in the flesh at the base of the ratfish's spine. Males have a club-shaped fleshy appendage that fits into a depression on the head. There are sharp structures on the appendage that we know from experience can quickly cut an ungloved hand. The male has a long clasper behind each pelvic fin. The body tapers into a long tail that is used more for steering than for propulsion. Spotted ratfish swim

Bill Pennell

with a motion of their pectoral fins that looks like flapping, giving the appearance of flying through the water, somewhat like a heron in the air.

LIFE HISTORY

Spotted ratfish are found from Baja California and the Gulf of California to southeast Alaska and range from shallow water down to 1,030 m. Their occurrence in shallow water over a muddy bottom usually happens at night. With large, emerald-green eyes, they are an attractive fish that is popular in commercial aquariums and a favourite image of the artist Ray Troll. The maximum reported size is 97 cm and 1.8 kg. The body is mostly silver, with iridescent bronze shades providing a darker background for the round white spots. In a year, a female produces 20 to 29 eggs, which are probably fertilized by sperm that was stored after mating. The fertilized eggs are released by the female with each embryo in a rigid brown capsule about 13 cm long. The filamentous attachments to the capsule may anchor the egg case to the bottom. Estimates of hatching time, before the young ratfish emerges from the egg capsule, vary from about 9 to 12 months. Eggs are probably released between May and October in the ocean, but may be released throughout the year in commercial aquariums. This is apparently a slow-growing fish with low reproductive capacity. The maximum age for the spotted ratfish has recently been reported to be about 16 years for males and 21 years for females by fisheries scientist Jackie King, who used the ridges on the vomerine tooth plate to age the fish. She determined that the age at maturity was between six and eight years. Spotted ratfish feed primarily on crabs, other crustaceans, polychaetes, molluscs, echinoderms and fish, including other spotted ratfish. They are eaten by bluntnose sixgill and sevengill sharks, Pacific spiny dogfish, halibut and some marine mammals. Oil from the livers of spotted ratfish was stored by First Nations and had medicinal value.

The spotted ratfish is often seen swimming near the shore on nighttime scuba dives, but also on occasion during daylight. Ratfish swim by flapping their broad pectoral fins like wings. The underslung mouth is adapted to snagging small critters off the bottom or snipping clam siphon tips. Bill Pennell

Sharks

The recent taxonomy of the class Elasmobranchii, generally called elasmobranchs, contains eight orders of sharks and three orders of rays and skates. The number of shark species in the world changes as the taxonomy changes, but it appears to be between 450 and 500. Sharks are marine fishes, but two species live in fresh water. There are 15 species of sharks in British Columbia, and 11 of these reside in or visit the Strait of Georgia. Most sharks are wider in the middle and taper at each end—an elliptical, or fusiform, shape. There are five to seven paired gill openings on each side of a head that has paired nostrils and no bony plates. Teeth are frequently replaced. The tail fin is large, and there are one or two rigid dorsal fins with or without immovable spines. The moderately sized pectoral fin is not attached to the head. There is no swim bladder, and an anal fin may or may not be present. There is a large liver with a high concentration of oil that provides buoyancy. The skin is tough and has denticles. Sharks are predators that have small eyes and large olfactory capsules, resulting in a reliance on smell more than eyesight to find prey. Fertilization is internal.

MACKEREL SHARKS, THRESHER SHARKS AND BASKING SHARKS

The order Lamniformes has eight families of sharks, with one species from each of the three families Lamnidae (mackerel sharks), Alopiidae (thresher sharks) and Cetorhinidae (basking sharks) that occasionally visit the Strait of Georgia. There are 16 to 17 species in the order. They have a worldwide distribution in tropical and temperate oceans, from the surface to about 1,200 m. This is a diverse group of species that includes some of the best-known sharks, such as the great white shark and the basking shark. They have two dorsal fins, an anal fin, five long gill slits and a large mouth that extends beyond the eyes. The second dorsal and anal fins are small. They have an extended snout and a torpedo-like body, and they swim continuously. They all lack a nictitating membrane.

SALMON SHARK

Lamna ditropis VISITOR ▼

IN THE STRAIT OF GEORGIA

Salmon shark have been reported in the Strait of Georgia, but there were no recorded sightings for over two decades prior to May 2013, when a recreational angler photographed one on his cellphone 1.5 km off Shelter Point, near Campbell River.

DESCRIPTION

The salmon shark is a robust, stout-bodied shark that is sometimes mistaken for a small great white shark. It is blue-grey to brown or black dorsally and white ventrally with grey blotches. The teeth are well developed on both jaws, with each tooth having two small cusplets.

LIFE HISTORY

Salmon shark occur singly or in schools in the northern Pacific Ocean from the Sea of Japan to the Bering Sea, off Alaska to Southern California and Mexico. These are pelagic fish that are near the surface, though they have been reported to dive to over 700 m. The size of an adult is up to 300 cm with a maximum weight of 450 kg. Most, however, are 200 to 260 cm and approximately 220 kg. Females mature at about age 10 and males perhaps as early as about age 5.

Andy Murch

Andy Murch

The maximum age is about 30. Recently it was discovered that most of the salmon shark on the Asian side of the North Pacific are males, while those off North America are mostly females. Mating occurs in the late summer or early fall in the Gulf of Alaska. The two to six pups develop inside the female and are born after 12 months around the late spring and at a length of 65 to 85 cm. Juveniles tend to remain in the southern areas. This shark is a major predator of Pacific salmon, though it feeds on other fishes, including other sharks. They are eaten by killer whales. This is one of the few fishes that can regulate their body temperature, but it does not maintain a constant body temperature. Fishing for salmon shark is popular off Alaska, particularly in the Prince William Sound area.

COMMON THRESHER SHARK

Alopias vulpinus

VISITOR ▼

IN THE STRAIT OF GEORGIA

There are no recent reports of the common thresher shark in the Strait of Georgia. There is only a report from Saanich Inlet in J.L. Hart's *Pacific Fishes of Canada*. There is no mistaking the identification of this fish because of its long tail, so the record is probably valid. The species is common off the west coast of Vancouver Island in the summer, which may result in future summer sightings within the Strait of Georgia. Hart also recorded a specimen from Johnstone Strait, which may indicate future visits from the north.

DESCRIPTION

The extremely long, curved tail makes the common thresher shark easily recognizable. The upper lobe of the tail is about as long as the body anterior to the tail. There are a variety of colours dorsally, ranging from black to dark blue and purple to brown and grey. The sides can be silver blue or golden, and the belly is whitish.

Thomas Alexander / Creative Commons

LIFE HISTORY

Common thresher shark occur around the globe in warm water. In the North Pacific, they are reported from the southeast Gulf of Alaska to Baja California and from the Gulf of California to Hawaii and Japan, from the surface to 572 m. In the summer, seasonal migrations within warm water bring the shark off the west coast of Vancouver Island, where they are occasionally caught in commercial fisheries. With the recent ocean-warming events, they could be more common in the summers off the west coast. There are reports of a maximum size of about 7.6 m and 420 kg, with most in the range of 4 to 5 m. There is a large range in the reported maximum ages. Estimates of 20 to 50 years may not be correct; an age of 15 years is more likely closest to the oldest. Males and females mature at about age three to seven and at about 3.3 m for males and 4.2 m for females. Mating occurs from July to August, with the young born from March to June after a gestation period of nine months. The pups are large, ranging from 1 to 1.6 m. It is reported that the large, scythe-shaped tail is used in feeding on schooling prey, such as squid or small fishes.

BASKING SHARK

Cetorhinus maximus VISITOR ▼

IN THE STRAIT OF GEORGIA

Basking shark historically were observed in the Strait of Georgia, with one sighted in Hammond Bay just a few kilometres from the Pacific Biological Station in Nanaimo. Any sightings now are newsworthy. The most recent sighting was in July 2009 off the San Juan Islands. They are now officially listed as endangered.

DESCRIPTION

There is no mistaking these enormous fish, black to brown dorsally and white ventrally, which swim slowly near the surface. In addition to their large size, they are distinguished by the big, prominent gill slits that almost circle the body.

Andy Murch

LIFE HISTORY

The basking shark is the second-largest fish in the world (the whale shark is the largest), reaching sizes of up to 15 m and 4 t, and is found around the globe in cooler water. There are reports from the Gulf of Alaska to Baja California and in the South Pacific off Chile. Basking shark also are sighted off Russia, China, Australia and New Zealand. No one has discovered how to determine their age, but an educated guess is that they could live up to 50 years. Once mature, a female will produce one to six pups, which are not released from the female for about 2.5 to 3.5 years. The pups are born at lengths of about 1.5 to 2 m. This is the longest known gestation period of any vertebrate. These are plankton-eating, filter-feeding, harmless giants. They continuously filter out plankton, mostly near the surface, but are known to dive to 1,200 m. There used to be aggregations in the spring and summer off the coast of British Columbia, where they are legally listed as endangered under the Species at Risk Act. Tragically, they were classified as a destructive pest in 1949 because they could damage commercial gillnets fishing for salmon, and hundreds were killed between 1955 and 1969. They are extremely rare in waters around British Columbia, but with protection, they may rebuild to the aggregations reported in the past. If you observe a basking shark anywhere in the ocean around British Columbia, you should report it to the Canadian Pacific Shark Research Lab by calling 1-877-507-4275.

CAT SHARKS, HOUND SHARKS AND REQUIEM SHARKS

The order Carcharhiniformes has seven families, and three of the families, Scyliorhinidae (cat sharks), Triakidae (hound sharks) and Carcharhinidae (requiem sharks), each have one species in the Strait of Georgia. The tope shark in the family Triakidae and the blue shark in the family Carcharhinidae are rare visitors to the strait, while the brown cat shark is a common species found in the deeper water everywhere. Carcharhiniformes is the largest order of sharks, with about 270 species, representing about 60 percent of all shark species. The cat shark family has about 115 species and is the largest of all shark families. Many of the shark attacks on humans are from species in this order. These species are distributed worldwide in temperate and tropical oceans, from intertidal zones to deep on the continental slope, with a few in fresh water and estuaries. Species range in size from about 45 cm to 600 cm and have two dorsal fins, an anal fin and five gill slits. They also have a nictitating membrane that can protect the eye.

BROWN CAT SHARK

Apristurus brunneus COMMON ●

IN THE STRAIT OF GEORGIA

Brown cat shark are found in deep water throughout the Strait of Georgia. They are not abundant, but they are common, particularly in the deep areas around the entrance to Malaspina Strait. In the Strait of Georgia, the average size is 41 cm and the maximum is 57 cm. In a two-day research survey in September 2012, most of the 75 brown cat shark caught were in water deeper than 300 m, either on the bottom or just off it. Lengths ranged from 17 cm to 56 cm and averaged 33 cm, with the number of males and females about equal.

DESCRIPTION

The brown cat shark is easily recognized by its small size, uniformly dark-brown body and shark-like appearance. Cat sharks typically have elongated eyes, which is probably where they get their common name. Their teeth are very small and positioned in three rows. The head is flat and long. Each tooth has three to five cusps, with a large central cusp. This is the smallest shark in British Columbia.

Bill Pennell

LIFE HISTORY

Brown cat shark are found from southeast Alaska to Southern California and the Gulf of California, though they are rare in Alaska. The distribution extends to central and southern South America through to Chile. Worldwide, cat sharks are the most abundant of all sharks, with about 134 species. They are typically found in warmer oceans, but the brown cat shark occurs in colder oceans, including in the Strait of Georgia. These deepwater residents have been found at depths from 33 m to 1,306 m throughout their distribution. The maximum length is 71 cm, with most ranging from 40 to 50 cm. In one study, the length at 50 percent maturity for males was 51.4 cm and all males were mature at 54.7 cm. In general, all appear to be mature at about 58 cm. Maturity for males is determined by the length and degree of calcification of the clasper. Brown cat shark are oviparous, which means that the eggs hatch outside the body, as for most fishes. The female deposits the eggs in a leather-like egg case with a distinctive translucent appearance. Egg cases are 5 cm in length, smooth, and amber-coloured and have a long, twisted tendril at one end. It is reported that it may take about two years for the egg to hatch, with the newly hatched individual measuring 7 to 9 cm. There may be increased reproductive activity in April and May, but there is no indication that reproduction occurs only in a particular season as gravid (pregnant) cat sharks have been reported in all months of the year. It is possible that mating occurs from April to June, with females storing some sperm that allows them to reproduce throughout the year. There are no reliable age estimates of their age at maturity or their maximum age.

Over the years, we have looked at the stomach contents of a number of individuals. In the Strait of Georgia, the contents routinely appear as a black, dense liquid with some unrecognizable chunks and some shimmering specks. On rare occasions, there are parts of glass shrimp, and perhaps squid and lanternfish parts. In one survey about half of the gut contents could be identified, with shrimp being the most common food item. Fish and squid remains were also found. The few fish remains that could be identified were northern smoothtongue and northern lampfish. It is probable that brown cat shark are slow-growing, and if they are long-lived, it is likely that they feed only occasionally and digestion may be slow. Bluntnose sixgill shark are known to feed on spiny dogfish and it is therefore possible that they also feed on the brown cat shark.

TOPE SHARK

Galeorhinus galeus VISITOR ▼

IN THE STRAIT OF GEORGIA

The only record of a tope shark in the Strait of Georgia is from the groundfish database at the Pacific Biological Station in Nanaimo. The specimen was caught in a commercial bottom trawl fishery just northeast of Saturna Island in September 2013. There is a second report from a commercial midwater trawl fishery in Queen Charlotte Strait just north of the Strait of Georgia. The common name "tope" or "tope shark" is the official name used by organizations such as the American Fisheries Society, but many old-timers prefer the name "soupfin shark." However, this name may no longer be socially acceptable because of an international effort to not use shark fins for soup.

DESCRIPTION

Tope shark are a solid dark blue to grey dorsally and white ventrally. They have a long, pointed snout, and their dorsal fin originates behind a relatively long pectoral fin. The second dorsal fin and anal fin are about the same size. Tope

Andy Murch

shark are readily distinguished from salmon sharks by the body shape, the position of the dorsal fin, the enlarged lobe at the top of the tail and the lack of a keel on the side of the caudal peduncle.

LIFE HISTORY

Tope shark are found worldwide but do not have a continuous distribution; they are not found in the western Pacific. They occur in the eastern Pacific from northern British Columbia to the Gulf of California and then from Peru to Chile and off the coast of Argentina. They are common off the west coast of Vancouver Island, in Queen Charlotte Sound and in Hecate Strait. The biology of this highly migratory shark is not well known. They have a maximum length of about 2 m and maximum weight of 45.5 kg for females and 1.8 m and 28.2 kg for males. Maximum age estimates range from 40 to 45 years, with females maturing at 1.5 m and 12 to 17 years and males at 1.3 m and 13 to 15 years. Mature females take about one year to produce 6 to 52 pups, which are released alive at lengths of about 30 to 37 cm, mostly from March to July. The tope shark feeds on a variety of fishes and species, such as squid and crabs, from the surface to near the bottom at 470 m. They are prey for larger sharks and marine mammals.

Andy Murch

BLUE SHARK

Prionace glauca VISITOR ▼

IN THE STRAIT OF GEORGIA

The first record of a blue shark in the Strait of Georgia was a report in the groundfish database at the Pacific Biological Station in Nanaimo. The report was of a specimen caught in a commercial fishery in May 2014. This blue shark was caught on the bottom at 150 m just northeast of Saturna Island. According to marine naturalist Andy Lamb, his colleague, commercial fisherman Sandy Matheson, reported that a blue shark was caught on a longline off Bowen Island, near Vancouver, in the 1970s.

DESCRIPTION

The shape and colour easily distinguish a blue shark from other sharks. The body is long and slender with a long snout and a very distinctive long pectoral fin. It is dark blue dorsally with a lighter solid blue side and a white belly. Pectoral and pelvic fins can be dusky.

Andy Murch

LIFE HISTORY

Blue shark are found throughout the temperate and tropical waters of the world's oceans. They have the widest range of all sharks. Off the west coast of the Americas, they occur from the Gulf of Alaska to Chile. Blue shark are migratory, commonly occurring off British Columbia, including in Queen Charlotte Sound and Hecate Strait in the summer. These are surface-dwelling offshore fish that are commonly in the top 50 m and rarely as deep as 100 m. A maximum age of 20 years has been published, along with a maximum length of 3.83 m. Most mature fish are 2.5 to 3 m. Males and females mature at the age of four to six years and at lengths of about 2 m. Females retain the pups for 9 to 12 months. The average litter size is 25, and the pups have an average length of 35 to 50 cm. Mating occurs from March to August. Blue shark are nocturnal, feeding on squid, pelagic invertebrates and small fishes, including Pacific salmon. There are reports of predators such as sperm whales, larger sharks and marine mammals, but the abundance of potential predators of larger blue shark is sufficiently small that it is mostly fishing that impacts their survival.

COW SHARKS

The order Hexanchiformes has two families, with recent reports identifying a total of seven species. Two species in the family Hexanchidae (cow sharks) occur in the Strait of Georgia: the bluntnose sixgill shark and the broadnose sevengill shark. Species in this order are considered to be the most primitive of the shark species. These species have one dorsal fin, six to seven gill slits (not five), and no membrane over the eye. These can be stout sharks and are found worldwide in deeper tropical and temperate waters along the continental slope, though their distribution can be patchy. They may, at some times of the year, move into shallower water.

BLUNTNOSE SIXGILL SHARK

Hexanchus griseus COMMON ●

IN THE STRAIT OF GEORGIA

Bluntnose sixgill shark are regularly found in the Strait of Georgia, and of those caught, most are believed to be immature. However, a dead 4.05 m female with five pups was found on a beach near Comox in 2001, indicating that mature individuals do inhabit the Strait of Georgia. It is possible that the strait is a nursery ground for subadults, with adults occasionally moving into the strait or Puget Sound to give birth. In the Flora Islet area, just south of Hornby Island, subadult shark averaging 2.4 m in length were observed via surveillance camera from the end of May to mid-October. Another study using satellite tags showed that most activity in the Strait of Georgia occurred deeper than 100 m. Bluntnose sixgill shark are also present in Puget Sound, where they have recently been protected from all fishing.

We list bluntnose sixgill shark as common because of the results of a study by the International Pacific Halibut Commission (IPHC) that showed a common distribution. In 2018, from July 31 to August 10, IPHC staff set baited longlines on bottom at 23 stations throughout the Strait of Georgia. They caught 114 bluntnose sixgill shark at 18 of the 23 locations throughout the strait.

DESCRIPTION

The bluntnose sixgill shark is the largest predatory shark found off the coast of British Columbia. It may be one of the most widespread of all shark species, occurring in temperate and tropical waters throughout the world. It is one of the largest predatory sharks, growing to lengths of up to 4.8 m. The world record sport-caught bluntnose sixgill shark was landed off the coast of Ireland in 2009 at a weight of 480 kg. The name "sixgill" comes from its six gill slits, compared to the five found in most other shark species. It also has only one dorsal fin, while other species of sharks that occur in the ocean around British Columbia have two.

LIFE HISTORY

Bluntnose sixgill shark occur globally. Off the west coast, they have been reported from California to Alaska. This is considered to be a deepwater species, found at depths greater than 90 m and perhaps as deep as 2,500 m. However, it can also be found in shallow areas, as observed in the Strait of Georgia. Some age estimates indicate that immature females were approximately 1.3 m at age 4 and 3.4 m at age 10. Females may mature at a length of about 4.2 m

Andy Murch

and an age of 18 to 35 years, while males mature at a smaller size, about 3.1 m and 11 to 14 years. Eggs hatch inside of the female after a development period that may be from one to two years. The pups are large when born, measuring approximately 60 to 70 cm in length. The litter size is also large, ranging from 22 to 108 pups. Females grow to bigger sizes than males, and while maximum ages are unknown, they are speculated to be about 80 years. Fisheries scientist Jackie King reported that females may give birth in the Strait of Georgia and that the juveniles may remain in the strait until they mature and migrate offshore.

Studies in Puget Sound and observations made in the Strait of Georgia suggest that the movements of subadults in the summer months are restricted, with individuals remaining within relatively small areas. It is interesting that studies in Puget Sound show that bluntnose sixgill shark appear to be readily attracted to bait. If this is true—keeping in mind the fish's limited movement—it is reasonable to assume that these large shark are highly susceptible to fishing. An examination of the stomach contents of 56 specimens off the west coast of Vancouver Island found that most stomachs were empty, while the others contained Pacific salmon and one squid. In the Strait of Georgia, it is possible that Pacific hake, Pacific herring, squid and even crustaceans are common prey. In a hook-and-line survey in the Strait of Georgia, bluntnose sixgill shark avoided squid and herring bait and ate only the Pacific spiny dogfish that were on the hooks. They are known to feed on marine mammals in other areas.

BROADNOSE SEVENGILL SHARK

Notorynchus cepedianus RARE ★

IN THE STRAIT OF GEORGIA

Broadnose sevengill shark are occasionally reported in the Strait of Georgia. The few records are from the southern end of the strait. In 2006, a 1.4 m specimen was caught in a research trawl that was fishing near the surface in the Gulf Islands off Crofton. Specimens have also been reported from commercial bottom trawls fishing in the southern Strait of Georgia in the early days of the groundfish fishery, from the 1940s to 1960s.

DESCRIPTION

Broadnose sevengill shark are easily recognized by the seven gill slits on each side of the head. The dorsal is silver-grey to red-brown with small dark spots, and the ventral is lighter or whitish. There is one dorsal fin and the nose is broad and blunt.

Andy Murch

LIFE HISTORY

This species occurs globally, mostly in temperate waters in the North and South Pacific as well as in the South Atlantic. In the North Pacific they are found off Japan, China and along the North American coast to the Gulf of California. This is a bottom-dwelling fish that prefers shallow water of about 50 m but can be found as deep as 510 m, as well as near the surface in bays and nearshore areas. Males are reported to mature at the age of 4 to 5 years and length of about 1.5 m, while females are reported to mature at the age of 11 to 21 years and length of 1.9 to 2.5 m. There is a two-year reproductive period with up to 86 pups released after a gestation period of 12 months. The pups range from 35 to 45 cm when released from the female. Maximum ages are uncertain, but experts guess that they could live to about 50 years old, growing to about 3 m and 180 kg. Smaller individuals feed on fishes, but larger broadnose sevengill shark can feed on marine mammals and even other sharks. This is not a friendly shark, as evidenced by a finding of human remains in the stomach of one individual.

DOGFISH SHARKS AND SLEEPER SHARKS

The order Squaliformes has six families and about 120 species. One species in each of the families Squalidae (dogfish sharks) and Somniosidae (sleeper sharks) is found in the Strait of Georgia. Members of this large group of sharks are found worldwide from estuaries to very deep marine waters, with a record depth of 3,675 m. They are even found in fresh water. Species in these two families have five gill slits, two dorsal fins (some with a spine in front of one or both), no anal fin and no membrane over the eye. This diverse order includes the smallest of all sharks, which is under 20 cm, and one of the largest, which is up to 7.3 m.

PACIFIC SPINY DOGFISH

Squalus suckleyi ABUNDANT ◉

IN THE STRAIT OF GEORGIA

The Pacific spiny dogfish is one of the key species in the Strait of Georgia ecosystem. This species is found at all depths and in virtually all areas of the strait. Female dogfish appear to aggregate in the winter in the vicinity of Mitlenatch Island to give birth. Thousands of newborn pups are found near the surface in January and February in this area. The species name of Pacific spiny dogfish has recently been changed from *acanthias* to *suckleyi* as biologists realized that it is distinct from the species in the Atlantic. Because this change is recent, the Pacific species may still be called the spiny dogfish in most publications. Going forward, however, the common name "spiny dogfish" applies only to the Atlantic species.

DESCRIPTION

Pacific spiny dogfish are mostly grey and perhaps brownish dorsally and whitish ventrally. There are prominent white spots on the sides that gradually disappear with age. They are elongate with a pointed snout and distinguishing sharp spines in front of each dorsal fin.

LIFE HISTORY

This species is found from Korea to Russia, to the Bering and Chukchi Seas and south to the Gulf of California. They are found from intertidal depths to 1,244 m. In the Strait of Georgia, feeding individuals will swim a few hundred metres up the Cowichan River. The maximum age reached by Pacific spiny dogfish is about 105 years, and the maximum size is 1.6 m. Ages are estimated from rings on the spine in front of the dorsal fin. These rings are deposits from an enamel gland just below the skin. The average age of a female at maturity is about 35 years at a length of 95 cm; males reach maturity when younger and smaller, at 20 years and 70 cm. It is not known where or when males and females copulate in the Strait of Georgia, but it is most likely late in the year. From the amount of wear on the spines, it could be speculated that large numbers of males compete to be able to fertilize the eggs. Females give birth to 2 to 16 live pups, ranging in size from about 23 to 30 cm, after a pregnancy of almost two years. These juveniles remain in the water column for 15 to 20 years, feeding on plankton and jellyfish. At a length of about 50 to 60 cm, they begin to migrate closer to the bottom and closer to shore. Pacific spiny dogfish larger than 60 cm begin to be captured in the commercial longline fishery, with most fish measuring about 80 cm. A staggering 51,000 Pacific spiny dogfish were tagged by researchers in the Strait of Georgia from 1978 to 1988. Approximately 2,500 were recaptured, with almost 98 percent recaptured within the Strait of Georgia, indicating that they are virtually resident within the strait. A few migrated into Puget Sound and off the west coast, and one took a trip to Japan.

Andy Murch

It is possible that Pacific spiny dogfish fulfill an important role in the ecosystem by removing dead or dying organisms. They are also an important animal in First Nations cultures. A representation of this fish is an honoured family crest among many Haida clans, and it is the family crest of Haida royalty. The crest is associated with feminine qualities of persistence, strength and innate leadership. It symbolizes a woman who could transform into a dogfish and enter the undersea world. The dogfish woman is the most prominent shark in the Haida legends.

Fisheries for dogfish were among the first commercial fisheries in British Columbia. The fishery in the late 1800s exploited them for oil used for lubrication, lighting and fertilizer. The oil was preferred for coal miners' lanterns because it produced a clear, bright light. After the late 1800s, dogfish were extensively fished for vitamin A, fish meal and eventually food. From 1870 to 1950 the total weight of dogfish caught in the strait exceeded the total weight of the Pacific salmon catch. It is ironic that today dogfish can be worth more per kilogram than some species of Pacific salmon.

PACIFIC SLEEPER SHARK

Somniosus pacificus VISITOR ▼

IN THE STRAIT OF GEORGIA

The Pacific sleeper shark has one sighting in the Strait of Georgia, near Comox, that is not dated. There was a recent report in Johnstone Strait in 2011. It is a large deepwater shark that could occasionally visit the Strait of Georgia, but it is unlikely to remain for any length of time.

DESCRIPTION

The Pacific sleeper shark has a large, flaccid, cylindrical body with a rounded snout. The colour is black-brown, grey or dark green. The low dorsal fins are about equal in size and there is no anal fin.

LIFE HISTORY

Pacific sleeper shark occur only in the North Pacific off Japan, in the Sea of Okhotsk, the Bering Sea and from Alaska to Baja California. Pacific sleeper

Andy Murch

shark may be relatively common offshore near the bottom in 240 to 2,000 m of water, but they do swim to the surface. Most range in size from about 3.7 to 4.3 m. There are reports that individuals can reach a length of up to 7 m and over 1 t, but these remain to be confirmed. Females mature at about 3.7 m and males at 4 m. Females retain large numbers of eggs, which hatch inside the body at a length of about 38 to 42 cm. The shark appears to be a mostly bottom-dwelling species that preys or scavenges on a variety of living and dead animals found on or near the ocean bottom. However, it is reported to have one of the most powerful bites of all sharks and is known to feed on giant Pacific octopus, harbour porpoise and even juvenile Steller sea lions. As the name implies, this shark is generally considered to be sleepy and sluggish, but it is capable of bursts of speed that catch fast-swimming species such as Pacific salmon. The only confirmed predator of sleeper shark is the offshore killer whale, which has been observed feeding on sleeper shark in waters outside the Strait of Georgia.

ANGEL SHARKS

The order Squatiniformes has one family, Squatinidae (angel sharks), with about 19 to 24 species. There is one species in this family that visits the Strait of Georgia, but rarely. Angel sharks are highly specialized, and a taxonomic puzzle to some. The head and body are flattened, and there are large eyes and large spiracles on the top of the head. The mouth is at the tip of the very short snout. The big pectoral fins have anterior lobes that are not attached to the head. The lower lobe of the caudal fin is longer than the upper lobe. There are five gill slits, no anal fin and no membrane over the eye. These sharks are found worldwide, mostly from shallow marine water to about 200 m, though the deepest sighting is 1,300 m. They commonly burrow in soft sediments and range in size up to about 2.1 m and 35 kg.

PACIFIC ANGEL SHARK

Squatina californica VISITOR ▼

IN THE STRAIT OF GEORGIA

The first report of a Pacific angel shark in Canadian waters was in April 2016 off Vancouver Island, near Victoria. The location is within the Gulf Islands and sufficiently close to an unclear southern boundary of the Strait of Georgia that we are including it in our list of species in the strait. The sighting included a very clear photograph of a 1.1 to 1.2 m specimen on a muddy bottom in about 12 m of water, 30 m off Clover Point.

DESCRIPTION

The Pacific angel shark has a unique flat shape with large, fleshy pectoral fins. The eyes are on the top of the head like a flatfish, but the mouth, which has barbels along the front margin, is at the front and not underneath the head. It is red-brown to grey with mottled dark patches and a white ventral side. The powerful jaws contain sharp teeth, indicating that care is needed when handling or approaching one when diving.

Andy Murch

LIFE HISTORY

The range of the Pacific angel shark is from southeastern Alaska to California and then south to Chile and northern Peru. However, the species is most common off California and Mexico. A commercial fishery for the species continues in Mexico. Angel sharks are found throughout the world in warm, shallow areas with soft bottoms. They are called sharks, but the taxonomic relationship to other sharks is not clear. The maximum length of the Pacific angel shark is 1.52 m. This shark occurs in depths from 3 to 183 m, but mostly below 100 m. Females are mature at 1.12 m and males at 1.03 m, and both mature after about 10 years. Females retain young that take about 10 months to develop. The pups are born from March to June at 21 to 26 cm. The maximum age might be about 35 years. This is a nocturnal species that remains partially buried during the day. They feed on fish that pass by close enough to be ambushed.

Rays and Skates

There are about 550 species of rays and skates. Classifications vary and could be considered a work in progress. In one classification there are four orders with about 16 to 21 families and over 540 described species. These are flattened triangular or diamond-shaped cartilaginous fishes with pectoral fins fused to the side of the head over the five or six paired gill openings on the underside of the head. The body is short and the tail can vary from large and thick and shark-like to long and slender. The one or two dorsal fins always lack a spine and there is no anal fin. Males have external claspers for mating. Most skates have thorns on the back and tail. Skates swim using undulations of their large pectoral fins. The five species in the Strait of Georgia belong to two of the four orders, Torpediniformes (electric rays) and Rajiformes (skates). The Pacific electric ray that has been reported in the Strait of Georgia is in the family Torpedinidae (torpedo electric rays). We use the classification that places two skates in the Strait of Georgia in the family Rajidae (hardnose skates) and two in the family Arhynchobatidae (softnose skates). Species are distributed worldwide from intertidal depths to over 3,000 m. Most skates are on bottom and often buried in soft sediments.

PACIFIC ELECTRIC RAY

Tetronarce californica VISITOR ▼

IN THE STRAIT OF GEORGIA

Pacific electric rays have historically been observed in Nanoose Bay and in the southern Strait of Georgia. There have been no research-related reports for at least 45 years; this includes all of the various research trawl studies, beach seine studies and scuba observations. However, there are three recent records in the groundfish database at the Pacific Biological Station in Nanaimo from the commercial fisheries in the Strait of Georgia. The records are from the ocean bottom in the southern strait and Gulf Islands areas, one in February 2006 and two in May 2016. Pacific electric rays are also found in most of Puget Sound.

DESCRIPTION

This is a distinctive-looking species that would not be mistaken for any other fish. It has small eyes and is flat with smooth skin and rounded pectoral fins that, along with the body, create a wide oval disc shape. The thick tail has two dorsal fins, with the first at least two times larger than the second. The colour is grey-blue with black spots. There is an internal electric organ on each side of the body that can be identified externally by its lighter colour.

The dorsal view of the Pacific electric ray. Adobe Stock

The ventral view of the Pacific electric ray.
Adobe Stock

LIFE HISTORY

The Pacific electric ray occurs only in the North Pacific from off Japan and the Kuril Islands to Haida Gwaii to Baja California and south to off Peru. It is most common off Southern California. The maximum size is 1.37 m for females and 90 cm for males. This is a fish that can be found near sandy or muddy bottom in 60 to 200 m of water, around reefs or among kelp. It is known to be pelagic, making excursions out into the open ocean. Females retain and feed the pups with a uterine milk. Females give birth throughout the year to as many as 17 to 20 pups but apparently only reproduce every other year. The maximum age is 16 years in one report and at least 24 years in another. Females are reported to mature at age nine and a length of 73 cm, while males mature at age six and a length of 65 cm. There are two kidney-shaped electric organs that can produce up to 45 volts of electricity for defence and to capture other fishes. The shock is considered to be substantial if the body is touched. The behaviour is mostly nocturnal, and the fish tend to be solitary as they feed on fish and invertebrates.

SANDPAPER SKATE

Bathyraja kincaidii UNCOMMON ◆

IN THE STRAIT OF GEORGIA

The sandpaper skate and the Bering skate are both considered to be present in the Strait of Georgia. The recent recognition that they are two distinct species was determined using DNA. We have only a few records of the sandpaper skate, some of which we now know could be misidentifications of the Bering skate. We have an older record from the Courtenay estuary and a more recent record from a longline bottom survey in the southern strait. There also have been catches in the transboundary area of the strait. We propose that both the sandpaper skate and the Bering skate are not common in the Strait of Georgia and probably more abundant in the southern strait. Sandy McFarlane and Jackie King, in their book *Sharks, Skates, Rays and Chimeras of British Columbia*, considered this species to be common in the Strait of Georgia. J.L. Hart, in *Pacific Fishes of Canada*, refers to this species as the black skate, but the accepted common name according to the American Fisheries Society is sandpaper skate.

Jackson W.F. Chu

DESCRIPTION

This is a soft-nosed, broad-bodied skate that has tiny prickles on the dorsal surface, hence the name "sandpaper skate." There is a prominent continuous row of strong thorns of similar size along the mid-back and onto the tail. It is dark brown or grey-black dorsally and a mottled white ventrally. There remains uncertainty about the ability to distinguish sandpaper and Bering skates diagnostically without using genetic information.

LIFE HISTORY

Sandpaper skate are reported to occur from the western Bering Sea to southern Baja California; however, now that it is known that what was previously thought to be one species is in fact two, the northern distributions may need to be revised. Like many skates, they are found on flat, soft bottom areas at depths from 23 to 1,500 m, but since this is a deepwater species, they are most often found at 200 to 500 m. They can live about 18 years and grow to about 56 cm. Fifty percent are mature by about age seven to eight at lengths of 47 to 49 cm. Females produce between 2 and 22 eggs, which are released in leather-like egg cases that have extended stiff pointed corners called horns. The young skate develop in the egg cases and are about 12 cm long when they hatch. The development time is not known, but for some skate, it can be about nine months. Reproduction is continuous, meaning that there is no particular spawning season.

BERING SKATE

Bathyraja interrupta RARE ★

IN THE STRAIT OF GEORGIA

A Bering skate has been confirmed in the southern Strait of Georgia, and it is likely that the species is rare. This species has only recently been shown, using DNA, to be distinct from the sandpaper skate. Skates are identified using their colour and the pattern of thorns and denticles on the dorsal surface. Denticles are hooked scales, and thorns are enlarged denticles. There is great variation in the prominence of the thorns, making skates difficult to identify. The use of DNA to help distinguish species will help, but if species are distinguished genetically, there is still a need to find characters that can be observed or measured and are diagnostic. Thus, the relative abundances of the Bering and sandpaper skates in the Strait of Georgia remain to be worked out.

DESCRIPTION

The key to identifying the Bering skate is to distinguish it from the sandpaper skate. There is a space near the anterior of the beginning of the row of thorns down the middle of the body and tail. This differs from the continuous spacing of the thorns in the sandpaper skate. The outline of the anterior of the body is less rounded and more notched and disc-shaped. The skate is brown or grey-brown with a white belly and distinct dark spots dorsally.

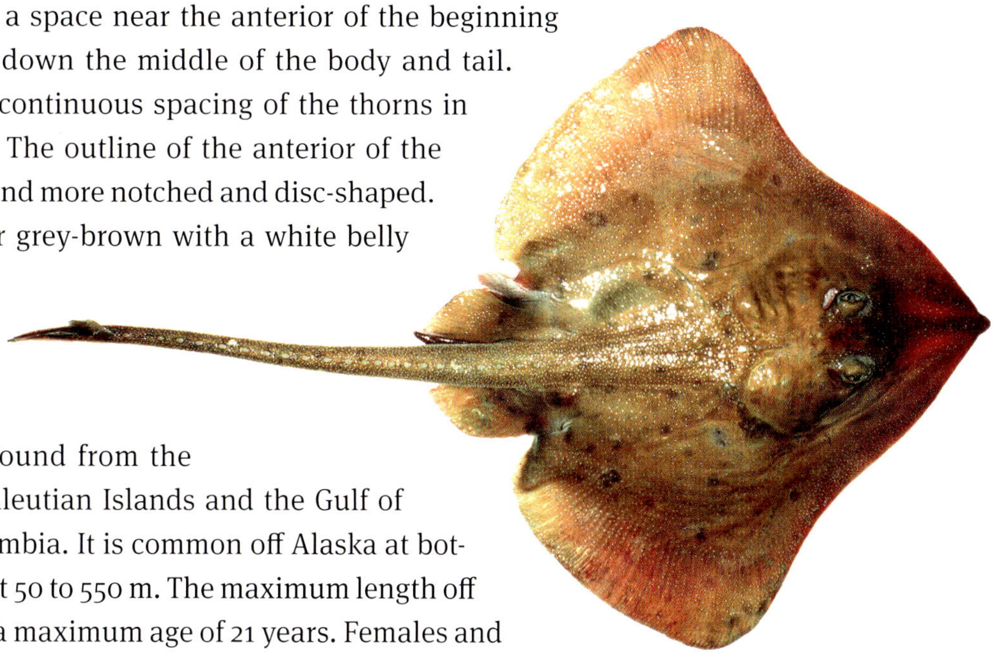

Alexei Orlov

LIFE HISTORY

The Bering skate is found from the Bering Sea, eastern Aleutian Islands and the Gulf of Alaska to British Columbia. It is common off Alaska at bottom depths from about 50 to 550 m. The maximum length off Alaska is 89 cm, with a maximum age of 21 years. Females and males mature at lengths of 67 to 72 cm and ages of seven to eight years. The production and release of egg cases from females is continuous, with the young hatching from the egg case at about 16 cm.

BIG SKATE

Beringraja binoculata

COMMON ●

IN THE STRAIT OF GEORGIA

The big skate is common in shallower waters throughout the Strait of Georgia, particularly off Cape Lazo, off Qualicum Beach and in the Gulf Islands. It is abundant in the southern strait but may be less abundant than the longnose skate throughout the strait.

DESCRIPTION

The scientific species name *binoculata* is derived from Latin words referring to the two eyes on the large, fleshy pectoral fins. These are not functional eyes but pigmentation patterns that have several rings, making them big and distinctive. The real eyes are small and anterior to large spiracles. The snout consists of thick, firm rostral cartilage, making it stiff and moderately long, but not as long as that of the longnose skate. The row of spines on the middle of the tail extends onto the body anterior to the pelvic fins, which are not deeply notched. The fish is mottled with reddish brown to light grey, with smaller, lighter cream-coloured spots. The ventral side is lighter white or cream, sometimes with brownish to mauve hues along the outer edges. The numerous teeth are small and in rows that look like bands.

Andy Murch

LIFE HISTORY

The big skate is the largest of the skate species. Maximum sizes are not known for the strait, but an unconfirmed length of 2.4 m and weight of 91 kg have been reported in another area. A recent survey of bottom fish in the southern Strait of Georgia indicated an average size of 60 to 80 cm and a maximum of 1.2 m, while a recent study in Hecate Strait found a maximum size of 2.04 m. Big skate are distributed from the Gulf of California and Baja California to the eastern Bering Sea and the Aleutian Islands in Alaska at depths from 3 to 800 m, but they are typically found from intertidal zones to about 110 m or shallower. This is a bottom fish, preferring flat sandy or muddy bottoms, where it preys on crabs, shrimps, molluscs, polychaete worms, decapods, crustaceans and smaller fishes, including flatfish and sculpins. As with other species of skate, the big skate swims with an undulating, wave-like movement of the pectoral fins, creating a gliding motion, rather than with the typical swimming movement of most fishes. Sexual maturity of males is signalled by an abrupt change in clasper length together with a hardening of the claspers. Male big skate copulate with females using the claspers, and eggs are fertilized internally. It is likely that sperm is retained, which results in eggs being fertilized and released over a prolonged period. Embryos are encased in an egg capsule that is one of the largest in all skate species. Olive-green egg capsules measure 23 to 31 cm long and 11 to 14 cm wide and may contain up to eight embryos in a single capsule, with three to four embryos being common. The female releases these capsules in pairs onto the bottom, where the embryos take about nine months to hatch. Concentrations of egg cases are found, suggesting that there are particular spawning areas. Captive skate have produced 350 egg cases a year. The young are 18 to 23 cm when hatched, but larger lengths of 30 to 34 cm have been recorded. Males and females grow quickly until about age five or six, when they begin to mature. In the Hecate Strait area, 50 percent of males mature at an age of six years and a length of 72 cm, while females mature at eight years and 90 cm. Sizes at 50 percent maturity have been reported to be larger in other areas. The maximum age has recently been reported to be at least 25 years for males and 26 years for females, but big skate may live as long as 30 years. There are commercial markets for the skate wings, but sustainable fisheries for these species are going to be difficult to manage until more is understood about the biology and population dynamics. In the Strait of Georgia, it is probably best to leave these magnificent creatures alone.

LONGNOSE SKATE

Beringraja rhina COMMON ●

IN THE STRAIT OF GEORGIA

The longnose skate is distributed throughout the strait, including around the Gulf Islands, and has been captured on bottom at depths of 400 m in experimental trawling in the central strait in the winter. It may be most abundant in Baynes Sound and the Hornby and Denman Islands areas. Although the abundance is not known, it may be the most abundant skate in the Strait of Georgia.

DESCRIPTION

The longnose skate is distinguished by its long nose, which extends as a stiff, pointed snout well in front of the eyes. There is a prominent but simple eye-spot on each pectoral fin. Unlike on the big skate, these eye-spots are not surrounded by a number of rings. Dorsally there are large spiracles behind each eye, and there are five gill openings ventrally. The ventral of the longnose skate is bluish, and the pelvic fin is deeply notched. Spines are almost all on the tail, with only a few extending onto the body.

Bill Pennell

LIFE HISTORY

Longnose skate are typically found in shallower water of less than 100 m, but are reported to depths of up to 1,069 m. They occur from Baja California, including the Gulf of California, to the Bering Sea. Skates are bottom-dwelling fish that may bury themselves in the soft sediment. As they undulate their large pectoral fins, these graceful swimmers appear to almost glide through the water. A maximum length of 1.36 m for longnose skate males and 1.45 m for females has been reported. A recent bottom trawl study in the southern Strait of Georgia found that longnose skate occupy deeper water than big skate. Most were 40 to 50 cm in length and the largest was 76 cm. The maximum age for males is 23 years, and for females, 26 years. Recent studies indicated that 50 percent of females mature at a length of 83 cm and an age of 10 years; 50 percent of males mature at a smaller size of 65 cm and an age of 7 years.

Maturity of males is evident when there is an abrupt increase in length and calcification of the claspers. The claspers form a stiff rod, suitable for internal fertilization. Once copulation occurs and the sperm is inserted into the female cloaca, the sperm appear to remain viable for an extended period. The fertilized embryo is encased in a smooth but tough leathery egg case with strands at the corners, commonly called a mermaid's purse. The egg case and enclosed single embryo are released on the ocean floor, where they continue to develop. The juveniles hatch from the egg cases in a similar way to the hatching of bird eggs and apparently as quickly. The longnose skate may produce about 140 egg cases per year at about 13 cm long. The size at hatching ranges from 12 to 17 cm. In one study the smallest free-swimming young was 18.1 cm, which is consistent with the estimated size at hatching. The reproductive period is prolonged, with estimates from six months to one year.

Skates have plate-like teeth arranged in diagonal rows. The flat teeth are used for catching and crushing prey such as molluscs. A finger carelessly placed in a skate's mouth experiences a crushing force but will not be lacerated. Longnose skate feed on benthic organisms, including crabs, shrimps, clams; smaller plankton, including euphausiids; and fish, including flatfish, rockfish, lanternfish and Pacific hake. Like many fishes, the diet changes as the fish increases in size. There are commercial fisheries off the west coast responding to a market for the fleshy pectoral fins, or wings, as advertised in seafood shops, but currently there is no commercial fishing for skates in the Strait of Georgia.

BONY FISHES

ALL FISHES THAT ARE LISTED FROM HERE TO THE END OF THE BOOK BELONG TO the class Actinopteri (bony fishes or ray-finned fishes). These are fishes with true bone—which provides supporting or protecting structures in the skull, jaws and paired fins—as well as a bony skeleton with vertebrae. Paired gills have bony arches with one opening and a bony covering (operculum).

Sturgeons

Sturgeons, in the order Acipenseriformes and family Acipenseridae, are primitive fishes that have remained relatively unchanged for about 175 million years. They occur only in rivers and lakes in the northern hemisphere. The reported number of species ranges from 23 to 29. All spawn in fresh water, but some are anadromous and enter the ocean for brief periods. The beluga sturgeon has been reported to be as large as 7.2 m and about 1.57 t, which would make it the world's largest freshwater fish. Sturgeon can also live very long, with the oldest thought to be about 150 years old. The skeleton is mostly cartilaginous, but this is a secondary development, as ancestors had bony skeletons. Their general shape is primitive, as is the shark-like tail with a large upper lobe that usually contains the terminal vertebrae. The snout is flattened and a protrusible mouth is under the head. In front of the mouth there are four barbels that have chemoreceptors to detect prey. There is a large swim bladder, a single dorsal fin far back on the body and no teeth in adults. There are rows of scutes on the body instead of scales. Sturgeon eggs are processed and sold as expensive black caviar. Sturgeon are also good to eat, resulting in conservation issues for all species. There are two species that are rare visitors in the Strait of Georgia.

GREEN STURGEON

Acipenser medirostris VISITOR ▼

IN THE STRAIT OF GEORGIA

The green sturgeon is an occasional transient into the Strait of Georgia, with very rare reports from the Fraser River and even rarer reports from bottom trawl catches in the strait. We have no records from any of the hundreds of off- and on-bottom research trawls from the mid-1990s until the present. There are occasional catches in commercial fisheries off the west coast of Vancouver Island.

DESCRIPTION

The green sturgeon is dark olive green with a white belly. It is distinguished from the more common and much less anadromous white sturgeon by the mid-ventral dark stripe on its belly, which is not present on white sturgeon. The vent is between the pelvic fins, rather than posterior to the pelvic fins as with the white sturgeon. There is a single row of one to four scutes between the pelvic and anal fins, compared to two rows of four to eight scutes between these fins on the white sturgeon.

Mark Conlin / Alamy

LIFE HISTORY

The green sturgeon is an anadromous species that is found in the ocean and rivers flowing into the ocean from the Bering Sea to northern Mexico. The three rivers where spawning is known to occur are the Rogue and Klamath Rivers in Oregon and the Sacramento River in California. Green sturgeon may live up to about 60 to 70 years, with a maximum length of 2.7 m and weight of 175 kg. Males mature at 15 years and females a little later, at 17 to 25 years. Juveniles spend their first 1 to 4 years in the lower river before entering the ocean at lengths of 30 to 75 cm, and then 6 to 10 years in the ocean before returning to their natal river to spawn. Subsequently, an individual will spawn every two to four years, entering a river from February to July and spawning mostly from mid-April to mid-June. Green sturgeon enter rivers only to spawn, and because they do not spawn every year, the majority of the population remains in the ocean. They tend to feed in estuaries, mostly on bottom organisms. All green sturgeon populations in the United States are listed as threatened and are protected. These are prehistoric fish that have survived for more than 200 million years. In Canada it is illegal to retain green sturgeon that are caught while fishing.

WHITE STURGEON

Acipenser transmontanus VISITOR ▼

IN THE STRAIT OF GEORGIA

White sturgeon are occasionally found in the Strait of Georgia, though they spend most of their lives in the Fraser River. In general, the older fish are not anadromous, but a few are known to move into the Strait of Georgia from the Fraser and other rivers. For example, two sturgeons around 12 to 18 years old were tagged in the Columbia River and recaptured upstream of Mission on the Fraser River about five to five and a half years after tagging. Juveniles in the lower river may be commonly found in the spring and summer in the estuary area up to Annacis Island. We have not caught white sturgeon in any of the hundreds of on-bottom and off-bottom research trawls in the Strait of Georgia from the mid-1990s until the present. The Squamish People once fished white sturgeon in the lower Fraser River estuary in the Point Grey area. They report that only a few men were sufficiently gifted to be able to catch sturgeon. Sturgeon flesh could be smoke-dried for storage for the winter, and the raw spinal cord was a sought-after meal.

DESCRIPTION

There is no mistaking a sturgeon for any other fish in the Strait of Georgia because of the prominent bony plates on the body and because of the mouth below the head with four barbels. There are 11 to 14 scutes dorsally and 9 to 12 ventrally. White sturgeon have 38 to 48 bony plates on their sides; this

Rick Swart

distinguishes them from green sturgeon, which have 23 to 30 plates. White sturgeon are usually grey with a white belly.

LIFE HISTORY

White sturgeon occur from the Gulf of Alaska to Northern California. Ages are determined from rings counted on cross-section slices taken from the pectoral fin. The oldest fish found so far seems to be a 138-year-old fish, found dead in 1992 at 4 m long. This fish was not weighed, but the maximum weight may be about 600 kg. There is a report of an individual that measured 6.1 m long and weighed 820 kg. Reported age and size at maturity differ and can be younger for specimens in captivity, but males generally begin to mature around age 15 and females around 20, at lengths for both sexes between 80 cm and 2 m. Individuals do not spawn every year, with intervals between spawning being about four years. White sturgeon have not been observed spawning in the Fraser River, but eggs have been found, indicating that spawning occurs in side channels of the river about mid-June. The number of eggs in a female is related to length, with the largest females producing three to four million eggs. White sturgeon are bottom oriented, with barbels ventrally around the protrusible mouth.

Beginning in 1994, white sturgeon could not legally be retained by any fishery or fishing method. White sturgeon have been listed as an endangered species in Canada since 2003 by the Committee on the Status of Endangered Wildlife in Canada. This was followed by a decision to make the designation a law in 2006 under the Species at Risk Act. In the Fraser River, juveniles larger than 48 cm and adults feed on migrating adult salmon, eulachon and other fishes, including lamprey. Adult Pacific lamprey migrating up the Fraser River in July to September are snagged by bait dealers and sold frozen to anglers catching and releasing sturgeon. Lamprey ammocoetes, known as "ditch eels" to anglers, are also common bait. It is possible that lamprey ammocoetes in the Fraser River are eaten by the smaller sturgeon. If this is true, the decline in sturgeon may explain the large abundances of western river lamprey ammocoetes in the lower Fraser River. Abundances of white sturgeon in the Fraser River declined as a result of the excessive fishing pressure in the late 1800s and early 1900s, when catches exceeded 200 t. Current numbers may be about 45,000 to 60,000, with most in the range of 50 to 80 cm or 4 to 10 years old. Many fish are under 12 years old. There may be about 7,500 fish that are mature but only several hundred that are larger than 2.6 m. Fish of this size were a daily occurrence in the fishery in the late 1800s.

Eels

True eels in the order Anguilliformes and family Nemichthyidae (snipe eels) are taxonomically distinct and differ from other fishes that have eel-like shapes. All lack pelvic fins and may lack pectoral fins. The dorsal and anal fins are long, with numerous soft rays. The head and gill area are elongate. True eels have a larval stage called a leptocephalus. This is a vertically flat, transparent larva that has no resemblance to the fish after metamorphosis. The leptocephalus period can be extended for up to about a year, depending on the species.

SLENDER SNIPE EEL

Nemichthys scolopaceus VISITOR ▼

IN THE STRAIT OF GEORGIA

This rare fish is found in the Strait of Georgia. Only two specimens have been found in our pelagic research trawls since the mid-1990s. Both were captured in July 2001, one week apart. The first specimen was northeast of Entrance Island in 225 m of water and had a length of 88.5 cm. The second was captured off Hornby Island in 210 m of water and was 72.6 cm. There are five specimens recorded in the groundfish research database at the Pacific Biological Station that are identified as the blackline snipe eel (*Avocettina infans*). One specimen was from south of Texada Island in January 1975, one from west of Texada in February 1983 and three from the middle of the strait off Gabriola Island in March 1981. There is also a record of three pale snipe eels (*Nemichthys larseni*) from June 1976, in the middle of the strait. There is no mistaking the identification of a snipe eel, but it can be difficult to identify the exact species. We are confident in our identification of the slender snipe eel from our surveys and note the possibility that the species recorded in the groundfish research database may include the pale snipe and blackline snipe eels, though all the species in the groundfish database may be slender snipe eels.

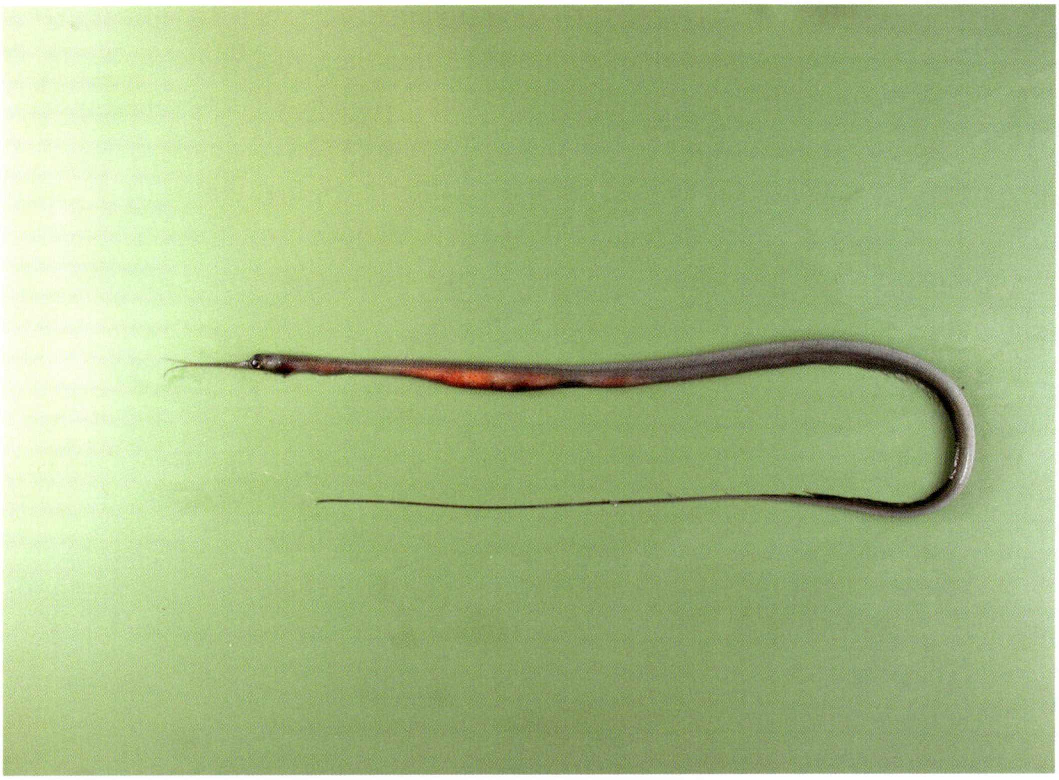

Alexei Orlov

DESCRIPTION

The colour is brown, but the dorsal fin and surface are pale with dark ventral areas. The anal fin and tips of the pectoral fin are black-brown. There is no pelvic fin. The head is bulbous and the eyes are large, a feature common to many deep-sea fishes. This is a long, slender fish, being 75 times longer than it is wide. The length results from the numerous vertebrae, which can number up to 750; this is the most vertebrae of any animal on the planet. The slender snipe eel gets its name in part from the long and slender upper and lower jaws that curve outward at the end. The jaws apparently reminded someone of the beak of a bird called a snipe. As male slender snipe eels mature, their jaws shorten, resulting in a mouth that is more typical of most fishes.

LIFE HISTORY

Around the world there are about 740 species of true eels. The slender snipe eel belongs to a family of snipe eels that has nine species. It is found in temperate and tropical seas throughout the world, from the surface to depths of about 3,700 m. Generally it is a deep-sea pelagic fish, but it may approach the surface at night, particularly if it is attracted by lights. The maximum length is 1.45 m, but specimens may be missing portions of the posterior body. Taxonomically, this is a true eel in the order Anguilliformes. True eels are elongate and can be either on bottom or pelagic. True eels have a larval form called a leptocephalus that is completely different in appearance from the adult. A leptocephalus is broad, flattened sideways and leaf-like, with a small head and a tapering body. Very little is known about the biology and ecology of leptocephali, except that they eventually metamorphose into a smaller individual that resembles the typical appearance of the adult. No leptocephali have been found in the Strait of Georgia, which probably indicates that the specimens in our catches arrived from the west coast of Vancouver Island.

The front ends of the jaws of slender snipe eels have thousands of small, backward-facing teeth. The fish is usually oriented vertically and remains almost motionless in the midwater. It feeds on shrimp-like prey, captured when the antennae become entangled in the eel's teeth. Eventually the struggling prey is close enough to the mouth to be swallowed. Very little is known about the life history of the slender snipe eel. The eggs are fertilized externally and float until they hatch. There are estimates that the fish may live for about 10 years, and some interpretations suggest that the fish spawns only once and then dies; however, if it lives for about 10 years, it seems logical that it may spawn more than once.

and Carps

Anchovies and herrings are in the order Clupeiformes and the families Engraulidae and Clupeidae. There are about 350 species of anchovies and herrings. These silvery fish occur in fresh water and the ocean worldwide. Fishes in these two families collectively account for about half of all fish caught commercially in the world. These are schooling fishes that mostly feed on plankton and are relatively short-lived (less than 20 years). Anchovies have a projecting snout and a long jaw that extends past the eye. Herrings have a terminal mouth and a shorter jaw. Herrings have scutes, which are thickened bony scales with a keel, along the belly. There are four species in the order Clupeiformes in the Strait of Georgia.

Common carp are in the order Cypriniformes and family Cyprinidae, along with a large number of minnows, chubs and shiners.

NORTHERN ANCHOVY

Engraulis mordax ABUNDANT

IN THE STRAIT OF GEORGIA

Abundances of the northern anchovy fluctuate in trends in the Strait of Georgia. Anchovy have been found in an archaeological site dating back as much as 3,000 years, and currently the northern anchovy is commonly caught or seen throughout the strait. In our pelagic research trawls from 1998 to 2009, it was an abundant species in the top 30 m. It was captured in trawls from the surface to 75 m and occasionally on bottom in shallower depths. It is found in estuaries and seen by scuba divers. Recent large abundances started again around 2014 and appear to be increasing. The largest abundances are found in the southern strait, along the mainland shore and in Howe Sound. There are reports of spawning occurring within the plume of the Fraser River from about March to the fall, with most spawning in June. Larval anchovy are currently captured in the Strait of Georgia; they grow to about 10 cm in their first year.

DESCRIPTION

The northern anchovy is easily identified by its elongate, herring-like body. It has a very short snout and large mouth, which is ventrally placed and underslung. It is a metallic blue-green dorsally and silvery ventrally. The origin of the anal fin is below the base of the last dorsal fin.

LIFE HISTORY

Northern anchovy occur from Magdalena Bay, Mexico, to Haida Gwaii, British Columbia, and very rarely off the southern coast of Alaska. They are widely distributed throughout the pelagic areas on the west coast of British Columbia. In the 1940s large schools were occasionally mixed with schools of sardine. There are three populations with overlapping distributions. Fish in the Strait of Georgia are in the northern population, from British Columbia to Northern California. The maximum age is seven years and the maximum length is 25 cm,

Bill Pennell

but fish older than four and larger than 16 cm are rare. This species forms large, tight schools that range from near the surface to near the bottom at depths of about 310 m. Adults avoid the surface waters during the day but migrate to near the surface at night as they disperse to feed. The large schools are more common in the fall and winter. Anchovy have commonly been found in archaeological sites and were the third most abundant fish in sites in Burrard Inlet.

About half of all females mature at the end of their second year at about 13 cm. All anchovy are mature at age four and 15 cm, though younger ages at maturity have been reported. A two-year-old anchovy has completed about 24 months of growth when it first spawns. Anchovy spawn throughout the year off Washington and Oregon, but most spawn from mid-June to mid-August. They can spawn several times a year, with females producing up to 30,000 eggs in a year, or about 17,000 eggs for an average individual. At the peak of spawning, females may spawn as often as once a week. Eggs are released into the water column and float near the surface at temperatures from about 13°C to 17.5°C. At these temperatures they hatch in two to four days at 2.5 to 3 mm long. The larvae also float near the surface and begin to school when they are about 1.2 cm in length. Metamorphosis occurs at about 2.5 cm. After one year of growth they are about 10 cm, at age two about 13 cm, and at age three about 14 cm. They have been periodically abundant in the Strait of Georgia, with the recent large abundance starting about 2014 appearing to be related to increasing ocean temperatures. Adult migration into the strait may have contributed to the recent abundance, but surveys over the past few decades have always found northern anchovy, indicating that some spawning probably has occurred in the strait. The current spawning that peaks in June and extends from March to September appears to be producing a large resident population that is affecting the ecology of the Strait of Georgia ecosystem. The current warming trend within the strait may result in a persistent population that significantly affects the production of predator populations, including fishes, seals, sea lions, birds and even whales.

The northern anchovy feeds on plankton and small fish, including their own eggs and larvae. There is an interesting oscillation in abundance between the northern anchovy and the Pacific sardine, with one increasing and the other decreasing, apparently in part a consequence of the type and abundance of plankton. Northern anchovy are an extremely important prey for a range of species, from fish to birds to mammals. There were seine fisheries for northern anchovy in the Strait of Georgia in the 1940s that apparently resulted in a tasty canned product.

AMERICAN SHAD

Alosa sapidissima UNCOMMON ◆

IN THE STRAIT OF GEORGIA

American shad are now found throughout the Strait of Georgia in both bottom and pelagic trawls, but most catches are from bottom trawls or near the bottom. In the 1990s they were very rare in our research trawl surveys, but they now occur regularly. In March 2012, 81 American shad, ranging in length from 20 to 43 cm and averaging 29 cm, were captured in the Strait of Georgia in a bottom trawl survey throughout the strait. They were most abundant in the southern strait, particularly in the Fraser River area. There was an abundance estimate of about 25 t for the survey area, which is an indication that they are becoming more common, possibly because the strait is getting warmer and American shad do best in temperatures from 13°C to 15°C. There has been no record of American shad spawning in the Fraser River or any other river around the Strait of Georgia; however, they have been reported in the Fraser River as far upstream as Hope. They are also found in Puget Sound. From 2018 to 2022, there was an annual spawning abundance of American shad returning to the Columbia River ranging from about 6 to 7 million fish. For context, historical returns of Pacific salmon in the Columbia River have ranged from 10 to 16 million fish. The increasing abundance may indicate that they will eventually spawn in the Fraser River. If this happens and abundances increase, as happened in the Columbia River, the American shad may become a common food fish within the Strait of Georgia, as well as a competitor of Pacific salmon.

DESCRIPTION

The American shad is a herring-like fish but is easily distinguished from Pacific herring by its deeper, compressed body, a single row of dark spots on each side, and sharp, saw-like scales, or scutes, on their ventral, keeled edge. It has a greenish-blue or green dorsal, silver sides and a white ventral, typical of herring-like countershading. The operculum has a series of radiating striations.

LIFE HISTORY

The American shad is native to the Atlantic coast of North America, where they spawn in rivers from Florida to Quebec. This is an anadromous species that is bony but good to eat. Because of its value as a food, it was transported to the Pacific coast and introduced into the Sacramento River in 1871 by the California Fish Commission. It became established and rapidly dispersed, reaching the Columbia River in 1876 and the Strait of Georgia in 1891. It now occurs south

American shad caught in a research study in the Strait of Georgia. Dick Beamish

to Baja California in Mexico and north to Alaska. It has occasionally been reported in Kamchatka, Russia. There is even a landlocked freshwater population in an area above a dam in the San Joaquin River.

These fish are found in schools from the surface to 250 m, and possibly as deep as 375 m. Most are caught in bottom trawls at depths of less than 150 m. In the Columbia River, the maximum length is 61 cm and the maximum age is seven years. In the Atlantic Ocean, the maximum length is 76 cm and the maximum age is 11 years. In the Pacific Ocean, males mature at three to four years old and at lengths of 36 to 43 cm. Females mature at four to five years old and 40 to 50 cm.

On the East Coast, American shad migrate long distances seasonally, swimming thousands of kilometres north in the summer and then south in the winter. Fish return to their natal river to spawn with a very high level of fidelity. The majority of fish at the southern limit of their distribution on the Atlantic coast die after spawning, while the shad at the northern limit may spawn more than once. On the West Coast it appears that fish spawn only once. They enter the Columbia River in April and migrate up the river from mid-April to August. Spawning occurs in the main river over sand and gravel bottoms. They spawn in groups, similar to Pacific herring. Females produce large numbers of eggs, between 200,000 and 600,000, which may result in large numbers of juveniles. The juvenile shad migrate to sea in the fall of their first year, at lengths of 6.5 to 11.5 cm. Their diet appears to be similar to that of Pacific herring: copepods, amphipods, shrimp and even small fishes. The American shad is eaten by marine mammals such as sea lions, as well as by sharks and most likely other predatory fishes.

PACIFIC HERRING

Clupea pallasii ABUNDANT

IN THE STRAIT OF GEORGIA

Pacific herring are one of the most abundant and important species in the Strait of Georgia, possibly numbering around two billion individuals in the summer. They are distributed throughout the surface waters of the strait, including in the Gulf Islands and the various inlets. Pacific herring are the most common fish found in middens around the Strait of Georgia. Their abundance in these middens generally exceeds that of Pacific salmon.

DESCRIPTION

Mature Pacific herring are elongate, blue-green dorsally and silvery ventrally. They have large scales that are easily displaced. There is a single dorsal fin and a strongly forked tail. Pacific herring are distinguished from northern anchovy, which have a large mouth at the bottom of the head and a blunt snout; from Pacific sardine, which have spots on their sides; and from American shad, which have a much deeper body that also may have spots.

A school of Pacific herring.
Janna Nichols

LIFE HISTORY

Pacific herring are distributed throughout the North Pacific and into the Arctic. They range from Baja California, Mexico, to Alaska and into the Beaufort Sea. Across the North Pacific they occur off Korea, Japan and Kamchatka, Russia. Until the late 1980s, scientists considered the Atlantic herring and Pacific herring to be so closely related as to be subspecies. However, the species are now considered distinct, with the Pacific herring retaining the name *pallasii* as the species name. The Atlantic herring is now *Clupea harengus*.

In the Strait of Georgia, the life cycle of the Pacific herring begins with the spawning period. Most adults migrate from offshore into the Strait of Georgia late in the fall in preparation for spawning. They return offshore in late March and early April after spawning. Most spawning occurs over a 42-day period, beginning in the north and ending in the south, with a peak at March 14. Young females produce about 19,000 eggs, and an eight-year-old can produce about 38,000 eggs. Spawning occurs in the intertidal areas in March. Over the years, the duration of the spawning has ranged from 11 to 87 days. Spawning is initiated when males eject milt due to environmental stimuli; however, the exact mechanisms remain to be discovered. Once spawning begins, the milt turns the water an attractive milky turquoise colour.

Eggs hatch almost simultaneously in a particular area within 10 to 21 days of spawning, depending on the temperature. After hatching, the translucent, thread-like larvae are about 9 mm long and found near the surface. The larval herring begin to feed, and if the timing of this aligns with the production of abundant copepod plankton, a large year class will result. After about 10 weeks, the larvae begin to look like the adults. In June to July, the juveniles are about 35 mm and 5 g. After a few months, the juvenile Pacific herring form large schools, typical of the behaviour of adults. These young herring are an important food source for many other species in the Strait of Georgia. An important and virtually unknown predator of the Pacific herring is the western river lamprey. One estimate is that these lamprey kill between 150 and 200 million immature Pacific herring in the several months that they are in the strait each year.

By fall, Pacific herring average 100 mm and 40 g. Beginning in October, the juveniles move offshore through Juan de Fuca Strait. Almost all juveniles migrate offshore, as abundances of adults have not been found in the winter in the Strait of Georgia except for the few resident populations. Once offshore, the juveniles feed and grow until they mature in their third year of life. Average sizes vary depending on the year, but in general, a one-year-old

TINRO

fish is about 15 cm long. The individual is about 30 months old when it begins to mature and returns to the Strait of Georgia from October to December to spawn for the first time, at an average of about 185 mm and 90 g. The longest Pacific herring in British Columbia was reported to be 31 cm and 15 years old.

There are resident or non-migratory stocks of Pacific herring around the Strait of Georgia. They occur in the mainland inlets, such as Bute Inlet, as well as in Johnstone Strait and the Gulf Islands. Old-timers tell of other resident stocks and their importance to the health of popular species such as lingcod and Chinook salmon. More samples are needed to be sure, but one study did show that there are some resident populations in the Strait of Georgia, with the vast majority being mixed genetically. In the Strait of Georgia, a population that spawned on the mainland side in Secret Cove and one that spawned near Victoria at Esquimalt Harbour and Portage Inlet were found to be resident in the 2001 genetic study. Resident stocks of herring in the strait are identified by the presence of adults in the winter. It is not known how many resident populations may have been lost in the early years of the commercial fishery, but now there is no choice but to deal with what remains—and the issue of what remains is of public concern.

There is always great concern about the welfare of Pacific herring, as there should be. Pacific herring remain as a major fishery in the Strait of Georgia, but the fishery continues to be controversial because of the unknown impact on the ecosystem. Herring stocks did rebuild into the 1990s and then in the early 2000s began to decline, and subsequently have increased again. Abundances in the Strait of Georgia generally remained much healthier than other major Pacific herring populations around the coast of British Columbia. We know that the species has survived commercial fishing in the Strait of Georgia for 130 years, but it is now time to find out why that is, and what impacts the fishing of Pacific herring has on the ecosystem.

PACIFIC SARDINE

Sardinops sagax RARE ★

IN THE STRAIT OF GEORGIA

Pacific sardine were very abundant on the west coast of Vancouver Island and occasionally in the Strait of Georgia until 1947, when there was a complete and unexplained collapse of the population and the fishery. For the next 45 years Pacific sardine, or pilchards, as they were commonly called, were virtually absent from British Columbia waters. In July 1992, Pacific sardine were found again off the west coast of Vancouver Island. Early the next year, a few were back in the Strait of Georgia. They occurred more frequently in research trawl catches later in the year, with the largest catches in November 1998. Pacific sardine were occasional residents in the strait over the next 10 years or so, but they have rarely been found in research trawl or purse seine catches since 2010.

DESCRIPTION

Pacific sardine are easily distinguished from Pacific herring by the large black spots on the side of the body. There is a series of almost straight lines, or striae, on the operculum. The dorsal body is a metallic blue or green, and the sides and belly are silver.

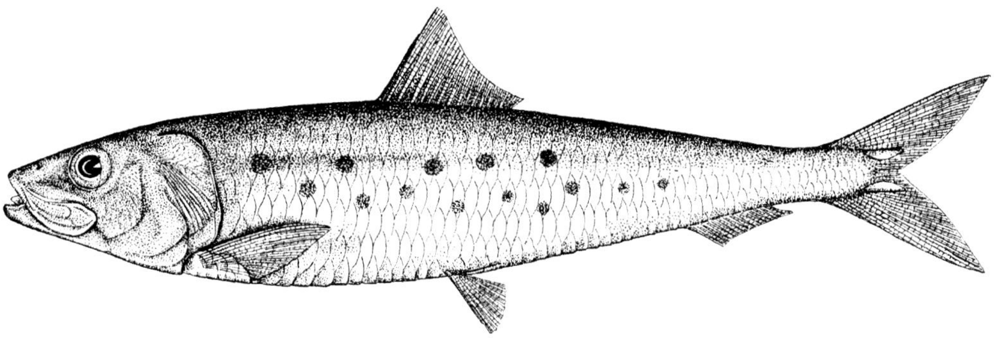

LIFE HISTORY

Pacific sardine occur off the coasts of Japan and Russia in the western Pacific and from Alaska to Baja California and the Gulf of California. They also occur in the eastern and western South Pacific and off South Africa. Off the west coast of North America, the population is highly migratory and has a history of major and abrupt changes in abundance. The recent population off the west coast of Vancouver Island continued to increase from the early 1990s and may have been as large as 200,000 t in the summer, with as many as several thousand tonnes remaining through the winter. Pacific sardine in spawning

TINRO

Jeremy Jodoin

condition were also observed off southern Vancouver Island, but the spawning abundances were relatively small. Abundances have since declined. The maximum length recorded is 43 cm, but the range in length off the west coast of Vancouver Island was 15 cm to 29 cm, with an average of 23 cm. The maximum age of Pacific sardine may be over 20 years. They were found singly or in small groups in the Strait of Georgia after 1993 but in larger schools on the west coast of British Columbia, where they are caught commercially by purse seines. Spawning occurs at night in the late winter to late spring, with females releasing eggs in batches. The coast-wide biomass is determined jointly with the United States, as the West Coast population mostly spawns off California; the larger, mature fish migrate into the Canadian zone as early as May and as late as November.

COMMON CARP

Cyprinus carpio VISITOR ▼

IN THE STRAIT OF GEORGIA

Common carp have been reported in the Strait of Georgia, but the records are not recent. One report was for 13 individuals, all smaller than 4 cm, caught in July 1967 in the Fraser River plume. They have also been found in the tidal sloughs of the Fraser River estuary. It is possible that juvenile common carp may occasionally be found in low-salinity water near rivers around the strait, but their survival in salt water would be limited to very low salinities; thus, they are likely to be only occasional visitors in the strait. We have never caught a common carp in any of the hundreds of research trawls around the Strait of Georgia since the mid-1990s.

DESCRIPTION

Common carp have a barbel attached to the posterior of the upper jaw, serrated spines at the anterior of the dorsal and anal fins, large scales and a long dorsal fin. They are distinguished from released goldfish, which are a different species, by the prominent barbel, which is absent on goldfish.

LIFE HISTORY

The common carp is a freshwater species that is common throughout the world, except at the poles, as a native, introduced or cultured species. The maximum size in fresh water is about 40 kg, and they live to at least 65 years old. Koi are an ornamental variety of the domesticated common carp, and some are believed to have lived for about 100 years. Common carp in British Columbia were first introduced in the Okanagan Valley in 1912 and reached the Fraser River drainage by 1928.

Edvard Ellric / Adobe Stock

Pencilsmelts, Smoothtongues and Smelts

Pencilsmelts, in the order Argentiniformes and family Microstomatidae, tend to be small, slender, silvery fishes with small mouths. Smoothtongues, in the family Bathylagidae, are small, smelt-like fishes with large eyes close to a small mouth. They can be very abundant in the midwater. The order Osmeriformes (family Osmeridae) contains smelts that occur in both fresh water and the ocean, but only in the northern hemisphere. There are 16 species of smelt, 6 of them in the Strait of Georgia.

BLUETHROAT PENCILSMELT

Nansenia candida VISITOR ▼

IN THE STRAIT OF GEORGIA

Two bluethroat pencilsmelt were caught in research trawls from 1997 to 2019. One specimen was caught at a depth of 212 m off Hornby Island in July 2001 and the other at a depth of 75 m off Mayne Island in September 2008. Thus, bluethroat pencilsmelt are extremely rare in the Strait of Georgia and are probably migrants from offshore.

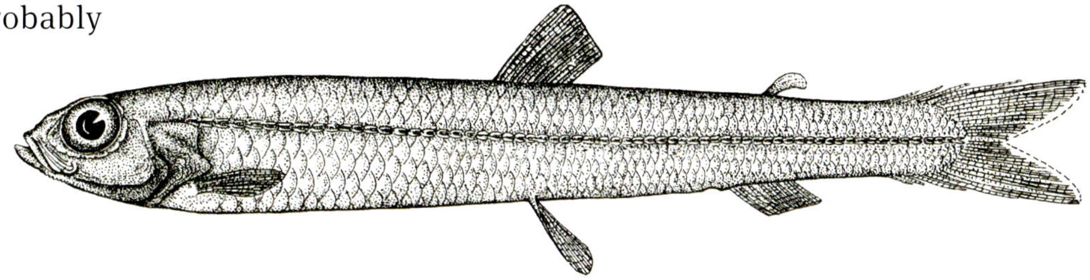

DESCRIPTION

This is an elongated fish, with a relatively small head and mouth but large eyes. The teeth in the lower jaw are long, slender and closely spaced. The bones around the upper posterior of the eye are transparent.

LIFE HISTORY

The species' common name was only recently changed from "bluethroat argentine" to "bluethroat pencilsmelt." The species name *candida* refers to its shining white appearance. This fish is found from Baja California to the southern Bering Sea, and across the Pacific off the southern Kuril Islands and northern Japan. The maximum length is 24 cm, and it is found from the surface to depths of about 1,500 m. The fish has been reported in the diet of birds feeding in the Queen Charlotte Sound area, indicating that it can be found near the surface. It is known to feed heavily on larvaceans, pteropods and euphausiids. The eggs and larval fish are planktonic, but virtually nothing is known about their life history. It is unlikely that this species spawns in the Strait of Georgia.

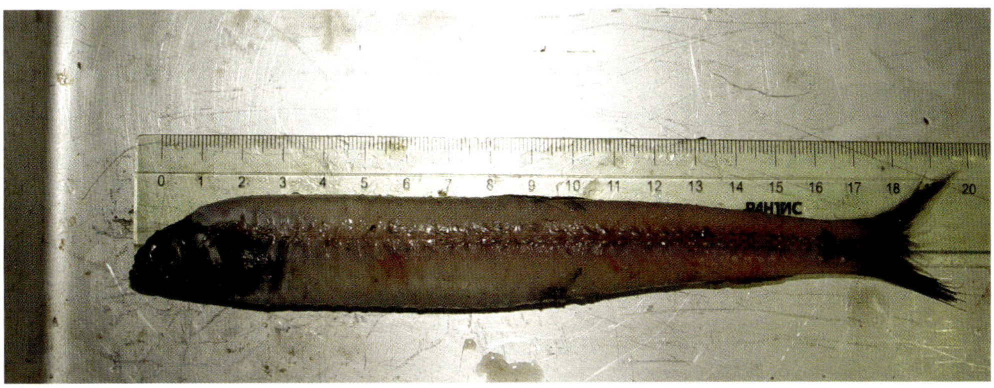

Typical appearance of specimens caught in the Strait of Georgia. TINRO

NORTHERN SMOOTHTONGUE

Leuroglossus schmidti ABUNDANT ⭘

IN THE STRAIT OF GEORGIA

The northern smoothtongue is a small, smelt-like fish that lives mostly in the midwater and deep water. It is one of the most abundant species in the Strait of Georgia. The northern smoothtongue are abundant, possibly numbering around two billion. In the Strait of Georgia, adults and juvenile northern smoothtongues are found wherever there is deeper water, but are most common in the open strait below about 100 m. Larger and mature northern smoothtongue are found at about 150 m and deeper. When not feeding, they may orient themselves vertically in the water column and remain nearly motionless with their heads down. In some areas, there can be a nighttime migration to the surface 50 m, but it is not clear if this happens in the Strait of Georgia.

DESCRIPTION

The northern smoothtongue is silver in colour with a slender body. The eyes are relatively large and the mouth is relatively small. The gill rakers on the front of each gill arch are long and fine, which is an adaptation to filter out plankton in the same way that the baleen of a whale does. The small mouth is adapted to sucking in prey.

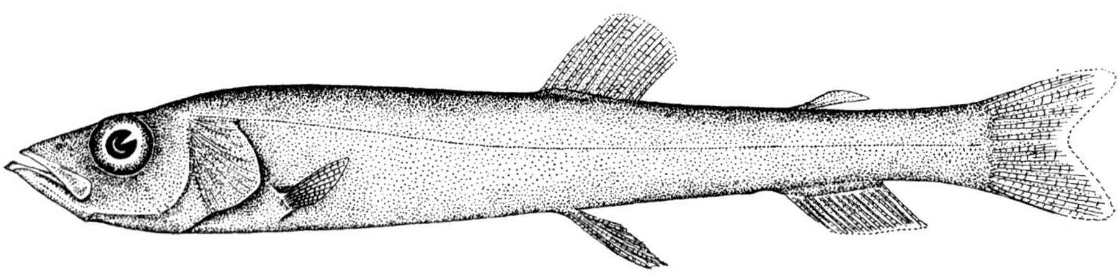

LIFE HISTORY

This deep-sea smelt is one of the most common fishes in the northern North Pacific and is thus well adapted to survive in a wide range of habitats. However, the northern smoothtongue is close to its southern limit in the Strait of Georgia; it occurs off Oregon, but is replaced by the related species California smoothtongue (*Leuroglossus stilbius*) off California and *Leuroglossus urotranus* (common name unknown) even farther south. This indicates that changes in climate may affect the ability of the species to survive. It is also possible that the more southern species would simply extend its distribution northward if the ocean continues to warm. The fish are generally 5 to 10 cm long, though they have been found at up to 20 cm.

Any species that is so wide-ranging is likely able to reproduce successfully in a variety of environments. Spawning begins in January, with most occurring in March and early April. The winter spawning of this species has evolved to match the timing of the reproduction of its prey. Spawning occurs in deep water, perhaps in the general area around Halibut Bank. Females produce a total of about 5,000 to 8,000 eggs and may spawn more than once in the same season. This batch spawning seems to be restricted to the larger females. Individuals mature after two years, or at about 5 cm in length. Males apparently do not live longer than about four years, which indicates that they would spawn for only one or two years. (It is possible that the age estimates for older males in the Strait of Georgia are not correct.) Females can live to age five or perhaps as old as seven. Fertilized eggs are common in the plankton at all depths, but most abundant at depths from about 250 to 350 m. After eggs hatch there is a movement of the newly hatched larvae to the surface and, generally, out of the spawning area. Larvae are about 5 mm long after hatching and have specific physical and morphological characteristics that are adaptations to facilitate feeding and growing quickly while avoiding predators. They retain these features until they metamorphose into their adult-like appearance in about 130 days and at about 3 cm in length. We observed that the eggs of key species of plankton are commonly found in the guts of other larval fish species, possibly indicating that larval northern smoothtongues also engage in this type of feeding. As the larvae grow, they eat the nauplii of copepods. The common food tends to be gelatinous plankton. The large eye is an indication that this fish looks for its prey, which it eats by sucking individuals into its mouth. There is a double stomach and a long gut, which is characteristic of fish that consume gelatinous prey and of fish that feed frequently. In the Strait of Georgia, the northern smoothtongue is eaten by the Pacific herring, eulachon, Pacific sand lance, Chinook salmon, walleye pollock, Pacific hake and Pacific spiny dogfish, as well as by its own species.

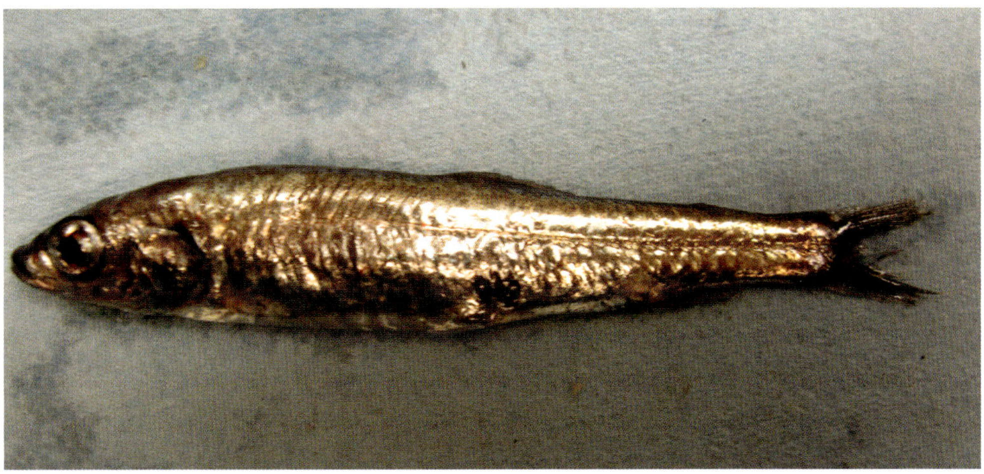

Dick Beamish

WHITEBAIT SMELT

Allosmerus elongatus UNCOMMON ◆

IN THE STRAIT OF GEORGIA

We have a number of records of the whitebait smelt throughout the Strait of Georgia from various surveys. In our pelagic research trawl surveys from 1998 to 2019, we captured about 200 specimens; almost all were caught in the top 75 m, with a few from deeper sets. Catches ranged from off Comox to the Sunshine Coast, but most were in the southern Strait of Georgia. There were also catches at the mouth of the Fraser River in a shrimp survey and in pelagic trawls in Bute Inlet. A survey for Pacific herring in the Strait of Georgia from 2004 to 2008 also recorded catching specimens. Thus, we consider whitebait smelt to be reproducing in the Strait of Georgia, but uncommon.

DESCRIPTION

The fish has big eyes, a large mouth and a long, pointed snout. It is greenish-grey on the back and has a distinct silver band on the body. There is a large tooth called a vomerine tooth on the roof of the mouth.

LIFE HISTORY

The maximum length of this species is 23 cm. It is reported in schools from the surface to 103 m, but our catches include three from depths below 150 m. It is distributed from Vancouver Island to San Francisco, California, and should be found in bays and nearshore areas. Spawning occurs in the fall in subtidal areas, but very little seems to be known about this species.

TINRO

PACIFIC CAPELIN

Mallotus catervarius RARE ★

IN THE STRAIT OF GEORGIA

There appear to be very few Pacific capelin caught in the Strait of Georgia today. In all of our pelagic trawls from 1995 to 2019, we captured only 23 capelin and all were in the top 45 m between July and November. Our last catches were in early November 2004 off the mouth of the Fraser River. It appears that the warming in the last 40 years may have greatly diminished capelin numbers in the strait.

Pacific capelin may not have been common in the Strait of Georgia, but they were an attraction during their spawning period in late September or early October. Catching them used to be a popular annual activity. Friends would collect chest waders, nets, and bottles of cheer to keep warm, and head out on a chilly evening to Pipers Lagoon. Spawning areas with specific bottom types were also known in Burrard Inlet, Departure Bay, Hammond Bay and off Lantzville. These were fall-spawning fish, but there are no recent records of their spawning. However, there are records of their spawning at the head of Bute and Knight Inlets in March in the mid-1990s. It is not known if these spring-spawning Pacific capelin are still present. There were reports of Pacific capelin spawning in the northern Strait of Georgia in the late 1990s, but in March, not the fall. It is not known if the fish continue to spawn in March in the northern strait. It is important to determine if they are disappearing from the Strait of Georgia, and one way to do so is for readers of this book to check the known spawning areas when there is a high tide and a full moon in late September or early October, remembering to bring something to keep warm on both the outside and the inside.

DESCRIPTION

Pacific capelin have a typical smelt-like form: they have an elongate, compressed body, a big head and a large mouth with small teeth situated in a slightly upturned jaw. They have a pointy snout, but the snout is more prominent in females. There is one dorsal fin, with a short base, situated midway down the body, and one bluntly shaped adipose fin posterior to the dorsal fin. The caudal fin, or tail, is highly forked and pointed at both ends. They have large fan-like pectoral fins immediately behind the operculum, as well as large pelvic fins just anterior to the long, curved anal fin. There is a distinguishing ridge between the main body of the fish and the anal fin; this area becomes prominent in males during spawning season. The dorsal surface of the capelin

TINRO

is olive green, with silver on the sides and ventral surface. The fish is readily recognized by the very small scales, large number of pectoral fin rays (17 to 20) and an adipose fin that has a long base and squared corners.

LIFE HISTORY

The Pacific capelin belongs to the smelt family. It is an Arctic species, with the limit to the southern distribution in the Strait of Georgia and Juan de Fuca Strait, and a centre of abundance in the Bering Sea and Beaufort Sea. It is found in the western Pacific off the coast of Russia, Japan and Korea. Their maximum length differs significantly within a geographical area. Fish from southern British Columbia reach a maximum length of 13.2 cm to 13.5 cm; those from the North Pacific are significantly longer, reaching 25 cm. Although there is not a significant fishery for Pacific capelin on the Pacific coast of Canada, it is often fished for recreationally. The maximum age appears to be six years.

Recent taxonomy separates Pacific and Atlantic species, with the Atlantic species (*Mallotus villosus*) having the common name "capelin." In the Atlantic, capelin can be found in Iceland, Norway, Greenland and the Arctic and throughout Canada's eastern coastline, extending into Maine. In Atlantic

Canada, capelin have high cultural value in many communities and are fished for food.

The Pacific capelin exhibits sexual dimorphism during spawning season. The males have highly prominent raised, hairy bands on their sides that are absent in females. Males also have a pronounced anal fin during spawning season, unlike females. Males and females are believed to reach sexual maturity at about three years of age. Both ovaries are completely functional and release viable eggs during spawning. However, the left ovary is several times larger than the right ovary. The fecundity of Pacific capelin from British Columbia is estimated to range from 3,020 to 6,670 eggs, which is substantially fewer than for capelin found elsewhere in their range, such as in the Sea of Okhotsk (6,000 to 53,200). Spawning occurs after dark, usually when there is a full moon and a high tide. Spawning fish can be captured with beach seines or can even be dip-netted as the spawning fish approach the water's edge in breaking waves. Spawning occurs quickly as the waves spread out onshore, with the adhesive eggs sticking to the sandy gravel bottom. Throughout their range, they spawn in two different types of habitats: intertidal beaches and subtidal sediments offshore. Intertidal-spawning fish may even enter rivers to spawn. Beach spawners prefer substrate made of fine gravel, and spawn at night on a high tide. During spawning, females deposit their eggs and males swarm in to fertilize over the fine gravel. The eggs are 1 mm in diameter, are adhesive and stick to the rocks and hard substrate. The fertilized eggs are covered over with fine sediment up to 20 cm deep through wind and tidal action until they hatch, which takes up to three weeks, depending on temperature. Prey are mostly copepods and euphausiids, but shrimp, amphipods, larval fishes and polychaetes are also eaten. There is a wide range of predators, consisting of mostly many species of fishes but also birds and marine mammals.

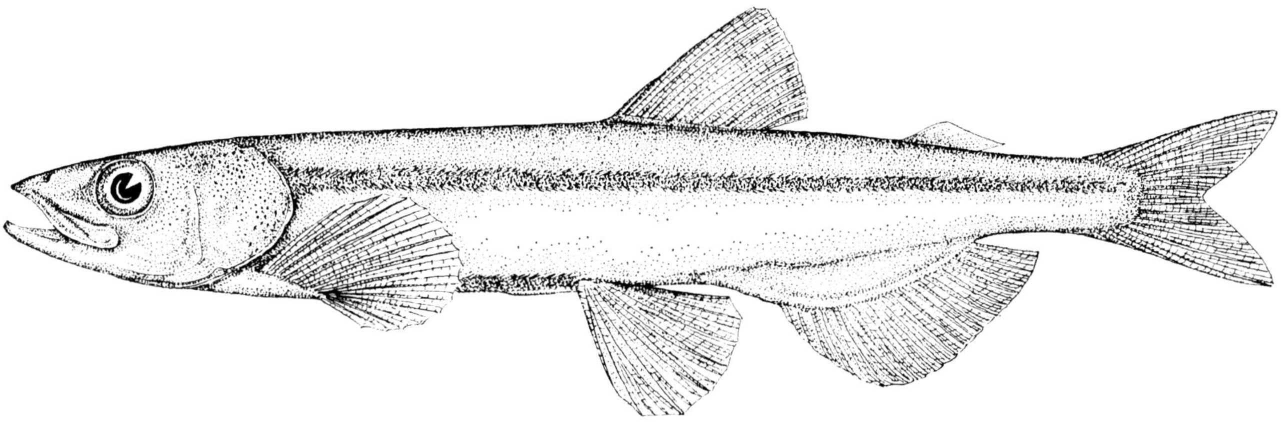

SURF SMELT

Hypomesus pretiosus UNCOMMON ◆

IN THE STRAIT OF GEORGIA

The surf smelt was once commonly found in the Strait of Georgia; however, in recent years their abundance appears to have declined at an alarming rate, particularly in Burrard Inlet. It is a marine species that has been observed moving into the Fraser River as far as 30 to 80 km from the mouth, though there is no evidence that it spawns in the Fraser River. In addition to the Squamish River estuary, it has been reported in the Nanaimo River and Cowichan River estuaries as well as around Stanley Park. They also are found on Wreck Beach on Vancouver's west side as the weather warms. In the summer on the beaches around Jericho Beach, the Squamish People traditionally collected spawning smelt that they smoke-dried for the winter.

DESCRIPTION

Surf smelt have the smelt family's typical slender body and silver appearance with a bright reflecting band along the side. They are distinguished from other smelts by their small mouth, pelvic fins that have the anterior ray behind the origin of the dorsal fin, the short pectoral fin, the bright strip along the side that turns green in preserved specimens, a small curved adipose fin and a short lateral line. There are no large teeth on the roof of the mouth. Spawning males have tubercles on the head, fins and sides and are brown dorsally.

LIFE HISTORY

Surf smelt occur from California to the Gulf of Alaska and along the north side of the Alaska Peninsula. The maximum length is 30.5 cm, but most are smaller than 20 cm. The maximum recorded length in British Columbia is 22.2 cm. In the Strait of Georgia, ages estimated from length frequencies indicate that the fish are about 12 cm after one year and 14 to 15 cm after two years. The maximum age is five years, but it appears that there are no detailed age and growth studies for fish from the strait. Such a study would be an excellent and important student project. This pelagic schooling species was fished commercially in the early 1900s and recently has been fished recreationally on beaches during spawning. It provided a popular summer evening recreational fishery on beaches in Burrard Inlet until the 1980s, when abundances declined. Most mature surf smelt in British Columbia are two years old, with average lengths of 14 cm in males and 15 cm in females. Fish may begin to spawn as early as the latter part of their first year, but most begin to spawn at age two to three. There is an equal ratio of males to females, with males aggregating in the spawning areas before females. In Puget Sound, most spawning occurs from June to September, but spawning continues throughout the year. In the Strait of Georgia, spawning has been observed only between early May and September.

Spawning is influenced by high evening tides. Spawning fish aggregate immediately off the spawning beaches, possibly selected for particular features involving freshwater seepage that reduces the drying of the eggs at low tide. It is thought that there is some homing ability, which would enable fish to return to the beach where they hatched. Spawning fish appear to prefer spring tides (which are extreme) as opposed to neap tides (which are moderate), and when daily tides are decreasing and the sea is calm. Spawning occurs at intervals of approximately two weeks. About one to two hours before high tide, the spawning female moves onto the beach, followed by several males. The female releases a portion of the eggs and these are immediately fertilized by the males. The males and females then return to the schools of fish in the deeper water. This behaviour is repeated for several days until each female has released all of her eggs. A large female of 17.5 cm and age four may release about 30,000 eggs. Eggs are laid singly and adhere to the fine bottom sediments by means of an upright short stalk. The small pedestal that attaches the egg to grains of sand is formed when the outer membrane of the egg ruptures and turns inside out. Currents then bury the weighted egg several millimetres down, providing protection from predators. Spawning occurs quickly so that the fish are carried back out with the receding wave. This sort of spawning results in the common name "surf smelt." The eggs have some tolerance to warm temperatures at low tides, but moist, cool conditions are required for maximum survival. Very little is known about the larval period, but it is believed that larval fish remain in the nearshore areas.

Surf smelt feed on zooplankton such as copepods, amphipods and jellyfish. They also consume crab larvae, shrimp and aquatic insects. Surf smelt are considered to be a forage fish and thus important prey for a number of other species, including salmon, harbour seals and birds. Forage fish are typically very abundant as they must survive large amounts of mortality. Surf smelt spawning habitat may have specific qualities that restrict the areas suitable for spawning. These spawning areas need to be mapped in the Strait of Georgia. If the surf smelt's abundance is diminishing, it will be important to ensure that the spawning beaches are protected.

Vicky Okimura

NIGHT SMELT

Spirinchus starksi RARE ★

IN THE STRAIT OF GEORGIA

Night smelt have been reported in studies of the Fraser River estuary and in collections of recent bottom trawl research surveys in the central and southern Strait of Georgia. These surveys captured one specimen in 2003 and 17 in 2005. Thus, the species is present but would be rare.

DESCRIPTION

The night smelt is a silvery-gold fish that is difficult to distinguish from the longfin smelt. The pectoral fin on the longfin smelt almost reaches the insertion of the pelvic fin, making it longer than the pectoral fin of the "night smelt," which does not reach the origin of the pelvic fin. Night smelt may also have a distinctive cucumber-like smell.

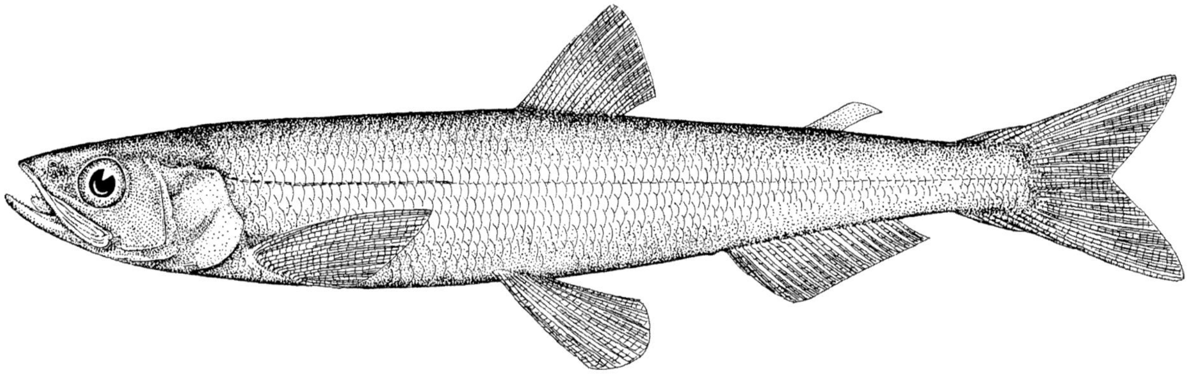

LIFE HISTORY

Night smelt are found from the southeastern Gulf of Alaska to Southern California at depths from the surface to 128 m. They are called "night smelt" because they spawn at night on sand and gravel beaches. Spawning occurs from January to October, but mostly from about February to August. During spawning, males will form large schools and follow females to shore as the surf spills onto the beach. Females are shorter than males, with maximum lengths of 23 cm, but most males and females are considerably smaller. Night smelt may live for two to three years, but more detailed studies on age and growth are needed. This species is an important forage fish in the diets of birds, fish and marine mammals. This is a good-to-eat, delicate-fleshed fish that has a history of being important to First Nations. In 1996, the commercial catch off California was a little over 500 t.

LONGFIN SMELT

Spirinchus thaleichthys COMMON ●

IN THE STRAIT OF GEORGIA

The longfin smelt is both anadromous and a resident freshwater species. It is common in coastal areas and estuaries throughout British Columbia and is reported from the Cowichan and Fraser River estuaries, as well as being abundant in pelagic trawl surveys in the Fraser River plume from 1989 to 1994. They were very common in the southern part of the Strait of Georgia in the early 1970s. The anadromous form has been reported to swim as far as 130 km up the Fraser River, but in most areas it migrates only a few kilometres up the river, prior to spawning. There are forms resident in Pitt Lake and Harrison Lake, British Columbia, and Lake Washington in Washington State. In Pitt Lake there are both the resident fish and larger anadromous forms that enter the lake in late summer and early fall. It is believed that these populations generally do not mix but are not genetically distinct.

DESCRIPTION

The longfin smelt is distinguished from other smelts by its canine teeth on the tongue, large pectoral fins that may reach or almost reach the insertion of the pelvic fins, an incomplete lateral line, a large anal fin, a lower jaw forming an angle of 68 to 90 degrees with the top of the head, and no concentric ridges or striations on the operculum. The species is pale olive to iridescent pink on the back, laterally a translucent silver, and white ventrally. Mature males are darker, with extensive stippling dorsally and along the margins of the scales. They have breeding tubercles on the paired fins and scales.

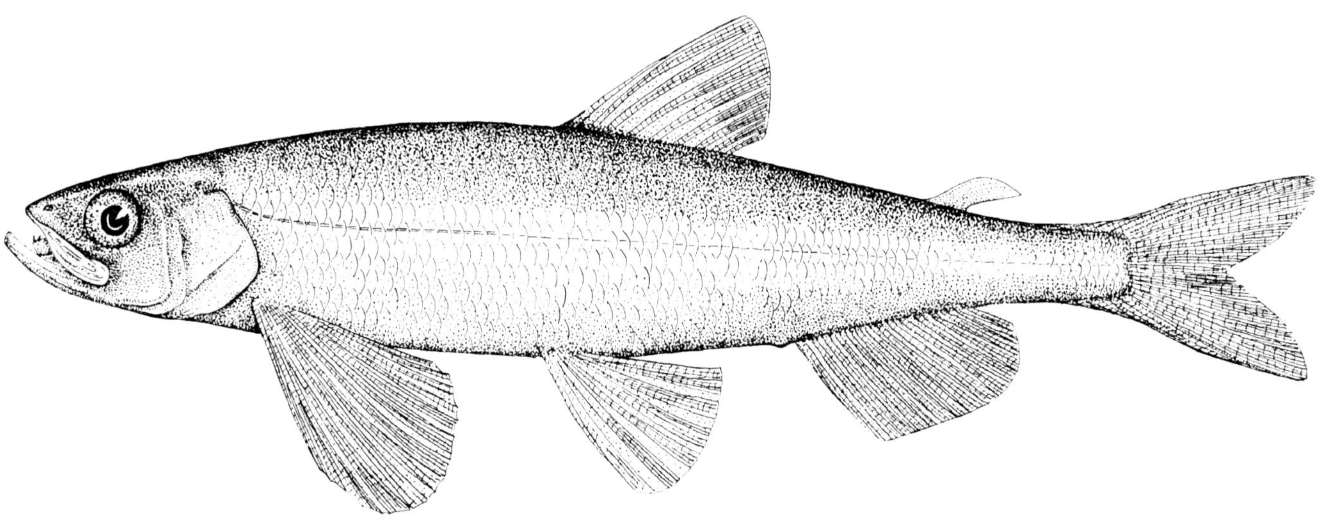

LIFE HISTORY

The longfin smelt is pelagic, occurring in the nearshore areas in bays and estuaries. It is distributed from Monterey Bay in Central California to the Shelikof Strait in southwest Alaska. The species currently has patchy distribution throughout its range, whereas abundances used to be large and more continuous. The maximum length is 20 cm, though most measure about 12 cm. In the San Francisco estuary, they are generally 9 to 11 cm in length, with some as large as 14 cm. The spawning time of the anadromous forms in fresh water appears to be between October and December, but may extend to July in some areas. Based on studies in other areas, mature adults migrate from the outside estuary into the river. Males may enter the spawning area in the rivers after dark, followed by females. Females release all eggs at once into the gravel. The eggs are fertilized by the males, and each fertilized egg has special adaptations that result in the egg being secured to the gravel by a short stalk. Most adults die after spawning. Females produce between 5,000 and 24,000 eggs, which hatch in about 40 days at 7°C. The larval fish are 5 to 8 mm when they hatch. Larvae that are 7.2 mm have been found in the Fraser River. The newly hatched fish move down the river with the tides and probably remain within the estuary, where they metamorphose after 30 to 60 days. There is very little age information available, but most spawning fish are believed to be two years old. There are reports of the resident form living to eight years old, which, if correct, would indicate that the landlocked form spawns more than once.

Dick Beamish

There is concern that the abundances are declining around the Strait of Georgia. There also have been major declines in the San Francisco Bay area, and longfin smelt were recently listed as threatened under the California Endangered Species Act, which means that they cannot be fished. The reasons for the declines are not known, but if there are some coast-wide causes, it is likely that changes in climate are one of them. Unfortunately, very little is known about the biology and ecology of the anadromous form. As with many non-commercial species, there are excellent opportunities for students to study the basic biology of longfin smelt and use this information to improve our understanding of the general ecology of the Strait of Georgia.

EULACHON

Thaleichthys pacificus UNCOMMON ◆

IN THE STRAIT OF GEORGIA

Eulachon are routinely found in research-related bottom trawl catches in the southern Strait of Georgia, often in deeper water outside of the mouth of the Fraser River. The Fraser River is the only area in the Strait of Georgia where regular spawning occurs. Spawning also occurs in the Squamish River, but not regularly. No one knows why eulachon spawn only in the Fraser River or how millions of fish find their way back to the spawning areas, but historical abundances returning to the Fraser River may have been in the thousands of tonnes. The sandy river bottoms between Mission and Chilliwack, 60 to 120 km from the ocean, are traditional spawning areas. There are reports that eulachon do not remain in the Strait of Georgia after their larval period, but we have routinely caught juvenile eulachon near the bottom in the southern strait in the past 15 to 20 years. Eulachon were extremely important to First Nations. The arrival of the fish in April was a highlight after a long winter, and oil from the rendered fat of larger individuals was eaten, used as a preservative and traded. Squamish people believed that the species that spawned in the Fraser River was distinct from the one that spawned in the Squamish River, which had more oil.

Archipelago

DESCRIPTION

The eulachon is a small, elongate fish belonging to the smelt family. There are other spellings of the name, such as "oolichan," resulting in several pronunciations. Eulachon have a relatively large mouth with small, pointed teeth that are good for catching euphausiids. The silver fish is easily recognized by striations on the operculum. The pelvic fin base is anterior to the origin of the dorsal fin, and there are four to six gill rakers on the upper limb of the first arch.

LIFE HISTORY

Eulachon occur from the Bering Sea and central Aleutian Islands to the Gulf of Alaska, and south to Central California. There is a small number of populations in British Columbia; only about 14 spawn regularly. Eulachon is an anadromous species spawning in large rivers that receive large amounts of water in the spring from melting snowfields. Eulachon are estimated to be three years old when they return to spawn. Ages of up to five to six years have been reported, but the accuracy of existing aging methods is uncertain. Females can produce an average of 25,000 eggs that hatch in two to three weeks depending on the temperature. The larvae move quickly into the Strait of Georgia. Most adults die after spawning, but there are reports that some are repeat spawners. Spawning occurs in large masses at night, and people living near a spawning area traditionally catch the spawning fish with dip nets. The time for spawning and for eggs to hatch is less than eight weeks, meaning that about 95 percent of a eulachon's life is spent in the ocean. Interestingly, the spawning time in the Fraser River in April to May is later than in the Columbia River to the south and the Nass River to the north. Eulachon larvae are less than 1 cm in length when they hatch, and as juveniles they grow to about 6 cm in the first year. Although lengths of up to 25 cm have been recorded, it is rare for individuals to reach 20 cm. Spawning fish appear to average about 16 cm. The uncertainty of how to age the fish makes it difficult to confirm the amount of growth after the first year. Solving the problem of age determination would be a useful student project.

Very little has been reported about larval and juvenile development of eulachon in the Strait of Georgia. However, there is extensive information about the distribution of larvae in May and June. We also have records of catches from trawl surveys in the strait. It is not possible to report all of the data here, but a brief summary follows. Larvae are 3 to 4 mm long when they leave the rivers and are mostly transparent. The larvae develop into the juvenile stage, when they begin to look more like the adult eulachon, with scales and pigmentation. Ichthyoplankton surveys in the early 1980s collected larvae beginning in late April. Larvae were common at a variety of depths through May and June, with the largest concentrations off Sechelt and Halibut Bank in the lower strait. In one oblique tow of the plankton net from 152 m near Sechelt, it was estimated there were about 1,600 eulachon larvae per 1,000 m^3 of water. Juveniles were first observed in late May, but were more common in June, especially off Sechelt, in the McCall Bank area, and in the open Strait of Georgia off Saturna Island.

In the spring of 1994, larger eulachon were captured throughout the central strait near the surface. It is possible that these fish were returning to spawn in the Fraser River. In late November 2008, we captured 17 eulachon, with an average length of 14.8 cm, just off the mouth of the Fraser, perhaps indicating that these fish had recently returned to the Strait of Georgia. Studies in the 1960s identified eulachon in the deep plankton layer, and in September 2009 we fished a pelagic trawl near the bottom in the deep areas of the central Strait of Georgia, where a total of 89 eulachon were caught, ranging from 9.7 cm to 21.2 cm, with two distinct size classes. It would be most interesting if these catches indicate that some eulachon have always resided in the deepest areas of the strait.

This anadromous behaviour resulted in the late 1800s in a Fraser River commercial fishery, which was only recently closed. Eulachon are highly prized by many First Nations for food and oil. With an oil content of about 20 percent of its weight, the eulachon can be dried and ignited, which explains one of their common names, "candlefish." In the area of the lower Fraser River, these fish are smoked and eaten, particularly as a potlatch delicacy. First Nations value the oil as a trade good, and in early days transported it inland along traditional routes known as "grease trails."

Eulachon are prey for a variety of fish, mammals and birds. However, except for the spawning period, their apparent low abundance as juveniles would indicate that they currently are not a major forage species in the Strait of Georgia ecosystem. In the 1950s to 1960s, there were large concentrations of Pacific spiny dogfish in the Fraser River estuary; these were probably feeding on the spawning eulachon. It appears that eulachon abundances have been declining in the strait since 1994, which was a year when abundances of a number of other species changed. Eulachon populations in other areas have also declined, and if this pattern is related to a general warming of the freshwater and marine habitat due to a changing climate, then it is possible that the declining trend, unfortunately, may continue.

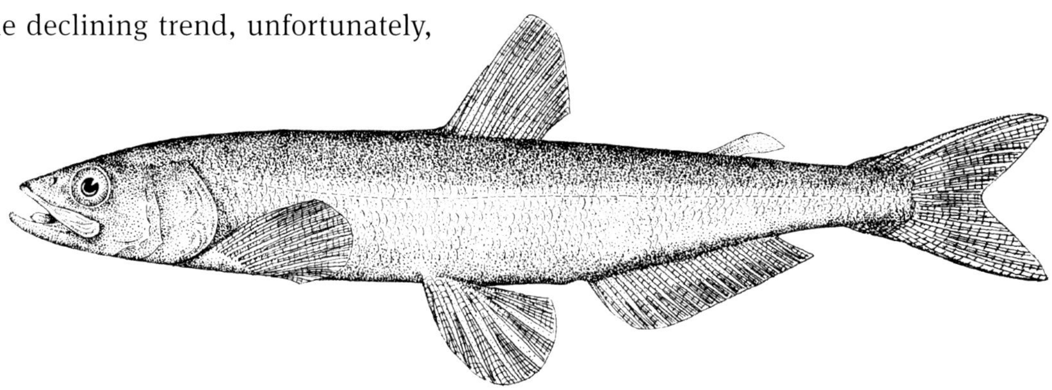

Trouts and Salmons

The order Salmoniformes and the family Salmonidae include trouts and salmons. It is estimated there are about 30 species. These are fishes that are familiar to many, and are characterized by a dorsal adipose fin. They are native to the northern hemisphere, but some, such as rainbow trout, have been introduced throughout the world. A few species are actively cultured; for example, Norway now has an annual production of farmed Atlantic salmon that exceeds all countries' total annual commercial catch of Pacific salmon. All trout and salmon spawn in fresh water and some are anadromous. Pacific salmon are in the genus *Oncorhynchus*, which has 12 species; 6 of these have common names ending with "salmon," and 6 names end with "trout." One of the salmon species, the cherry salmon (*Oncorhynchus masou*), is found in Japan and Russia but not in North America. The five major species of Pacific salmon—Chinook, coho, chum, pink and sockeye—are the dominant groups of fishes in the daytime in the subsurface waters of the subarctic Pacific, an area of about 15,000,000 km^2. The Strait of Georgia has seven species that rear as juveniles from the spring to the fall, when almost all of the five major salmon species leave. A few Chinook and coho may be resident. Juvenile Pacific salmon are the dominant fish in the offshore surface waters in the Strait of Georgia during the daytime in the spring, summer and fall. Pacific salmon have a life span of two to eight years, and all die after spawning.

Pacific salmon in the Strait of Georgia are probably the most familiar and popular species of fish for British Columbians. Our reports for these species are longer than others because of this importance.

COASTAL CUTTHROAT TROUT

Oncorhynchus clarkii RARE ★

IN THE STRAIT OF GEORGIA

The coastal cutthroat trout is a popular sport fish that is found in virtually all major estuaries around the strait but is generally uncommon in the open waters of the strait itself. There have been several thousand research-related pelagic trawl sets in the open waters of the Strait of Georgia since the mid-1990s, with only one catch of a coastal cutthroat trout, off Nanaimo in 2014, confirming that in the strait they remain in estuaries and close to shore. Knowledgeable anglers know where and when this species can be caught in the estuaries and nearshore areas around the strait in the late summer and early fall.

DESCRIPTION

The coastal cutthroat trout is recognized by the reddish streak along the underside of the lower jaw and by the rather intense spotting compared to other Pacific salmon. The many densely spaced black or brown spots are on the body, fins and head. There are small teeth at the back of the tongue.

Eiko Jones

LIFE HISTORY

A review of the taxonomy of this fish in 1989 found that it was closely related to Pacific salmon, resulting in the genus name being changed from *Salmo* to *Oncorhynchus*. Even though we still consider this fish to be a trout, it is now counted as a Pacific salmon. Coastal cutthroat trout are one of the numerous subspecies of cutthroat trout and considered to be the most abundant of all of them. This subspecies is unique to North America, occurring from Northern California to Prince William Sound in Alaska. The biology of the fish is complicated and not well studied. It is the anadromous form of the coastal cutthroat trout that moves into the Strait of Georgia at age one to six, but mostly at age two to four. Once in the ocean, most fish remain close to shore. The ocean residence time is short for most individuals, from about mid-May to August and September. Most return to fresh water in the fall, but some will remain in the strait over the winter. Up to 40 percent of these trout may spawn in fresh water in February and March more than once.

The maximum reported age for the coastal cutthroat trout is 10 years old, and they mature between the ages of 2 and 5. Adults grow to a length of about 68 cm and weight of 3.6 kg, though there are reports of much larger sizes. Homing is precise, and individuals have been recorded repeating their migrations into the ocean and back to spawn for up to four years. Spawning in fresh water is from December to May with a peak in February. In the ocean, coastal cutthroat trout feed on small fishes, including juvenile salmon.

There is evidence that declining survival during the ocean residence in some areas off the coast of Oregon is related to climate and ocean conditions. Unfortunately, abundance data are generally not available in the Strait of Georgia, so trends in abundance are speculative. There is concern that freshwater habitat loss threatens the survival of the subspecies and the life history type. There also is concern that hybridization with rainbow trout affects the abundance of the anadromous form. This is a fish that needs the attention of more than one graduate student.

PINK SALMON

Oncorhynchus gorbuscha ABUNDANT ○

IN THE STRAIT OF GEORGIA

Around the Strait of Georgia there are about 150 rivers that support pink salmon populations, and about 14 of these rivers also have hatchery operations. Pink salmon are labelled as odd-year and even-year populations; these labels refer to whether the year that they spawn ends in an odd or even number. The major populations are in the Fraser River, where they are virtually all odd-year pinks that return as adults to the Strait of Georgia and the Fraser in odd-numbered years, with the juveniles entering the strait the next year, an even-numbered year. After emerging from the gravel from February to June, the fry move quickly downstream into the strait, with a peak migration from mid-April to early May. Literally hundreds of millions of pink salmon fry enter the Strait of Georgia in April and May. They are most abundant in the Fraser River estuary area in April but they quickly find their way to the nearshore areas throughout the strait. Large abundances of these Fraser River fingerlings can occur across the strait in Hammond Bay and Pipers Lagoon at the end of April. They are not abundant in the open waters of the northern Strait of Georgia until mid-June. At that point they are leaving the strait, but relatively large abundances continue through to September. In the strait virtually all juvenile pink salmon are in the top 30 m and most are in the top 15 m.

TINRO

In the Strait of Georgia, about 95 percent of all juvenile pink salmon originate from the Fraser River. This means that the very large abundances of juveniles in the strait occur only in years ending in an even number. In the northern areas of the strait, there are some populations of pink salmon that return to spawn in years ending in even numbers, but their abundance is very small compared to those in the Fraser River. There are some pink salmon that remain in Puget Sound over the winter, but by late fall none are left in the

Strait of Georgia. Adults return from August to September and spawn in early to late fall. Adult abundances begin to increase in Johnstone Strait in late July to early August, with a peak in mid- to late August and finishing in mid- to late September. It is during this migration that most commercial and recreational fishing occurs. Pink salmon that spawn the farthest up the Fraser River leave the strait about two weeks earlier than the adults that spawn in areas closer to the ocean.

DESCRIPTION

Adult pink salmon are readily distinguishable from other species of Pacific salmon by their smaller size, small scales, large numbers of slender, closely spaced gill rakers on the first gill arch, and the large, dark oval spots on the dorsal body and tail. Juvenile pink salmon in the ocean are a metallic blue or blue-green dorsally, silvery on the side and white ventrally. In addition to the small scales and numerous gill rakers, juveniles can be distinguished from other juvenile Pacific salmon in the ocean by the small, oval-shaped otolith and absence of spots. Mature male pink salmon have a prominent hump dorsally and an elongate, hooked lower jaw.

LIFE HISTORY

Pink salmon occur along the Pacific Rim from Central California to North Korea and are the most abundant of all Pacific salmon, with the simplest and least variable life history. They also may be increasing in abundance in the Arctic. Pink salmon are the most evolved of the Pacific salmon, meaning they were the last of the various types to become a distinct species. Returning adult pink salmon have the highest straying rate of all Pacific salmon, as they more frequently venture into other rivers near to their natal river. There are some odd-year pink salmon in Puget Sound and one even-year population, but there are very few persistently spawning populations south of Washington State, indicating that the populations around the Strait of Georgia represent the largest abundances near the southern limit of their range off the west coast of North America. There are 2,200 distinct populations of pink salmon in British Columbia. Their maximum size is 76 cm.

The average size of juvenile pink salmon in early July varies, but generally they are about 11.5 cm. By September those juveniles still in the Strait of Georgia are an average of 18 cm. The movement of juveniles out of the strait is relatively unstudied but it appears that most now leave through Johnstone Strait. Previously the exit was thought to be through Juan de Fuca Strait. It is

Andy Murch

believed that pink salmon from the Fraser River spend the winter and second ocean year in the Gulf of Alaska. There is evidence that some pink salmon populations migrate south in the winter and then north into the Gulf of Alaska after the winter. The small pink salmon are heavily preyed upon by birds and other fishes, such as sculpins and coho salmon. Studies have shown that juvenile pink salmon eat some of the same food as other salmon. This competition for food has been shown to affect the survival and growth of other species in the Strait of Georgia, such as coho salmon.

Pink salmon feed about 30 percent on amphipods, 40 percent on decapods (mostly crab larvae) and 14 percent on euphausiids. They invariably mature and die at age two. This means that the odd- and even-year lines of pink salmon do not interbreed. Indeed, recent studies show very distinct genetic separation between the two different life-history types.

In the open ocean pink salmon have the widest distribution and are the most abundant. It is remarkable that in this relatively short life, an individual migrates about 4,000 km and grows to an average of about 45 cm and 1.5 kg. Pink salmon are currently increasing in abundance in the North Pacific, and catches by all countries are at record levels. The increasing abundance is partly related to increased hatchery production, which accounts for the release of 1.4 billion fry worldwide annually.

Pink salmon have the simplest life history of the Pacific salmon, but there is a complexity that is one of the great unsolved mysteries in fisheries science. The distinct broodlines and fixed two-year life span prevent interbreeding, resulting in the genetic differences between broodlines. For example, the individuals in the odd-year line are larger and collectively more abundant. Some rivers have only one broodline, but many rivers that support pink salmon will have both lines. It is common for one line to be consistently more abundant than the other, without any apparent explanation for its dominance. The balance between the two lines can also shift quickly. This unexplained dominance of one line over the other puzzles scientists; almost all attempts to artificially introduce the missing broodline in a river have failed, to the point where these experiments are no longer attempted. Surprisingly, an accidental release of a small number of pink salmon into Lake Superior in 1956 resulted in the fish surviving and reproducing in fresh water. Eventually we will solve the mystery of the regulation of pink salmon abundance in the ocean, perhaps by using the talents of a team of international researchers.

CHUM SALMON

Oncorhynchus keta ABUNDANT ⭘

IN THE STRAIT OF GEORGIA

Chum salmon occur in the Strait of Georgia as juveniles in their early marine residence and as adults when they return to spawn in their natal rivers. There is no evidence of them remaining in the strait from ocean entry through to maturity. Juvenile chum salmon used to be the most abundant species of juvenile salmon in the strait, on average, but the abundance is declining. The numbers of chum fry entering the strait are not known, but in 2008 it was estimated the number could be between 650 million and 1 billion. Initially after entering the strait, the fry remain close to shore. Mortality during this early marine period is extremely high, with the smaller, slower-growing individuals having the highest mortality. Large numbers of juvenile chum salmon are captured in trawls in early July in the surface 30 m throughout the Strait of Georgia, including the Gulf Islands. Surprisingly, in nearly half of the years, there were more juveniles in our research trawl surveys in September than in July. It is difficult to interpret these results without knowing when and how many juveniles leave the strait; however, the results do show that some remain in the Strait of Georgia for about five months. They can be abundant in the strait as late as mid-November, when they average about 22 cm. Juveniles are not found in the strait in the winter.

TINRO

Chum salmon are the latest of all Pacific salmon to spawn, with the returning adults spending very little time in the strait. After entering from Johnstone Strait, they may take only a few weeks to reach their natal river. Most adult chum salmon migrate through the Strait of Georgia in October and November, with the maximum number in mid-November. A few return in December and occasionally in January.

DESCRIPTION

Juvenile chum salmon are distinguished by the absence of distinct black spots on the back and tail. However, it is not always easy to differentiate these juveniles from other species shortly after they enter the Strait of Georgia from the river. At this time, chum salmon are distinguished by their small size and the presence of parr marks dorsally, but it is wise to check the gill rakers if parr marks are not found. The gill rakers on the first gill arch are short, stubby, smooth and widely spaced. Very small chum in their first weeks in the ocean have a dark edge along the posterior of the caudal fin, which is distinguishable against a white background. Larger juvenile chum salmon are most easily identified by the absence of black spots and by the distinctive gill rakers. The otolith shape is also diagnostic, but an image is needed for comparison. Adults are metallic blue dorsally and have silver bellies. Otherwise, mature chum salmon are largely dark olive or black, with a reddish colour on the sides and blotches of grey, resulting in a calico pattern. Spawning males have the typical secondary sexual characteristics of mature salmon: an enlarged head, hooked nose, jaws with prominent teeth and a dorsal hump.

LIFE HISTORY

Chum salmon are distributed throughout the northern Pacific Ocean and along the coastal areas of the Arctic Ocean. They range from Central California to the Bering Sea and into the Arctic off North America and Russia. Their range of spawning distributions is greater than any other species of Pacific salmon, and they used to have the greatest biomass in the ocean: about 60 percent of the total weight of all Pacific salmon. However, over the past decade chum salmon abundances have declined just about everywhere.

Most chum salmon mature and return to spawn at age three or four, which means they were in the ocean for about 30 to 42 months, including three or four winters. For example, of the chum salmon that return to the Harrison River, on average 15 percent are age three, 74 percent are age four, 10 percent are age five and fewer than 1 percent are age six. The percentages of the various ages vary considerably between years, likely related to growth rates. The maximum age is seven years. After spawning in rivers, it takes about 8 to 12 weeks for the eggs to hatch in the gravel, depending on the temperature. The alevins remain in the gravel for about 4 to 10 weeks, after which they emerge as fry and migrate almost immediately into the Strait of Georgia. The peak fry migration to the ocean is mostly at night between mid-March and the end of April, but the full range is from February to June. Juveniles migrate into the open ocean in summer and fall. At the end of their first winter they are about 28 cm long; the maximum length is 1.09 m. Chum salmon increase their weight by between 60 and 90 percent in their final marine year. Thus, any competition for food on the high seas could affect growth and, ultimately, the number of eggs produced.

We know that chum salmon from British Columbia mostly rear in the Gulf of Alaska, with some rearing in the Bering Sea. During the open-ocean residence, chum salmon are eaten by a variety of predators, such as longnose lancetfish, North Pacific daggertooths and salmon sharks. Predators of the adult fish in Johnstone Strait and the Strait of Georgia are killer whales, seals and California sea lions. In general, their survival rate in the ocean is between 0.3 percent and 3.2 percent, with an average of 1.8 percent.

Adult chum salmon populations are categorized based on the time of year they return to their spawning rivers. Almost all chum that return to the Strait of Georgia are fall chum. These fish return to the rivers in early October, with peak spawning in November, and continue to spawn as late as mid-December. The last to arrive are the Big Qualicum River fish. There also are summer-run and winter-run populations, but these are rare. On average it takes about 22 days for a fish to swim from Johnstone Strait to the mouth of the Fraser River. The fishery for chum salmon returning to the Strait of Georgia is mostly in Johnstone Strait.

In the Strait of Georgia, which includes Bute and Jervis Inlets, most chum salmon are produced in the Fraser River, downstream of Hope. The returns to these spawning areas are the largest in all of British Columbia. The largest returns are to the Harrison River area, Stave River and the Chilliwack area. Collectively, these three areas account for about 90 percent of all chum production in 124 spawning sites in about 110 streams and rivers in the lower Fraser River drainage. The numbers of chum salmon produced outside of the Fraser is quite variable. There are about 186 other streams and rivers around the Strait of Georgia that provide chum salmon fry, for a total of about 300 spawning populations that contribute fry to the Strait of Georgia.

Spawning salmon are iconic to British Columbians, attracting crowds and busloads of schoolchildren. One of the most frequented areas is along the Goldstream River, near Victoria. When chum salmon are abundant, the fish arrive about mid-October and attract visitors for over two months. As the males mature in the river, they develop their large, hooked jaws and the big teeth that give them the common name “dog salmon.” Visitors can readily recognize the grey-and-green males with the dull red-purple bars. Females are less colourful and greyer with a dark bar along the length in the middle. Males are aggressive and do not hesitate to take on whatever offends them.

Shigehiko Urawa

Beginning in 2019, there was a poor return of chum salmon coast-wide in British Columbia, possibly resulting from the warm ocean years beginning in late 2014. Chum salmon fisheries in British Columbia have now collapsed. Spawning abundances have also collapsed in most areas. The fisheries are well managed, so any major declines are not from overfishing. There is an urgent need for an assessment of why ocean survival of British Columbia chum salmon is declining.

COHO SALMON

Oncorhynchus kisutch ABUNDANT ○

IN THE STRAIT OF GEORGIA

Coho salmon are an anadromous species that is abundant in the surface waters throughout the Strait of Georgia as fry, smolts, juveniles and adults. The colour of smolts becomes silver as they enter the strait from April to late June, mostly about mid-May. The yearling smolts (one year spent in fresh water after emerging from gravel) are about 7 to 12 cm, and the two-year-old smolts are about 10 to 15 cm. Most smolts entering the Strait of Georgia are yearlings. There are also fry that enter the strait shortly after they emerge from the gravel during their first year in fresh water. The numbers of fry entering the strait can be large.

Juvenile coho salmon spend less time in estuaries than other salmon. These juveniles rear throughout the Strait of Georgia at depths mostly above 45 m, with larger densities in the north central area. A variable percentage of juveniles overwinter in the strait and the fish mature into adults in their second ocean year. The remaining percentage of juveniles migrates in the late fall to the west coast of Vancouver Island, where the fish mature in their second ocean year and then return to the Strait of Georgia and their natal rivers in the summer to early fall. The reason that one group migrates to the west coast of Vancouver Island appears to be related to the condition of the individual fish, which in turn may be related to feeding conditions in the early ocean residence. There was a period from about 1994 to 2013 when almost all juveniles left the Strait of Georgia and did not return until late summer of the next year. Beginning about 2022, a large percentage of the juvenile population remained in the strait over winter.

Wyatte Keltie with a 10 lb coho caught off Schooner Cove. Shaughn Keltie

DESCRIPTION

Larger coho salmon have the familiar silver colour and fusiform salmon shape. These larger individuals are distinguished from other salmon by the small black spots on the dorsal body and the upper lobe of the tail fin. The gums at the base of the teeth are white. Coho smolts that have recently entered the Strait of Georgia may have 8 to 12 parr marks that are long, narrow and widely spaced, with a lateral line through the middle of the parr mark. Both juvenile and adult coho salmon can generally be recognized by the small spots and white gums, but on occasion they may be difficult to distinguish from Chinook salmon because these features are not always a reliable characteristic. There are other characters that can be used to identify juvenile coho salmon, such as the number of pyloric caeca and branchiostegal rays. The otolith, or ear bone, of Chinook salmon is always about double the size of coho salmon of similar lengths.

Daniel Thornberg / Adobe Stock

LIFE HISTORY

Coho salmon in the Strait of Georgia are one small component of a very large population that extends from Monterey, California, to Alaska, along the Aleutian Islands, and from the Anadyr River in Russia south to Japan and North Korea. The maximum size of coho salmon is 1.08 m and 17.7 kg. Sportfishing records of size are not kept in British Columbia, but a recent record in Washington State for coho salmon is 11.5 kg. There are about 2,600 populations of coho salmon in British Columbia, and many of these have fewer than 500 spawning adults. Most spawn within 250 km of the ocean, and the longest migration in British Columbia, about 500 km, is in the Nechako River. Females 55 to 70 cm in length produce 2,500 to 5,000 eggs.

In general, coho salmon increase in size in the ocean more rapidly than other Pacific salmon. The diet of coho salmon changes as the fish grow, with juveniles feeding on amphipods, decapods, euphausiids and other fish. By early summer over one-half of their prey are fish, and about one-half of the fish are Pacific herring. Seals are a major predator of juvenile coho salmon, as are western river lamprey. Some coho salmon leave the Strait of Georgia in the fall and migrate to waters off the west coast of Vancouver Island. There is evidence that they tend to migrate to a particular location consistently. By August of the second ocean year most adults begin to return to spawn. Precocious males, or "jacks," can return after two to six months in the ocean when their growth is rapid, but much about their development remains to be discovered. Coho salmon enter their natal rivers from September to January, with most returning in the fall. In North America, most spawning also occurs from late September to January, but it can continue through to March.

In the 1970s and earlier there was relatively little management of coho salmon in the Strait of Georgia, as they were abundant and escapements (number of spawning fish) were at least adequate. Historically, coho salmon supported both a commercial and sport fishery in the strait. The sport fishery started in the spring with a so-called blueback fishery, which captured coho salmon that exceeded the size limit of 30 cm. In the late 1970s, a salmon enhancement program was introduced with the intent of doubling the total catch of Pacific salmon in British Columbia. As a result, a number of hatcheries were either built or enlarged to allow coho salmon to be produced artificially. The intent was to achieve a much larger abundance of the species. However, beginning in the 1980s, spawning escapements declined and there was concern that abundances were decreasing because of freshwater habitat loss and overfishing. By the mid-1990s it was accepted that major management decisions were needed to protect the coho salmon that reared in the Strait of Georgia. At the same time, it was discovered that the marine survival of juvenile coho salmon in the strait had declined from about 15 percent in the 1970s to about 1 or 2 percent in the 1990s. Thus, there needed to be a major reduction in the catch of coho. When it was realized that abundances continued to decline and that hatcheries could not replenish the stocks, particularly for stocks in the interior Fraser River watershed, managers and the public became alarmed to the point that commercial and recreational fisheries in British Columbia were greatly restricted beginning in 1998. The declining trend in marine survival beginning in the late 1980s was synchronous with declines in Puget Sound and off the coast of Oregon, indicating that large-scale climate and ocean events were involved.

There was another change in the capacity of the Strait of Georgia to produce coho salmon in about 2008 that increased their food supply, resulting in faster growth and better survival for juveniles. In 2023 and 2024, coho salmon were the most abundant they had been in the strait since the early 1990s. The recent increase occurred despite about a 70 percent reduction in hatchery releases over the past two decades.

STEELHEAD

Oncorhynchus mykiss RARE ★

IN THE STRAIT OF GEORGIA

Steelhead are found in the Strait of Georgia as juveniles that are migrating out to sea and as adults that are returning to spawn. In both cases, the time spent in the strait is relatively brief. From 1995 to 2019, 79 juvenile steelhead were captured in research trawl catches numbering in the hundreds in the Strait of Georgia. All were captured in the surface 15 m of water throughout the strait, but more were caught in the northern strait. They appear to be more abundant toward the end of May, but a few were captured through to about mid-July. The very low catches probably are a consequence of low abundances and a relatively rapid migration out of the Strait of Georgia prior to early July. The number of adults entering rivers could be about 2,000 on the east coast of Vancouver Island and perhaps about 5,000 in the Fraser River drainage. The largest populations occur in the Chilliwack and Cowichan Rivers. A smaller population of 200 to 300 adults in the Englishman River is used as an annual index of the health of steelhead around the Strait of Georgia. The Squamish People believed that steelhead came from coho salmon that remained in the river and therefore were a type of salmon. It was in the late 1980s that scientists first determined that steelhead were in the same genus as Pacific salmon and thus were indeed a salmon as the Squamish People had determined.

DESCRIPTION

Steelhead can be distinguished from most other Pacific salmon by the small dark-brown or black spots on the back, sides, dorsal area, adipose fin and the entire caudal fin. There are no teeth on the tongue, and the species has 9 to 12 anal fin rays. In the ocean, these silvery fish may have a reddish stripe.

LIFE HISTORY

Steelhead are sparsely distributed in the ocean, but they occupy the entire subarctic Pacific above latitude 38° north. They are found along the west coast of North America from Southern California to the north side of the Alaska Peninsula in the Bering Sea, and in Russia along the western shore of the Kamchatka Peninsula in the Sea of Okhotsk. A scientific survey in the Gulf of Alaska in the winter of 2022 showed that steelhead occur in the top few metres of the ocean. Tagging studies showed that steelhead are in the top few metres during virtually their entire ocean life. Their migrations are more extensive than any other Pacific salmon, with one report of a tagged

Liquid Art / Wikimedia Commons

steelhead travelling a point-to-point distance of 6,705 km across the North Pacific, from its hatchery site in Idaho to the Kuril Islands in Russia, in 17 months. Those originating in North America are distributed throughout the entire North Pacific, south to 40°58′ N and west to 151°16′ E.

In British Columbia, there are over 400 populations in about 400 watersheds. Abundances in more than one-half of the populations are generally declining, mainly because of poor survival in the ocean, though damage to the freshwater habitat is also an issue. The reasons for the poor survival are not known but believed to be related to large-scale ocean changes. The number of populations that enter the Strait of Georgia from all rivers, including the Fraser River drainage, depends on the interpretation of what a population is. Estimates range from 35 to 70 populations, but most of them are very small, and even the estimate of 35 populations includes about 70 percent that have fewer than 200 adults.

The return migration from distant waters to natal rivers is rapid and direct. Consistent with the complexity of steelhead's life history, the fish can enter fresh water throughout the year: in the summer and fall as immature fish, or in the winter and spring as mature adults. The summer run returns from May to September. These fish are immature, taking up to about eight months to mature and spawn in mid-May to late June. Most coastal populations return in the winter, and the interior populations in the summer. The coastal group matures in both the ocean and river, while the inland group matures only in the river.

Steelhead spawn in the late winter and spring and are capable of repeat spawning, which distinguishes them from Pacific salmon. In some rivers, rainbow trout and steelhead are produced from a common gene pool. In other rivers, the two types are separated. In general, steelhead tend to produce steelhead, but they can also produce rainbow trout. Rainbow trout can also produce steelhead, which is one reason they are considered to be the same species. Juveniles can rear in fresh water for one to seven years before smolting, and immature fish can remain in the ocean for one to three years before returning to spawn for the first time. Most steelhead spend two years in the ocean before returning to spawn. The percentage of fish that spawn more than once varies by area but appears to range from 6 to 12 percent. More females than males may be repeat spawners. Repeat spawners are called kelts; they typically spawn two times but up to five repeat spawnings have been reported in North America, and up to seven in Russia. Approximately 1 percent may spawn three times. The longest a steelhead has been reported to live in the ocean is five years, but this is very rare. In a study in Alaska, a large number of steelhead were held in a net pen in the ocean. Over time, these steelhead were used for various studies, but after four years a small percentage was still alive. The species' maximum age would be 11 years if the oldest fish in fresh water remained in the ocean for 5 to 6 years. Steelhead introduced into a river in Argentina have been reported to return to spawn up to eight times and survive for 11 years, but this is exceptional.

The maximum length is 1.22 m and the maximum weight is 23.6 kg. A fish of about 19 kg is mounted on the wall at the Smithers Airport. Steelhead are opportunistic feeders, with fish as a major prey. Of interest in their diet is the virtually complete absence of gelatinous zooplankton, which are common in the diets of chum salmon. Juvenile steelhead in the Strait of Georgia feed on euphausiids and small fish. Larger steelhead in Puget Sound consume mostly euphausiids, but also feed on amphipods and crab larvae. In the open ocean, squid and small fishes are the major diet items.

SOCKEYE SALMON

Oncorhynchus nerka ABUNDANT

IN THE STRAIT OF GEORGIA

Almost all sockeye salmon in the Strait of Georgia originate from the Fraser River. There are very small populations in Sakinaw Lake on the mainland and Village Bay Lake on Quadra Island. Anadromous sockeye salmon in the Fraser River have three life-history types. Almost all populations are lake-type sockeye. When these fish emerge from the gravel, they migrate into the adjoining lake, where they feed and grow for a year and sometimes longer. A few populations remain in the river and are called river types. The third type is represented by one major population from the Harrison River, known as the sea type, which means they enter the Strait of Georgia in the same year they emerge from the gravel. They also enter at a smaller size and much later in the year than the other sockeye smolts. In fact, after about mid-July, almost all juvenile sockeye in the strait are from the Harrison River.

The lake-type smolts migrate out of the Fraser River over a four- to five-week period, with a peak in early May and finishing by the end of that month. By early June, the young sockeye are spread throughout the Strait of Georgia, including the Gulf Islands. These lake-rearing or one-year-old smolts remain in the strait for an average of six weeks. By mid-July virtually all lake-type juveniles have left the strait. Some spend two years in fresh water and migrate faster out of the Strait of Georgia, but these represent only a few percent of all juveniles. River-type smolts are believed to behave similarly to the lake type.

About the time the lake-rearing juveniles are leaving the strait, the smolts from Harrison River are entering the Strait of Georgia. These sea-type smolts migrate to the lower part of the river and estuary area in late June and early July. In July, they are very common in Howe Sound. In recent years, these juvenile Harrison River sockeye salmon have been abundant in the Strait of Georgia

Alaska Seafood

from mid-July to October. They leave through Juan de Fuca Strait and Johnstone Strait. Occasionally a juvenile sockeye remains in the Strait of Georgia in the winter, but except for this rare occurrence, there are virtually no juvenile or adult sockeye in the strait in the winter. Adults return to the strait and into the Fraser River over a period of about three months, from late June to mid-September, with most adult fish from a typical Fraser River population moving through the strait in 7 to 12 days. These sockeye salmon return as adults in four general groups: the early Stuart River fish enter the river in early July, the early summer fish in late July, the summer fish in early August and the late group in late August. Adults returning to the Fraser River spend little time in the strait, except for the late-returning populations, which traditionally might remain off the mouth of the Fraser for several weeks.

DESCRIPTION

Sockeye salmon are recognized by their silver appearance without distinct black spots. If spots are present, they are a fine black speckling on the back. The gill rakers on the first arch are long, slender and closely spaced. Spawning fish have a characteristic bright-red body and green head. Adult sockeye salmon are known for their red flesh. Juvenile sockeye are readily distinguished by the long and numerous gill rakers, the absence of spots on the body, and scales with a dark margin to the scale pocket. The scales are larger than those of pink salmon.

LIFE HISTORY

Sockeye salmon are distributed from Northern California to the Bering Sea, with some strays found in the Canadian Arctic, specifically the Mackenzie River in the Yukon Territory. They occur in Russia from the Anadyr River, flowing into the Bering Sea, to Hokkaido, Japan. The Fraser River populations are at the southern edge of the range of the species and thus likely to be more susceptible to changes in the freshwater and marine environments that might be caused by a warming of the environment. The maximum size of sockeye salmon is 84 cm long. In the Strait of Georgia, juvenile sockeye salmon consume mostly amphipods, euphausiids and decapods. Larger sockeye salmon will also consume squid and small fishes. In turn, they are eaten by seals, western river lamprey and Pacific lamprey in the strait.

Virtually all lake-type juveniles now leave through Johnstone Strait to the north; previously, they also left through Juan de Fuca Strait to the south. The reason for the change is unknown. Once the juvenile sockeye salmon leave

Spawning sockeye salmon have bright-red bodies and green heads. Male and female sockeye can be distinguished by the hump in the back and the hook of the upper jaw in a male, in contrast to the streamlined contours of the female. Danny Kent

the Strait of Georgia, they migrate into the Gulf of Alaska and join the populations of other sockeye salmon and other species of Pacific salmon. Juveniles that leave the Strait of Georgia are widely spread by mid-July, occurring in Queen Charlotte Sound, Hecate Strait and even off Kodiak in Alaska. In 2019, an expedition to study the winter ecology of Pacific salmon found a juvenile sockeye from Chilko Lake, west of 180°. Other sockeye salmon from the Fraser River were far out in the central Pacific in their first and second ocean years. Sockeye salmon spend two winters in the Gulf of Alaska, with some migrating into the Bering Sea and south of the Aleutian Islands in the North Pacific, where they are mostly in the top 30 to 40 m. Other sockeye populations can remain in the ocean longer, and some fish return after only one year. Most fish returning after one year are males, called "jacks," but a very small percentage are females, called "jills." The combination of freshwater and ocean residence can result in a maximum age of seven years (most adults are age four). On average, a female produces about 3,720 eggs.

Fernando Lessa / Alamy

Production of sockeye salmon in the Fraser River occurs in cycles and trends. There is a remarkable and generally unexplained consistency to the pattern of abundances and to the timing of the adult return. In the Fraser River, the four-year cycle typically includes one large run (dominant cycle), one moderate run (subdominant cycle) and two low-production years (off cycle). There are several possible explanations for these cycles, but no consensus has been reached. A complication to forecasting returns is that the cycle of dominance has changed over the past 100 years. Perhaps the best-known dominant run is the Adams River sockeye salmon run, which can return in huge abundances every four years, to the delight of thousands of visitors to the spawning grounds. The Adams River sockeye are some of the last to migrate into the Fraser River; they then travel 385 km to an area 1,200 m above sea level. Equally remarkable is that the fish reached this point, after starting from the Gulf of Alaska, using their own biological global positioning system. Various populations of Fraser River sockeye also need to begin their return migration at different times, and how this is done is another mystery. It has been estimated the migrations exceed 3,700 km, with an individual fish finding its way back to the exact area where it was hatched.

CHINOOK SALMON

Oncorhynchus tshawytscha ABUNDANT ●

IN THE STRAIT OF GEORGIA

There are Chinook salmon in the Strait of Georgia throughout the year. Most originate from rivers around the strait, but a small percentage are from Puget Sound and other areas. The number of juveniles entering the Strait of Georgia from all sources, including hatcheries, varies year to year, but in a good year could be as high as 85 million. Larger Chinook salmon are caught throughout the strait, and maturing adults return throughout the year. In recent years, an average of about 240,000 adults enter the Fraser River.

The species has two main life-history types: yearlings and subyearlings. Yearling Chinook salmon spend at least one winter in fresh water after they emerge from the gravel. Subyearling juveniles leave fresh water in the year that they emerge from the gravel, often within three months. In the past it was proposed that most of the yearling juveniles, which are larger than subyearling fish, migrated out of the Strait of Georgia in June and July, leaving the subyearling juveniles resident until about November. However, this interpretation is now known to be oversimplified, as both types remain in the strait. In mid-July, the juvenile Chinook salmon caught in trawl surveys were found to be about one-half yearlings and one-half subyearlings. By September, about 75 percent of the juvenile Chinook in the strait were subyearlings and mostly in the surface 30 m. Recent studies also showed there are distinct juvenile rearing areas for particular populations within the Strait of Georgia. For example, juvenile Chinook salmon from the Cowichan River remain within the Gulf Islands until

Beth Swanson / Alamy

Clyde Wicks

the late summer. Thus, in addition to specific life-history types in fresh water, there are genetically distinct ocean-rearing types that occur in specific areas of the Strait of Georgia.

DESCRIPTION

Most fishermen recognize a Chinook salmon by its large size. However, the diagnostic characters are the small black spots on the upper and lower lobes of the caudal fin and back, the black gum on the lower jaw and the short, rough, widely spaced gill rakers on the first gill arch. Large Chinook salmon in the ocean are a metallic greenish-blue to black dorsally, with silver sides and white ventrally. They also have a variable flesh colour that is genetically controlled; for example, Harrison River Chinook salmon are well known for their white flesh. Juvenile Chinook can be more difficult to identify than adults. Smolts that recently entered the ocean have parr marks (long vertical bars) located fairly evenly above and below the lateral line, and the gums are black. Experienced fishermen can distinguish older juveniles by their shape, black gums and slippery skin. When doubt remains, one can look at the otolith, which is always about two times larger in Chinook salmon than in coho salmon of the same size.

LIFE HISTORY

Chinook salmon, which are also commonly known as "spring salmon" or "king salmon," are the least abundant of the five major species of Pacific salmon. Chinook occur from the Sacramento River in California to Kotzebue Island in Alaska, and occasionally in the Mackenzie River in the Arctic. They are not abundant in Asia, where commercial catches account for only about 10 percent of the total commercial catches by all countries. They are rare in northern Japan but more common in Russia in the Kamchatka Peninsula, with some occurring as far north as the Anadyr River. At sea, Chinook are found deeper than other salmon, at about 40 to 70 m, and as deep as 150 m, particularly during the winter. In North America, the major drainages producing Chinook are the Kuskokwim and Nushagak Rivers in Alaska, the Yukon River in Alaska and Canada, and the Fraser River in British Columbia. Chinook are the largest of all Pacific salmon, reaching a maximum size of 1.6 m and 61 kg. Catching a tyee (14 kg) around Campbell River is a mark of a true sportsman. In North America there has been a gradual decline in abundance that appears to have resulted from climate-related ecosystem changes. The size of Chinook salmon has also generally declined, which may result from genetic changes relating to the past catching of the larger individuals, from earlier maturation or from reduced ocean growth, or from a combination of all of these factors.

In the mid-1970s, the commercial and recreational catch of Chinook in the Strait of Georgia was about 800,000 fish. However, catches started to decline, and in the mid-1980s it was discovered that about 80 percent of the fish being produced in all rivers were being caught in fisheries, and this was too many. At the same time, the ocean survival of juveniles was declining, but this was not discovered until the early 1990s. There also were other problems, such as freshwater habitat loss and interceptions of Fraser River fish in Alaska, that affected abundances. Marine survival continues to decline and is generally about 1 percent, with fish from hatcheries having poorer survival than wild fish. In one study, only 0.1 percent of the fish from the Cowichan River Hatchery survived long enough to be caught in a fishery or return to spawn. Chinook salmon are still caught in the sport fishery, but the catches are now 20,000 to 30,000 a year, compared to about 800,000 a year in the early 1980s.

Most Chinook salmon return to spawn at age four or five, but some return at age three (jacks) or as old as age eight. Thus, the time in the ocean ranges from one to five years. Fish about four years old are the most common in the Fraser River. Older fish tend to be females. Usually, the fish returning early to the Strait of Georgia enter the rivers in May and June, hence the common name "spring salmon." However, these fish spawn at about the same time as those that enter the river later in the season. Later-returning adults, or fall Chinook salmon, return from September to October. Most spawning occurs in the Fraser River between November and January.

In the Strait of Georgia, juvenile Chinook eat amphipods, decapods and euphausiids, but mostly small fish, of which the greater part are Pacific herring. Large Chinook are preyed upon by Pacific lamprey, seals, sea lions, white-sided dolphins and orcas. Smaller Chinook are also eaten by some species of rockfish, coho salmon, other Chinook, seals, mergansers, lingcod and lamprey.

ATLANTIC SALMON

Salmo salar VISITOR ▼

IN THE STRAIT OF GEORGIA

Atlantic salmon now occur in the Strait of Georgia as a result of fish escaping from salmon farms in the strait. Farming of Atlantic salmon started in Puget Sound in 1972 and in the Strait of Georgia in the mid-1980s. Atlantic salmon are rarely caught in the strait or associated rivers. There were 29 reports from the Fraser River between 1991 and 1998 and two from 2000 to 2007. We did not catch any Atlantic salmon in over 3,000 scientific pelagic trawl surveys in the Strait of Georgia from the mid-1990s to the present. Atlantic salmon are occasionally reported being caught in the recreational fishery in the strait.

DESCRIPTION

The large black spots on the gill cover (operculum) readily distinguish Atlantic salmon from Pacific salmon. Atlantic salmon also have small heads with the upper jaw not extending beyond the back of the eye. They have 12 or fewer anal fin rays, while Pacific salmon have more than 12.

LIFE HISTORY

There are no recent reports of escaped Atlantic salmon successfully reproducing in rivers flowing into the strait. Similarly, from 1905 to 1935, approximately 275,000 Atlantic salmon were introduced into rivers in British Columbia, but there was no evidence that these fish reproduced and colonized any of the rivers. There is also no evidence that the escaped Atlantic salmon have successfully hybridized with Pacific salmon. Laboratory attempts to produce viable hybrids have also been unsuccessful. Thus, it appears that a small percentage of the escaped Atlantic salmon can survive in the Strait of Georgia and enter rivers such as the Cowichan, Quinsam and Fraser, but the reports in rivers are rare.

Alexander Raths / Adobe Stock

BROWN TROUT

Salmo trutta VISITOR ▼

IN THE STRAIT OF GEORGIA

Brown trout occur occasionally in the Cowichan estuary and along the near-shore areas around Courtenay. They are freshwater fish that move into the tidal areas of rivers down to the estuary. They have not been caught in any of the several thousand pelagic research trawl sets in the Strait of Georgia that began in the mid-1990s.

DESCRIPTION

Brown trout are a golden-yellow colour with large dark spots that can be red or black and can be surrounded by pale halos. There are no teeth on the back of the tongue and the tail is almost square.

Daniel Thornberg / Adobe Stock

LIFE HISTORY

Brown trout mature in their third or fourth year and can live for about 20 years, though the maximum age for the populations around the Strait of Georgia may be about 10 years. Brown trout were originally introduced into British Columbia as eggs in 1932–34 from Wisconsin and Montana. The hatched young were put into the Cowichan and Little Qualicum Rivers to provide a fish for summer fishing. They are major predators of other fishes and feed on juvenile Pacific salmon in the Cowichan River, including both hatchery and wild Chinook juveniles.

BULL TROUT

Salvelinus confluentus

VISITOR ▼

IN THE STRAIT OF GEORGIA

Bull trout are occasionally found in Burrard Inlet and Howe Sound, often in the spring when young Pacific salmon are entering the ocean. At this time of the year, anglers will target what they consider to be sea-run populations, particularly in the Squamish River area. No bull trout were identified in any of the several thousand research pelagic trawls made in the Strait of Georgia beginning in the mid-1990s, indicating that the sea-run individuals remain in the area around the river mouth and probably for a limited time.

Creative Commons

DESCRIPTION

The bull trout is difficult to distinguish from the Dolly Varden. Bull trout have a total number of branchiostegal rays (counting both sides) between 26 and 28, while Dolly Varden have 21 to 23. Bull trout also have more robust gill rakers that have prominent teeth on the medial margin. Dolly Varden gill rakers are longer, less robust and flexible, with no teeth on the medial margin. Bull trout have prominent oval white to pink and red to pink spots on their sides, but the sea-run life-history types are more silver and the spots may be difficult to see. The pelvic and anal fins have a white anterior margin.

LIFE HISTORY

These are large fish that can grow to 55 to 85 cm. A typical size is about 40 to 70 cm and 0.5 to 2.5 kg. They can live for at least 24 years and mature at age 3 to 4. The distribution is not clear because of the difficulty in identification, but one report places them between about 41°N and 61°N on the west coast of North America. Migration into the ocean first occurs at age 3 to 6, and the individuals that migrate are larger than those of the same age that remain in fresh water. It is known that the sea-run form is a distinct life-history type; however, very little is known about its life history and behaviour. A fish tagged in the Squamish River was recovered about 150 km away in the lower Skagit River in Washington State. It is also known that individuals tagged above Pitt Lake are caught in the Fraser River delta area. On the west side of the Olympic Peninsula in the United States, studies have shown that there are mature, anadromous bull trout that migrate into rivers to spawn and then return to the ocean in the winter to feed. The large mouth allows the fish to capture prey up to 50 percent of their own length. Bull trout are voracious predators, as well as major competitors for species such as juvenile Pacific salmon.

DOLLY VARDEN

Salvelinus malma VISITOR ▼

IN THE STRAIT OF GEORGIA

Dolly Varden have been found in the Squamish, Cowichan, Courtenay and Campbell River estuaries. However, there have been no catches in several thousand pelagic research trawl sets in the Strait of Georgia since the mid-1990s, indicating that Dolly Varden remain close to the mouths of their natal rivers and are not found in the open waters of the strait.

DESCRIPTION

Dolly Varden in the ocean have light-orange spots on the side and can be difficult to distinguish from bull trout, as anglers fishing the Fraser River sandbars know. Known as "dollies," they can be distinguished from bull trout because they have a smaller number of branchiostegal rays, ranging from 21 to 23, and their gill rakers have no teeth on the medial margin. As they approach spawning, they become brightly coloured with bright-orange and black fins, with a prominent white leading edge.

tab62 / Adobe Stock

LIFE HISTORY

Dolly Varden have three life-history types: a river-resident form, a lake-resident form and a form that enters the ocean (an anadromous form). The common name "Dolly Varden" comes from a character in Charles Dickens's novel *Barnaby Rudge*. Dolly Varden occur from Puget Sound to the Aleutian Islands, in the Bering Sea, and from the Arctic Ocean to the Mackenzie River. On the western side of the Pacific, they occur along the coast of the Kamchatka Peninsula and south to Japan and Korea. Juveniles that enter the ocean remain in the river for about three years and are known as "parr." After about three to four years in the river, juveniles become smolts and enter the ocean, but return to the river for the winter. Most are mature by age five and at a length of about 25 to 45 cm. The maximum age is about 10 years. The largest reported is about 1 m and 8.6 kg. In the spring they are common along the beaches of southeast Alaska, where they feed heavily on the juvenile salmon that are entering the ocean.

Hatchetfishes, Dragonfishes, Lancetfishes and Barracudinas

Fishes in the order Stomiiformes, with families Sternoptychidae (hatchetfishes) and Stomiidae (dragonfishes), and in the order Aulopiformes, with families Alepisauridae (lancetfishes) and Paralepididae (barracudinas), have a worldwide distribution and are deeper-water residents that occasionally find their way to the surface. All are extremely rare in the Strait of Georgia, with records of only one species from each family. They probably are carried into the strait in the deeper, inflowing currents in Juan de Fuca Strait. Hatchetfishes have a hatchet-shaped, flat, thin body with prominent light-producing organs (photophores). There is no mistaking these species, and there are about 70 marine species of hatchetfish worldwide. Dragonfishes are represented by the Pacific viperfish, which also would never be confused with any other species in the Strait of Georgia. About 270 species of dragonfish occur throughout the world's oceans. Lancetfishes have a huge dorsal fin and large, prominent teeth. They are common in open areas of the world's oceans and an occasional visitor close to shore. They are a predator of Pacific salmon. There are about 50 to 60 species of lancetfish in the world, and some are known for their vertical swimming behaviour. Barracudinas are another worldwide-distributed species at depths from 50 to 2,000 m.

LOWCREST HATCHETFISH

Argyropelecus sladeni VISITOR ▼

IN THE STRAIT OF GEORGIA

No specimens of the lowcrest hatchetfish have been reported in the Strait of Georgia in over 50 years, and only two specimens have ever been recorded. Both were captured near the mouth of the Fraser River, and it is likely they were transported in the bottom current in Juan de Fuca Strait that flows into the Strait of Georgia. We have completed several thousand pelagic research trawls since the mid-1990s without catching any of these fish.

DESCRIPTION

This is a small, laterally compressed fish with a maximum length of 6.7 cm. The body is silvery and the mouth is nearly vertical. The eyes are telescopic, on tubes that are in grooves. The abdomen has bony plates with numerous light-producing photophores. There is no mistaking this fish once you have seen an image.

LIFE HISTORY

The species is commonly found throughout the Pacific and Atlantic Oceans, ranging from 55 to 600 m and migrating toward the surface at night. A different species of hatchetfish that lives in fresh water is commonly found in tropical fish stores.

Jeff Dubose

PACIFIC VIPERFISH

Chauliodus macouni VISITOR ▼

IN THE STRAIT OF GEORGIA

Only three Pacific viperfish have been reported in the Strait of Georgia. Two were caught over 60 years ago near the mouth of the Fraser River, and one was caught in a midwater trawl in June 1975 in the deep water in the middle of Malaspina Strait. No other specimens have been found, despite the extensive sampling over the past 25 years using research pelagic trawls and bottom trawls. Thus, the Pacific viperfish is another example of a species exotic to the Strait of Georgia that is occasionally transported into the strait.

DESCRIPTION

The Pacific viperfish is a most unusual and ferocious-looking fish. Its mouth, eyes, and overlapping needle-like teeth are all large. The first dorsal fin ray is greatly extended and dangles over the front of the head, as seen in some angler fishes. It is speculated that the dangling may attract prey, and fish remains have been found in their stomachs. This is another species that is readily identifiable once an image is seen.

LIFE HISTORY

The Pacific viperfish is one of 273 species called dragonfish that are found in the deeper waters of the world's oceans. Pacific viperfish are found from 25 to 5,000 m, with some migrating to within 200 m of the surface at night. They may be green or silver, but most specimens that are landed on the deck of a ship are black. The maximum length is 27 cm, but the maximum age and age at maturity remain to be identified. There are photophores along the body, as its normal habitat would only have biological light. It is quite amazing how biological light from numerous organisms actually lights up the deep ocean.

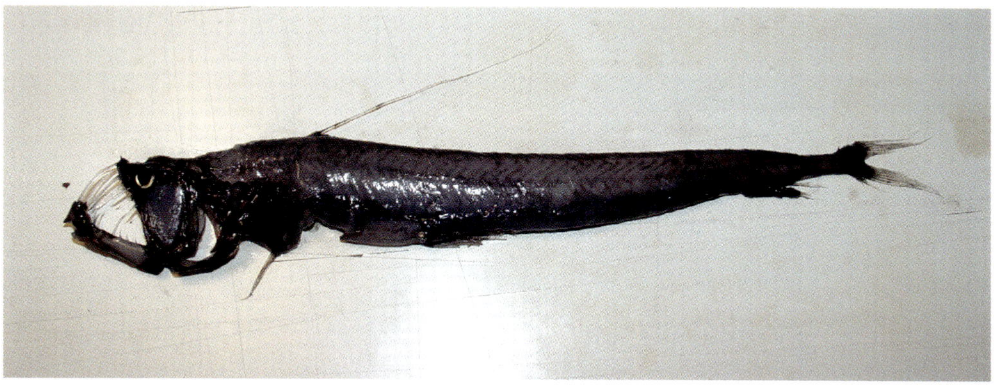

TINRO

LONGNOSE LANCETFISH

Alepisaurus ferox VISITOR ▼

IN THE STRAIT OF GEORGIA

There is one record of the longnose lancetfish in the Strait of Georgia, but nothing in over 50 years. We have no catches in any of several thousand pelagic research trawl sets in the Strait of Georgia since the mid-1990s. If spotted by an observant beachcomber, it is worth reporting to a local newspaper.

DESCRIPTION

There is no mistaking this fish, and if found, it would most likely be washed up on a beach. This is a skinny fish up to 2.3 m long with a silvery colour and an exceptionally high, sail-like dorsal fin. The longest rays in the dorsal fin can be about half the length of the body. The body is flaccid and translucent or iridescent with loose outer skin. There are large dagger-like teeth on the upper jaw, and a few on the lower jaw. Because the flesh is delicate, specimens washed up on a beach would not be in good shape, but the dagger-like teeth would most likely still be present in the head and the sail-like dorsal fin would be evident.

Adobe Stock

LIFE HISTORY

The longnose lancetfish is found in all warm ocean areas but not in polar regions. It is reported as a deep-sea fish, down to 1,830 m; however, it is caught on the surface in commercial hook-and-line fisheries. It is a carnivorous species but not a strong swimmer, so it must ambush its prey. There are studies that identify attacks on Pacific salmon based on the salmon's wounds, but it's the North Pacific daggertooth (*Anotopterus nikparini*), not longnose lancetfish, that appears to be the major cause of the backward-slanting slash marks found on one side at the posterior of some salmon.

TINRO

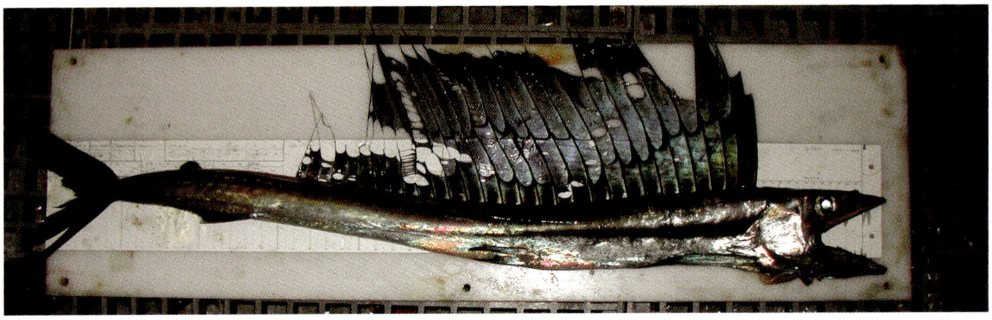

WHITE BARRACUDINA

Arctozenus risso VISITOR ▼

IN THE STRAIT OF GEORGIA

Several white barracudinas have been recorded in the Strait of Georgia. Two specimens have been captured in the research pelagic trawls from the mid-1990s to the present. One was caught at a depth of 300 m south of Texada Island in September 2000 with a length of 26.7 cm, and the other at a depth of 375 m in Malaspina Strait in September 2011 with a length of 25.2 cm. There is one record in the groundfish research database at the Pacific Biological Station from deep water in the central strait in August 2000. There is another record in the same database of the slender barracudina (*Lestidiops ringens*) from deep water in the southern strait in March 2002. The two species of barracudina should be readily distinguishable by the position of the pelvic fin; we are listing only the white barracudina but recognize that this second species, which is abundant offshore, could also be an occasional visitor to the strait.

DESCRIPTION

The white barracudina is a long, slender, bright-silver fish that can occur singly or in schools. It is an excellent swimmer and probably orients itself vertically in the water column, where it darts up and down in search of prey. It is distinguished by the pectoral fin being shorter than the anal fin, a small dorsal fin well back on the body just anterior to the insertion of the pelvic fin, and the teeth on the lower jaw that have smooth edges. The scales on the body are easily lost when the fish is captured, leaving obvious scale pockets. The pelvic fin

Alexei Orlov

is mostly behind the base of the dorsal fin, and there is a black area at the base of the anterior rays of the anal fin. There is also a band of dark pigment dorsally.

LIFE HISTORY

The common name used by the American Fisheries Society and recent publications is "white barracudina," but the species is also reported as the "spotted" or "ribbon" barracudina. Although the species is not abundant, it is frequently reported in fish surveys throughout the world, including in the Canadian Arctic. It is most unusual for any species to have such a broad range in its distribution. The maximum length of the white barracudina is reported to be 31 cm, though there is a record of 33 cm. There are estimates that the fish is about 5 years old at 23 cm and has a possible maximum age of 11 years, but there are no detailed published studies. It is primarily a deepwater species that occurs mostly from about 200 to 2,200 m throughout the world, from the Arctic to the Antarctic. The white barracudina feeds on crustaceans and fishes. Throughout the distribution, it can be a prey item for some fish species and a major prey item for squid. There are records from the stomachs of Chinook salmon. In the North Atlantic, it is eaten by Atlantic cod, redfish, seals and swordfish. It is known that some juveniles migrate close to the surface at night, which would make them available as prey to animals such as seals. Very little is known about the spawning behaviour of the white barracudina other than that it spawns in the midwater and the larvae are pelagic. It is probable that the specimens in the Strait of Georgia were originally on a foraging migration from offshore and ended up in the strait by accident.

Lanternfishes

Lanternfishes are a diverse group of fishes in the family Myctophidae and commonly called myctophids. The common name "lanternfish" comes from their light-producing organs, or photophores, which have a light gland, reflecting tissue and a lens. Much of the identification of myctophids depends on the presence and arrangement of the photophores. In the open ocean, lanternfish live at depths of about 75 m to 1,000 m during the day, but many migrate right to the surface at night to feed on plankton. Myctophids are abundant and include a large number of species worldwide. They are extremely important in the ocean as a form of prey, and because they move nutrients from the surface into deeper waters and back again. The high fat content of some myctophids makes them a source of high-quality oil of commercial interest in some areas.

We list four species of myctophids in the Strait of Georgia. The recent book *Fishes of the Salish Sea* lists the California flashlightfish (*Protomyctophum crockeri*), which was found on a beach on Pender Island. We do not include this species as we have no listing of it in any of the research records that include examinations of thousands of myctophids.

CALIFORNIA HEADLIGHTFISH

Diaphus theta COMMON ●

IN THE STRAIT OF GEORGIA

California headlightfish are moderately common throughout the Strait of Georgia at depths below about 60 m during the day. This species was abundant in the deeper pelagic research trawl catches from the mid-1990s to the early 2000s. In those years it was always among the 10 most abundant pelagic species. However, beginning about 2005, the species became less abundant.

DESCRIPTION

The California headlightfish is easily recognized because of the large head that has a pair of large luminous photophores positioned between the eyes, like a headlight. There is a luminous organ below the eye and one above the base of the pectoral fin. There are no luminous glands dorsally or ventrally on the caudal peduncle. The California headlightfish is also distinguished from other lanternfish in the Strait of Georgia by the soft pink flesh. It is rare to catch specimens of this species with all their scales, but when fresh specimens are found, the small photophores can appear almost blue.

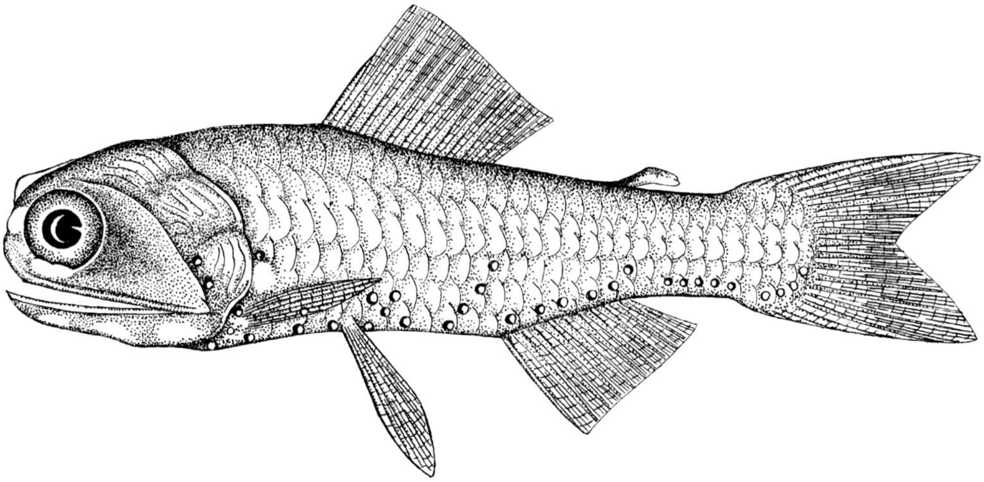

LIFE HISTORY

The species is found around the North and South Pacific from Japan, the Sea of Okhotsk and the southern Bering Sea to the Gulf of Alaska and south to Baja California and Chile. The largest specimens in the Strait of Georgia are about 80 to 90 mm. Dick Beamish used otoliths from specimens of this size to estimate an age range from four to seven years. In the open ocean, this species is commonly found between 300 m and 500 m during the day but can

Alexei Orlov

be as deep as 800 m. The larval stage is found in some areas from late March through to September, indicating that spawning occurs from late winter through to the summer. Larval fish will stay in the surface 50 m and migrate into deeper waters as they develop. This is a faster-growing species, and this enhanced growth is associated with the frequent nightly migrations into the surface waters to feed. The large mouth, big eye, large stomach and short intestine indicate that this is an active predator of larger plankton. The bigger, older fish also replace the gas in their swim bladders with fat, which helps maintain buoyancy. This fish is found in the stomachs of Chinook salmon in the Strait of Georgia and in the stomachs of a variety of fishes and marine mammals, such as dolphins, in the open ocean. This is another species in the midwater ecosystem of the Strait of Georgia that is virtually unstudied.

PINPOINT LAMPFISH

Lampanyctus regalis VISITOR ▼

IN THE STRAIT OF GEORGIA

The groundfish database at the Pacific Biological Station in Nanaimo includes two records of the pinpoint lampfish. One specimen was collected in deep water near Halibut Bank in September 2010, and two specimens were captured in Jervis Inlet in January 2010, also in deep water. The only other report is of a specimen collected off Point Roberts in 1968 and recorded in the book *Fishes of the Salish Sea*.

Alexei Orlov

DESCRIPTION

Pinpoint lampfish caught in trawls are dark and have a mostly scaleless body with prominent photophores. If the scales are not disturbed, the fish is silver laterally, darker dorsally and lighter ventrally. The location and number of the distinct photophores provide species-specific characters. In the Strait of Georgia, it is important to distinguish the pinpoint lampfish from the northern lampfish. The pinpoint lampfish has a much smaller eye and smaller photophores. The pectoral fin of the pinpoint lampfish extends to the base of the pelvic fin, compared to the northern lampfish's much shorter pectoral fin, which terminates well before the origin of the pelvic fin. The supracaudal and infracaudal luminous glands are large and prominent on the northern lampfish. There also are specific differences in the number and sizes of photophores. The pinpoint lampfish is one of the largest myctophids, reaching a length of 18.3 cm.

LIFE HISTORY

Pinpoint lampfish range from northern Japan to the Bering Sea and south to Baja California. It is relatively common off the west coast of British Columbia. Specimens in the Strait of Georgia would most likely be occasional visitors transported into the strait by deep currents. Like many myctophids, this fish will migrate to near the surface at night and has been found as deep as 1,500 m. Very little is known about the age at maturity or the maximum ages. Spawning is believed to occur early in the year. Myctophids commonly feed on small plankton.

NORTHERN LAMPFISH

Stenobrachius leucopsarus ABUNDANT ○

IN THE STRAIT OF GEORGIA

The northern lampfish is one of the most abundant species throughout the Strait of Georgia and North Pacific Ocean. It is the fourth most abundant pelagic species in our research-related pelagic trawl catches in the Strait of Georgia at depths below 75 m during the day.

DESCRIPTION

Northern lampfish are silvery, but most will have lost their scales when captured. They have a small pectoral fin and a thick caudal peduncle, which has luminous glands on its top and bottom. They are also distinguished by the pattern of photophores, but we look for the size of the luminous glands on the caudal peduncle as the distinguishing character.

TINRO

LIFE HISTORY

Northern lampfish are common off the coast of Japan, in the Sea of Okhotsk and north off Russia, and in the Bering Sea and south to Baja California. In the Strait of Georgia, juvenile northern lampfish are found in and below a layer of plankton that occurs at about 75 m during the day. Mature individuals occupy deeper water in the strait, but some will migrate to the surface very quickly after sunset. There appear to be vertical migratory and non-migratory groups, but it is not known if they are persistently distinct from each other. Fish mature at about 6 to 7 cm and age four. Dick Beamish aged fish in the 9 to 10 cm range as being four to five years old, and there are estimates that the maximum age is about eight years. Maximum lengths are reported to be from 11.2 to 12.7 cm. Spawning occurs in the winter in the Strait of Georgia and in winter and spring throughout this fish's distribution. Their large mouth, relatively few gill rakers, a well-developed stomach and short intestine identify the fish as a predator of fast-moving plankton such as copepods and euphausiids. Vertical migratory fish feed on euphausiids, copepods and amphipods. The northern lampfish is a common food item for walleye pollock, tuna, Pacific salmon, sharks, porpoises, whales and seabirds. In the Sea of Okhotsk, walleye pollock are estimated to eat 1,500 t of this fish each day in the summer. There is no commercial value, but one of the authors has eaten a few and they taste similar to smelt. Fishermen might find them in the stomachs of Chinook salmon. The northern lampfish is part of the midwater community of small fishes that are relatively short-lived and very poorly studied.

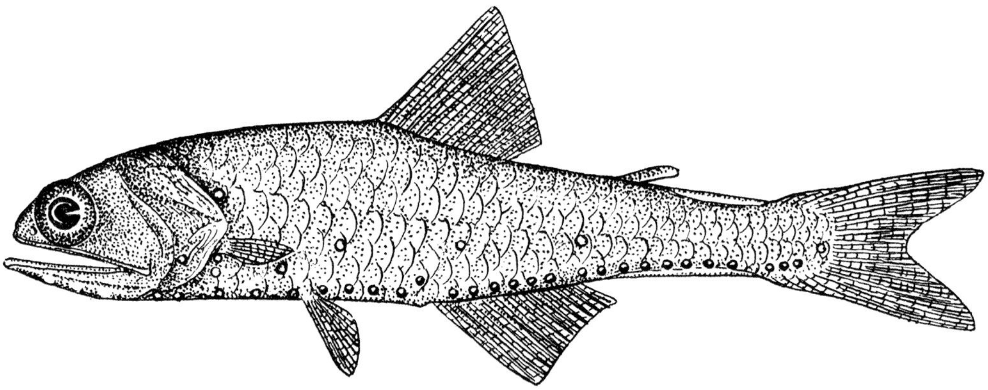

BLUE LANTERNFISH

Tarletonbeania crenularis RARE ★

IN THE STRAIT OF GEORGIA

The blue lanternfish is a pretty little lanternfish that is found throughout the Strait of Georgia, but it is rare. These fish are caught in our deepest sets in the strait. One or two dozen are routinely caught throughout the strait in a two-week trawl survey. The frequency and abundance have not changed over the past 20 years.

DESCRIPTION

The very narrow tail region (peduncle) easily distinguishes this fish from other species of myctophid. It has a longer body with no photophores above the angle of the operculum. When caught in a trawl net, it retains most of the scales, whereas the other species of myctophids in the Strait of Georgia lose most of their scales when captured in trawls. The wonderful scientific name originates from Tarleton Hoffman Bean, a well-known ichthyologist in the United States.

Bill Pennell

LIFE HISTORY

Blue lanternfish are one of the most abundant mesopelagic fishes in the North Pacific, and possibly one of the most abundant fishes in the entire subarctic Pacific. It is distributed from off the coast of Japan, in the Sea of Okhotsk, around the rim of the North Pacific and south to Mexico. Specimens we obtained in our trawl catches in the Strait of Georgia were 3 to 6 cm in length. Maximum lengths may be between 8 and 9 cm, though some reports indicate lengths up to about 12 cm. It is known that they live to be about three years old, but it is suspected they live longer. Offshore, where they are common, they undertake daily vertical migrations from very deep water during the day to near the surface at night. The blue lanternfish is known to spawn throughout the year, possibly with a peak in the winter. Very small specimens of about 2 cm in length have been captured in the strait, indicating that, though rare, they are reproducing there. Larger individuals feed on euphausiids and amphipods, and in turn are eaten by Pacific hake and Pacific salmon. They were a common prey of steelhead that were captured in a winter survey in the Gulf of Alaska in 2022.

Opahs and Ribbonfishes

Opah and king-of-the-salmon are two species that very rarely find their way into the Strait of Georgia. Both have distinctive sizes, shapes and colours and are readily identifiable. Opahs are in the family Lampridae, which has two species that occur from the surface to about 500 m throughout the world in warmer water. King-of-the-salmon are ribbonfish in the family Trachipteridae, which contains about 10 species that occur throughout the northern oceans.

OPAH

Lampris guttatus VISITOR ▼

IN THE STRAIT OF GEORGIA

The first and only record of an opah in the Strait of Georgia was from a relatively recent report in the groundfish database at the Pacific Biological Station in Nanaimo. The opah was caught in a commercial bottom trawl—which was probably targeting the English sole—at a depth of 106 m near Saturna Island in September 2006. The opah weighed 43 kg but was not kept. There would be no mistaking the identity, particularly of a fish this large, and the commercial fisherman had a history of reporting catches of unusual species.

DESCRIPTION

NOAA

Opahs have a deep, laterally compressed or oval body that is brightly coloured, as are their fins. Individuals that have not lost their scales have a solid golden-orange to deep-red body with a rosy belly. The body is covered with white spots and the snout, lips and fins are bright red. There are other colours reported on dead fish that have lost their easily displaced scales. One description is of an iridescent vermilion and blue dorsally, shading to silver and red ventrally, with numerous silver spots. The low dorsal fin has a higher, sickle-shaped anterior. The large and long pectoral fins have a slightly hooked posterior edge. Pelvic fins are also long and large with a slightly sickle-shaped posterior edge. The caudal peduncle is small, but the tail is large and slightly forked. The dorsal and anal fins can be depressed into grooves in the body. There is no other fish this large that looks like an opah.

LIFE HISTORY

The common name "opah" is used in most publications. However, there are now three recognized species of opah and a recent DNA study proposes five species. We continue to use the name "opah," but "spotted opah" may eventually be the correct common name. Opahs are also known as "moonfish." Opahs are found in temperate to tropical waters in oceans around the world. In the Pacific off the Americas, they occur from the Gulf of Alaska to Chile. They are rare off British Columbia and farther north. A maximum size of 1.83 m and 74.1 kg has been reported. A larger reported size of 270 kg is likely incorrect. In one commercial fishery off Hawaii, where opahs are caught incidentally, most fish were about 13 kg, with the largest being 52.6 kg, and had lengths of 35 cm to 1.63 m, with an average of 98 cm.

This is a pelagic species that is mostly found at 50 to 400 m well offshore but that ranges from the surface to 736 m. Opahs are considered to be solitary once they mature, but very little is known about their biology and ecology. There are no age estimates or any understanding of growth rates and ages at maturity. There are reports of spawning throughout the year, with the slender larva metamorphosing into the deep-bodied shape, but no estimate of the timing. Opahs appear to be superior predators, feeding on fish, squid and crustaceans. It is believed that they are efficient swimmers when flapping their pectoral fins, perhaps like penguins. They are famous for being able to maintain high body temperatures. The flesh is highly esteemed, but the solitary behaviour restricts landings to incidental catches in other fisheries.

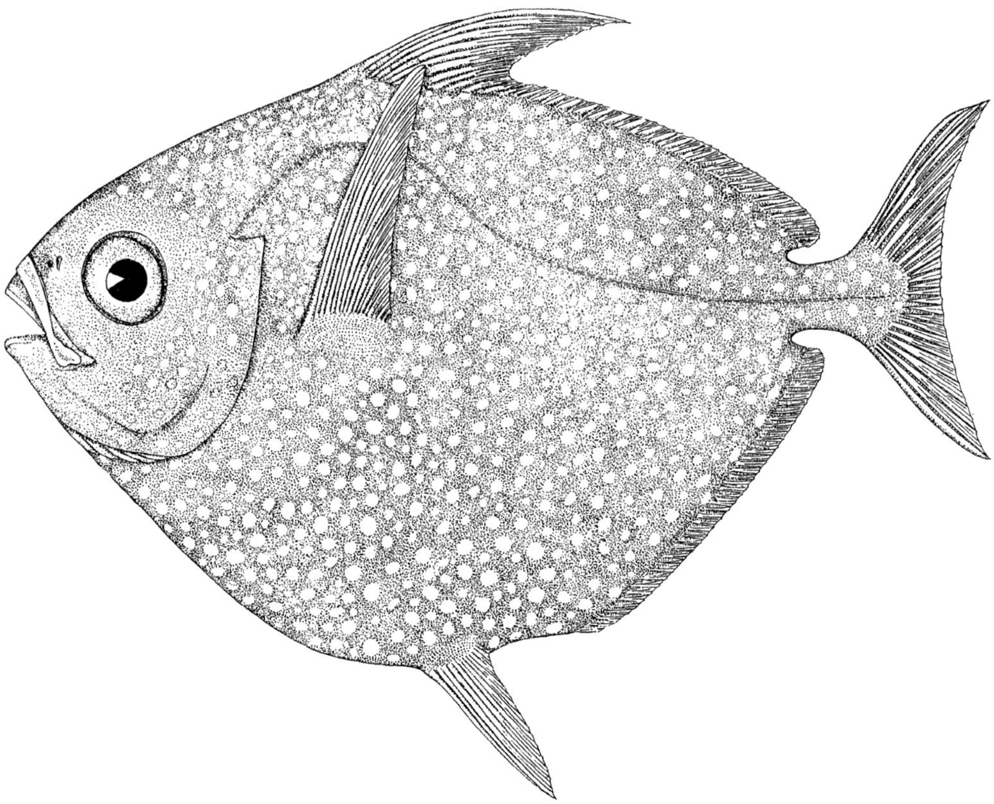

KING-OF-THE-SALMON

Trachipterus altivelis VISITOR ▼

IN THE STRAIT OF GEORGIA

One specimen of a king-of-the-salmon was recorded at a beach on Pender Island. A second record in September 2017 was by Ben Baker, who was walking his dog on Oak Bay's Rattenbury Beach and observed a washed-up, 2 m king-of-the-salmon. Two other records are from a beach in Tsawwassen in January 2021 and from a commercial fishery for Pacific hake in the central strait on March 2020; the latter is recorded in the groundfish database at the Pacific Biological Station in Nanaimo.

Tara Logan

DESCRIPTION

There is no mistaking this species, which is long, thin and ribbon-shaped, with the dorsal fin along the length of the body. It has large eyes, a small mouth, crimson-red fins and a rather delicate silver body.

LIFE HISTORY

King-of-the-salmon is a mesopelagic fish, meaning that it is found at mid-depths in the open ocean. It occurs in the eastern Pacific from the Bering Sea to Chile and out to the Hawaiian Islands at depths from the surface to 900 m. The maximum length is reported to be about 1.8 m. Very little is known about the biology except that the species feeds on plankton and small fishes and spawns throughout the year. It is prey for larger fish such as sharks, swordfish and even longnose lancetfish. The name comes from a Makah legend that holds that this fish brings the Pacific salmon back to their spawning rivers annually. Catching this fish was prohibited, as it was believed that doing so would stop the salmon return.

and Cods

Grenadiers, hakes and cods are in the order Gadiformes. Grenadiers are in the family Macrouridae; fish in this family are commonly called rattails because of the short body that tapers to a long, thin tail. There are about 300 to 350 species of grenadiers that commonly occur throughout the world in deeper waters, from about 200 to 2,000 m. The deepest record appears to be 5,825 m. Grenadiers are one of the dominant groups of fishes in the deep ocean, and the giant grenadiers are one of the most abundant.

Cods and hakes contribute about 25 percent of the world's commercial catch. There are two families represented in the Strait of Georgia, Merlucciidae (hakes) and Gadidae (cods). Merluccid hakes were recently placed in their own family, with one genus and between 13 and 22 species worldwide. Hakes are abundant along the continental shelves in the Atlantic and Pacific. They are schooling fishes that are found deeper than about 50 m. The one species in the Strait of Georgia was not considered to be abundant until the mid-1970s. Cods are some of the most recognized species because of their dominance in worldwide commercial fisheries. The Atlantic cod fishery is well known, but even before the collapse in the late 1980s, Pacific cod catches were approaching the equivalent of 50 percent of the Atlantic's. There are about 30 species of cod worldwide, and all but one are marine residents on continental shelves in northern areas, including the Arctic. Atlantic cod are the largest, with a record length of 1.83 m. Arctic cod dominate in the Arctic food web and have an amazingly short life span of only a few years. Walleye pollock support one of the world's largest fisheries in the subarctic Pacific. There are three species of cod in the Strait of Georgia: Pacific cod, walleye pollock and Pacific tomcod.

GIANT GRENADIER

Coryphaenoides pectoralis VISITOR ▼

IN THE STRAIT OF GEORGIA

There are no published reports of a giant grenadier—or any grenadiers, for that matter—in the Strait of Georgia. There are two records in the groundfish database at the Pacific Biological Station in Nanaimo from two experienced skippers with a history of reporting catches of unusual species. The first specimen was caught in a bottom trawl in January 2006 near Burrard Inlet at 94 m. The second record, in February 2013, was also caught in a bottom trawl, near Pender Island at 98 m. There are five more records in the database of catches from a commercial rockfish fishery just north of the Strait of Georgia in Queen Charlotte Strait.

It is possible that the species in the groundfish database may be another species of grenadier, as the specimens were not retained and as distinguishing characters among grenadiers may be subtle, such as the length of the barbel, the position of the anus and the length of the space between the two dorsal fins. However, because giant grenadiers are a common and abundant species in the eastern North Pacific, the two specimens may have been correctly identified.

NOAA

DESCRIPTION

Giant grenadiers are a pale- to grey-brown on the body, with a darker head, fins and lateral line. The mouth and gill cavity are black. Scales are easily lost when the species is caught in trawls, resulting in fish with a pale appearance. The head is large and the body tapers to a long, pointed tail, resulting in the common family name "rattails." The large mouth is subterminal and the barbel on the chin is short. The first dorsal fin and pelvic fins are small. The scales have a dark posterior border and are covered with short spinules, particularly on smaller specimens. The space between the two dorsal fins is smaller than the length of the first dorsal fin base.

LIFE HISTORY

Giant grenadiers occur in the Sea of Okhotsk, off Japan, in the Bering Sea and south to northern Baja California, Mexico. They are very abundant in deeper water, and of commercial importance in a few areas despite their very soft flesh. They are on or near a variety of bottom types in deep water, ranging from 140 to 3,021 m, with most at depths from 300 m to 1,500 m. Despite their abundance, relatively little is understood about their life history. A maximum age of 58 years has been reported, and their average age is about 30 years. This is the largest of all grenadiers, apparently growing to at least 2.2 m and 41.8 kg. Females grow larger than males and occur in shallower water. Off Russia, the maximum size for females is 1.35 m and 11.8 kg, and 97 cm and 3.6 kg for males. Spawning occurs in batches throughout the year. Young, immature fish are found well off the bottom. In the Gulf of Alaska grenadiers are abundant deeper than 400 m and often caught in sablefish (black cod) fisheries. They feed on a variety of organisms, from fishes to invertebrates, and are prey for halibut, sharks, sablefish and sperm whales.

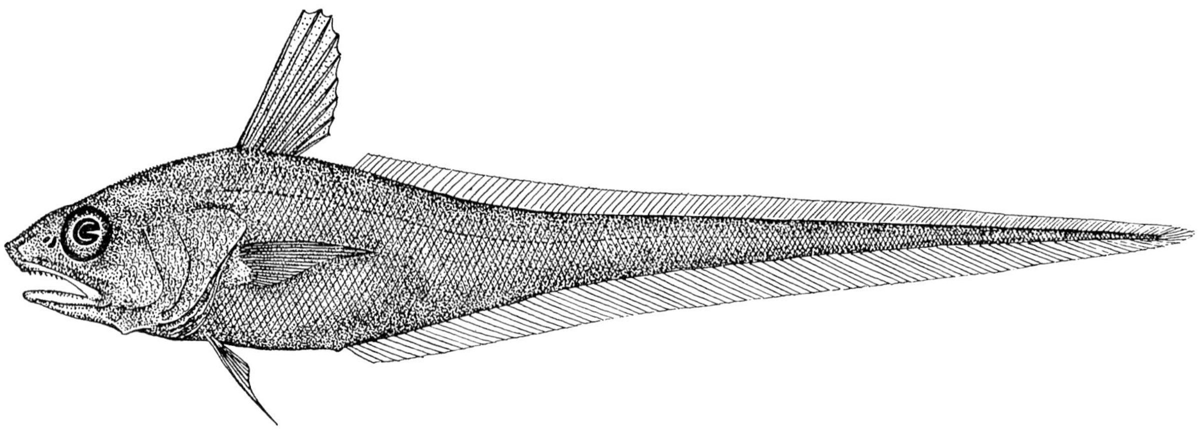

PACIFIC HAKE

Merluccius productus ABUNDANT ○

IN THE STRAIT OF GEORGIA

Pacific hake are one of the dominant resident species in the Strait of Georgia due to the abundance and weight of the population. Pacific hake are distributed throughout the deeper waters of the Strait of Georgia, including the inlets and Gulf Islands. They are a different population from the offshore population. The absence of a parasite commonly found in the offshore population proved that the strait's Pacific hake are distinct. Importantly, it also proved that the two populations became separated a long time ago. Groundfish fisheries in the strait in the 1940s and 1950s did not record Pacific hake as a major species in their catches, which may also indicate that the current abundances are recent. In the late 1990s, the number of Pacific hake in the strait was thought to range from about 50 million to 90 million.

DESCRIPTION

Maturing and mature Pacific hake are elongate, silvery dorsally and whitish ventrally. They are distinguished from walleye pollock and Pacific cod by their prominent and sharp teeth, their silvery colour and the continuous anal fin rather than two separate anal fins as in walleye pollock and Pacific cod. Larval Pacific hake in the Strait of Georgia have conspicuous dark pectoral fins that contrast with the clear pectoral fins on the larvae from the outer coast. Anyone catching a Pacific hake needs to be aware of their sharp, pointed teeth, which can quickly open the skin on a careless finger. Authorities in the United States changed the name to "Pacific whiting" as they thought this name would be more appealing to the public. There are publications claiming that "Pacific whiting" is the correct, scientifically accepted common name, but it is not.

CANFISCO

LIFE HISTORY

The coastal population of Pacific hake ranges from Southern California to southeast Alaska. Because the offshore population is distinct from the Strait of Georgia population, our focus is on the Strait of Georgia's. Pacific hake spawn from February to the end of April in the deep waters of the central Strait of Georgia. There is a small stock of much larger hake that spawns in the Gulf Islands off Yellow Point about four to six months earlier. In the central strait,

A research trawl catch of Pacific hake in the Strait of Georgia, showing the large abundance.
Dick Beamish

the major spawning aggregation is in the vicinity of Halibut Bank at about 150 to 350 m. It appears that fish that are ready to spawn will move deeper in the water column. After spawning, the hake move closer to shore while remaining in deeper waters. Over the summer they migrate north. Adult Pacific hake are less abundant in the strait in late fall as they move into the northern passages and inlets. In early winter, hake reappear and begin to migrate down Malaspina Strait and Sabine Channel toward the central strait in preparation for spawning.

Female Pacific hake do not mature all of the eggs in the ovary. The proportion of eggs that mature could be as low as 50 percent. This means there is always a capacity to produce more eggs and larvae if there is sufficient food for the adults to turn into the energy needed to allow the ova to develop into eggs that can be fertilized. Pacific hake eggs are pelagic and commonly found in the plankton at 170 to 220 m depth in early April. Larval Pacific hake are common throughout the midwater in April and May in the Strait of Georgia. They start their feeding on the eggs and early stages of common types of plankton. Older larvae eventually shift their distribution from surface waters to about 70 m, where they develop into juveniles. The same species of plankton are commonly eaten by many other species, including Pacific herring and Pacific salmon. Pacific hake, therefore, are a key species in the Strait of Georgia ecosystem. Someday we will understand how the ecosystem functions and how hake affect

NOAA

the ability of young fish of other species to survive. However, for that day to arrive there need to be more biologists on the water and looking through microscopes.

Otoliths, or ear bones, are used to estimate the age of Pacific hake, and it is relatively easy to determine ages for the younger fish up to about age four or five. Pacific hake in the Strait of Georgia experience reduced growth after about age five, resulting in difficulty determining the age of older fish. The largest Pacific hake we captured in the strait was 78 cm and 14 years old. Offshore hake on the west coast of Vancouver Island are easier to age. The oldest fish off the west coast that has been aged is about 23 years, but some could be older. Most fish in the commercial fishery offshore are 4 to 11 years old. In the strait, by age three 50 percent are mature, and at age four virtually all fish are mature. Pacific hake in the Strait of Georgia used to prey on larval and juvenile Pacific hake, as well as other fish such as the northern smoothtongue, young Pacific herring and young walleye pollock. However, when Pacific hake decreased in size in the 1990s, it was observed that most fed almost exclusively on plankton. Their sharp teeth caused biologists to wonder if Pacific hake were a major predator of juvenile Pacific salmon. After looking at thousands of stomachs' contents, they have found only one or two salmon, clearly showing that Pacific hake prefer other prey.

A commercial fishery started in 1979 in the Strait of Georgia, reaching a maximum of about 12,000 t in 1995. Landings have since declined. They are caught by recreational anglers in Puget Sound, but not intentionally in the Strait of Georgia. Most would not consider eating a Pacific hake that might be caught while fishing for Chinook salmon even though it is an ingredient in some common processed-fish products. Fresh Pacific hake is actually delicate and tastes good.

Predators of Pacific hake include lingcod, large rockfish, Pacific spiny dogfish and any larger fish that would prey on the juveniles. Seals are also a major predator of Pacific hake, which is an important item in the seal diet. One thing appears clear: hake species are one of the most successful groups of fish all over the planet, which means that the dominance that Pacific hake presently have in the Strait of Georgia is unlikely to change in the near future.

WALLEYE POLLOCK

Gadus chalcogrammus ABUNDANT

IN THE STRAIT OF GEORGIA

Walleye pollock are abundant throughout the midwater of the Strait of Georgia but may be more abundant in the northern areas around Texada Island, particularly in Sabine Channel. Adults occur in the midwater and on bottom, particularly in the Hornby and Denman Islands area, and in inlets such as Jervis Inlet. In midwater, adults occur in the same area as Pacific hake and Pacific spiny dogfish, except they are found more toward the top of a plankton layer at about 75 m to 150 m. Juveniles are more abundant at the top of the plankton layer and also abundant in the Gulf Islands.

DESCRIPTION

Walleye pollock are elongate, mottled green and brown dorsally and whiter ventrally. The mottled olive-brown blotches occur in imperfect rows. There are three separate dorsal fins and two separate anal fins. There may be a very short barbel on the chin. The flesh is moderately firm and the peritoneum is dark. Scuba divers distinguish small juvenile walleye pollock by the fish's silvery eyes.

LIFE HISTORY

Adult walleye pollock in the Strait of Georgia are generally 40 cm in length and three to four years old. An occasional female will reach lengths of about 65 cm and eight years old. There are small abundances of five-year-olds and an occasional six-year-old. Virtually all of the older and larger fish are females. However, the ages of walleye pollock in the Strait of Georgia are not easy to estimate after they mature. Walleye pollock in the Gulf of Alaska are generally thought to have a maximum age of about 28 years; however, Dick Beamish has aged pollock from the Bering Sea to 33 years old. Thus, it is possible that the maximum age estimates are not correct for older fish in the Strait of Georgia. The maximum length of walleye pollock in the Gulf of Alaska is 105 cm for females and 90 cm for males. There is a commercial fishery for pollock in the Strait of Georgia, but in recent years the catch has been very small.

Walleye pollock spawn in the deep water in the central Strait of Georgia in March, near Halibut Bank. Most fish spawn at age three, and by age four virtually all are mature. Males and females are not all mature at the same time during the spawning period. Males generally are in spawning condition first and appear to wait for females to mature. There are no observations, but catch

The colour pattern of a walleye pollock is some help in identification, but the real key to distinguishing it from a Pacific cod is the lack of a long barbel on the tip of the lower jaw, along with the precise size and location of the three dorsal fins relative to the two anal fins. Danny Kent

data suggest that when females are ready to spawn, they move deeper in the water column, probably attracting a number of males. Spawning appears to end by early April, after which adults disperse and the pelagic eggs mix with the eggs of Pacific hake and other species. In the Strait of Georgia, eggs float at depths of about 150 to 250 m. In the Bering Sea, a 75 to 80 cm female produces two million eggs. The young pollock in the strait migrate inshore, where they develop near the bottom and in midwater. The young age classes are mostly separate, probably because walleye pollock can be cannibalistic. Euphausiids are the major food item, but juveniles and adults also eat amphipods, glass shrimp, northern smoothtongue, Pacific hake, Pacific herring, eulachon and myctophids. Seals eat walleye pollock and so do the parasitic Pacific lamprey.

Walleye pollock are a species of cod, and the most abundant commercially fished species in the northern Pacific. This species is also one of the world's largest suppliers of edible fish. In recent years the total annual landings by all countries averaged 2.8 million tonnes. Sometimes called "Alaska pollock," walleye pollock are distributed from California around the Pacific Rim to the coasts of Korea and Japan. Much of the pollock eaten in North America is processed for fast food outlets. In Alaska, the annual catch is worth about $1.5 billion USD. The fishery in the Bering Sea is generally well managed, with fluctuations in abundance resulting from changes in climate and ocean conditions. There is a fish on the east coast of North America that is also called a pollock, but it is a completely different species and not even in the cod family.

PACIFIC COD

Gadus macrocephalus ABUNDANT

IN THE STRAIT OF GEORGIA

Pacific cod occur throughout the Strait of Georgia, including in the Gulf Islands and major inlets. They are genetically distinct from Pacific cod off the west coast of Vancouver Island and into the Bering Sea, indicating they have been isolated from the offshore populations for thousands of years. This is interesting, as tagging studies in the 1950s showed that of the almost 5,000 fish tagged, about 97 percent remained inside the strait, but 3 percent migrated offshore, where genetic analysis indicated the Pacific cod were distinct. This suggests there was some mixing of populations offshore but not inshore. Within the strait, there are two separate stocks, but they are not genetically distinct. One stock is in the northern central area on the west side of the strait, and another in the south and in the Gulf Islands. Tagging many years ago showed there was some exchange of Pacific cod between the Strait of Georgia and Puget Sound.

Andy Murch

DESCRIPTION

Pacific cod are easily recognized by their large head, three dorsal fins and brown to yellow spotted body. A closer look will reveal a long barbel that hangs from the chin. The length of this barbel can quickly distinguish the juveniles from some similar-looking walleye pollock, tomcod and Pacific hake.

LIFE HISTORY

Pacific cod are commonly found in the coastal areas of the North Pacific. They occur from California to the Bering Sea, and westward to the waters off the coast of Korea and the Yellow Sea off China. Pacific cod are most abundant in the Bering Sea, the Gulf of Alaska and the Sea of Okhotsk. In recent years the total commercial catch of Pacific cod by all countries has been equivalent to about 40 percent of the total catch of Atlantic cod. The Pacific cod is related to the Atlantic cod but does not live as long or grow as large.

The Pacific cod fishery is one of the oldest fisheries in British Columbia, having begun in the mid-1800s. The bottom trawl (or "otter trawl") fishery started about 1912 close to Vancouver. At that time, fishermen commented that the fish were plentiful. Up to about the mid-1950s the catches of Pacific cod in the strait represented about one-third to one-quarter of all species in the annual bottom trawl fishery in British Columbia. From 1945 to 1977, Pacific cod were about 50 percent of the total catch of all bottom fish. In some fish shops, Pacific cod is sold as "true" or "grey" cod. These names became necessary because the first settlers on the West Coast added the name "cod" to many of the species they found, such as rock cod (rockfish) or black cod (sablefish). Females may grow to about 1.45 m, but these larger fish are no longer found in the Strait of Georgia and would be rare in offshore populations. Males are smaller, with a maximum length of about 1.35 m. There are estimates of a maximum age of 25 years, but it is probable the maximum age is about 15 years.

Pacific cod spawn from January to March, beginning when they are about 50 to 55 cm in length and about two years old. Spawning can occur in Nanoose Bay, in the area from Hornby Island to Cape Lazo, and in Swanson Channel. Females produce a large number of eggs, ranging from over one million to over four million, depending on the size and age of the female. Fish produce large numbers of eggs for the same reason that plants produce large numbers of seeds: their survival is strongly related to environmental conditions when the eggs hatch. Female Pacific cod release all their eggs at once; the eggs are mildly adhesive and stick to the bottom after being fertilized. Larval fish are found near the surface or in the plankton layers at midwater depths during the day. Young fish become bottom dwellers and grow quickly, feeding initially on euphausiids, shrimp and other invertebrates. As they increase in size, adult Pacific cod feed on a variety of fishes, including their own juveniles.

During the 1950s to 1970s, the majority of historical catches for the trawl fishery occurred in the Winchelsea Islands area, from Qualicum Bay to Northwest Bay off eastern Vancouver Island and in Nanoose Bay. Catches declined significantly throughout the 1970s, particularly in Nanoose Bay. Biologists think the declines were the result of natural factors, including temperature changes in the strait. The best temperature for eggs to hatch is between 3.5°C and 4°C. Eggs will hatch at temperatures between 2.5°C and 8.5°C, but fewer than half of the eggs will hatch at temperatures above 8°C to 9°C. In fact, some biologists have reported that adults will not spawn at temperatures above 10°C. Since about 1980 the bottom temperatures in the strait in January have consistently been warmer than 9°C, and every year since 1993 the temperatures in January, February and March have been warmer than 9°C. In about half of these years the temperature exceeded 9.5°C. These temperatures most likely would reduce or eliminate the survival of eggs. However, since about 2010 there has been a period of cooler bottom-water temperatures and Pacific cod are now more common. Most biologists think that the cooling period is temporary and the warming will continue, which indicates that Pacific cod might eventually disappear from the Strait of Georgia. If this were to occur, the loss would not be from fishing but from oceanographic effects brought on by global warming.

PACIFIC TOMCOD

Microgadus proximus ABUNDANT ⭘

IN THE STRAIT OF GEORGIA

Pacific tomcod are abundant throughout the Strait of Georgia. Young-of-the-year occur in the top 75 m in our research trawl catches, but adults are mostly found on or near the bottom. Juveniles are common in the Gulf Islands area, and also appear to be common in the major estuaries around the Strait of Georgia. Tomcod traditionally were abundant in Burrard Inlet and considered by the Squamish People to be the best eating of all cods.

DESCRIPTION

Pacific tomcod are distinguished by the three spineless dorsal fins and the anus in line with the first dorsal fin, but mainly by their smaller length and the presence of a much smaller barbel relative to Pacific cod. This species resembles other cod-like fishes as it has three separate dorsal fins, a large, slightly

Andy Murch

forked tail fin and two anal fins. The body is elongate with a moderately sized head and upturned mouth. The coloration is olive green dorsally and creamy white ventrally. Juvenile tomcod can readily be distinguished from juvenile walleye pollock and Pacific hake as there is no barbel on Pacific hake and either a minute barbel or none at all on the chin of walleye pollock.

LIFE HISTORY

Pacific tomcod are schooling fish that are found from Central California to the southeastern Bering Sea. In British Columbia they are found at depths of 27 to 92 m. Their maximum depth is approximately 260 m, but they usually occupy water shallower than 100 to 120 m. Juvenile Pacific tomcod can be found in bays and estuaries, where they prefer a bottom substrate of sand, silt or mud. The maximum length is approximately 37 cm. Pacific tomcod in the Strait of Georgia have a substantially higher initial growth rate than the same species found elsewhere in their distribution; they also tend to mature earlier and have a shorter life span. These differences have been attributed to the warmer annual water temperatures in the area and to the species being near the southern end of its range in the eastern Pacific.

In British Columbia, age at maturity is between two and three years, with an average fecundity of 1,000 to 1,200 eggs. They spawn in the winter and early spring. Eggs are adhesive and stick to substrate during incubation. The larvae are pelagic and have been found in estuaries and nearshore marine environments, where they consume copepods, mysids and juvenile shrimp. As tomcod larvae approach metamorphosis to the juvenile form, they start to school. Juveniles eat shrimp, crab megalops, fish larvae, polychaetes, isopods, amphipods and copepods. Adults prey on many different bottom-dwelling invertebrates, shrimp, fish, amphipods, polychaetes and mysids.

Viviparous Brotulas, Cusk-Eels, Toadfishes, Tubesnouts, Sticklebacks and Pipefishes

There is one species in the Strait of Georgia in each of the following six families.

The red brotula is in the order Ophidiiformes and family Bythitidae (viviparous brotulas), which has about 100 species that are found throughout the world's oceans from the shallows to the very deep ocean. These are relatively small fishes of 15 to 30 cm, and the females release living young instead of eggs.

The spotted cusk-eel is in the order Ophidiiformes and family Ophidiidae (cusk-eels), which is also a group of fishes found in all oceans. One species has been reported at depths of over 8,000 m. There are about 220 species in the family Ophidiidae, and they are not very abundant.

The plainfin midshipman is in the order Batrachoidiformes and family Batrachoididae (toadfishes), which has 70 to 80 species. These are mostly bottom-dwelling coastal fishes that are found throughout the world's oceans, and a few in fresh water. There are about 14 species that have light-producing organs (photophores), which normally occur only in deep-sea fishes.

The tubesnout is in the order Perciformes and family Aulorhynchidae (tubesnouts). There are only two species in the family that are found in coastal marine areas, sometimes in large schools in the North Pacific. Males use pieces of plant material to build nests.

The threespine stickleback is in the order Perciformes and family Gasterosteidae (sticklebacks). The species in this family are marine and freshwater fishes that can also be anadromous around the northern hemisphere. Sticklebacks are frequently studied species; their constantly changing morphology attracts the curiosity of researchers studying evolution and behaviour. There are currently fewer than 12 species, but in the past up to about 60 were described.

The bay pipefish is in the order Syngnathiformes and family Syngnathidae (pipefishes and seahorses), which has about 225 species throughout the world's shallow marine areas and a few in fresh water. Pipefishes are noteworthy because the female transfers unfertilized eggs to a pouch on the male, where the eggs are fertilized and incubated to hatching.

RED BROTULA

Brosmophycis marginata UNCOMMON ◆

IN THE STRAIT OF GEORGIA

The red brotula is considered to be a reclusive fish and may be largely nocturnal, yet it has been reported by scuba divers throughout the Strait of Georgia, particularly in the Stuart Channel area. It has not been found in any of the pelagic or bottom research trawl studies and is probably uncommon. The fish's tendency to live in deep crevices may result in it being uncommonly observed or caught.

Red brotula larva.
Danny Kent

DESCRIPTION

The red brotula, like other members of this family, has an elongate, compressed, eelpout-shaped body. It has a large head with a very large mouth and small teeth. There is one long dorsal fin extending from behind the head to the end of the body and a similarly shaped anal fin on the underside of the fish extending from behind the anal pore to the end of the body. These fish have a very small rounded tail. The pelvic fins located below the operculum are nothing more than long, slender filamentous rays. The lateral line is in two sections. This species has loose, thick red- to brown-coloured skin covered with reddish mucus. The colour is darker dorsally, lighter below. The margins of both the dorsal and anal fins are coloured red while the caudal and pectoral fins are pink.

Janna Nichols

LIFE HISTORY

Red brotulas are believed to inhabit moderate depths (3 to 256 m) from Northern California to southeast Alaska, where they probably reside in rock crevices. The maximum size appears to be 45 to 50 cm. A 30 cm fish was aged by research scientist Milton Love as 12 years old. Very little is known about the red brotula's reproductive biology; however, we do know it is an ovoviviparous species, incubating eggs that are fertilized internally from sperm stored by the female and then giving live birth. Each female can give birth to hundreds of larvae. Red brotulas form monogamous mating pairs and the male remains close to the female during pregnancy. The live-born larvae swim near cliff walls to avoid being swept away by currents. Bonaparte's gulls, cormorants, seals and sea lions are known predators of this species.

SPOTTED CUSK-EEL

Chilara taylori VISITOR ▼

IN THE STRAIT OF GEORGIA

This species has only recently been reported in the Strait of Georgia, with several specimens collected in the southern strait, including from a research shrimp trawl in Howe Sound in the early summer of 2007. There is one record in the groundfish database at the Pacific Biological Station from a bottom trawl catch in the southern strait at 199 m in June 2018. The book *Fishes of the Salish Sea* records a specimen from Silva Bay near Gabriola Island that was collected in May 2004.

Charlotte Seid

DESCRIPTION

The spotted cusk-eel is readily recognized because of the long, tapering body and large dark spots on the head and body. It is a creamy brown, even orange colour with prominent dark margins along the dorsal and anal fins. There are small, paired, thread-like pelvic fins of unequal length at the bottom of the mouth anterior to the eye.

LIFE HISTORY

Spotted cusk-eels range from the Strait of Georgia to Southern California, with one record in Ecuador. The maximum length is about 40 cm, but most are less than 30 cm. There is limited information about their biology. It is known that the eggs are in a floating, gelatinous mass. This fish, like most cusk-eels, lives in burrows in soft sediments during the day. There is one video online that shows an alarmed spotted cusk-eel burrowing tail-first into the muddy bottom. This is another species that might be identified by night-diving scuba divers. The depth range for most individuals is from intertidal to about 90 m, but there is a record of up to 730 m. It feeds on invertebrates including polychaetes, crustaceans and small clams. Certainly, this species is rare, but it is possible that as waters warm in the Strait of Georgia, more individuals may move north into the strait.

PLAINFIN MIDSHIPMAN

Porichthys notatus

ABUNDANT

IN THE STRAIT OF GEORGIA

This unusual, relatively small fish is abundant in the Strait of Georgia, particularly in the Gulf Islands and between Texada and Vancouver Islands. It is common in estuaries and seen by scuba divers throughout the strait. It is one of the most abundant species in catches in bottom trawl studies in shallower water; for example, it is one of the five most abundant species in research-related shrimp trawl catches. Plainfin midshipmen are common but not abundant in midwater trawl catches below about 30 m and above about 75 m.

DESCRIPTION

The plainfin midshipman is dark brown with a large head and mouth. Its eyes are positioned on the top of the head, allowing an individual to find prey that swims over it when it is partially buried in the mud bottom. These fish do not have scales but do have rows of photophores. It is most unusual for fish anywhere in the world to have photophores when they live in shallow water. It is the pattern of the photophores along the bottom of the fish that produced the name "midshipman": the paired rows resemble neatly arranged buttons on a sailor's tunic.

Danny Kent

LIFE HISTORY

The plainfin midshipman is reported to occur from Alaska through to Baja California; however, catches off Alaska remain to be confirmed. Adults generally remain on bottom at depths of 70 to 100 m but can be found as deep as about 350 m. They migrate off the bottom at night, when we occasionally catch them in our midwater trawls.

The fish is also capable of making quite loud grunting and humming noises. The humming mating call of the males was so loud in an area of Sausalito, California, that the community organized the Sausalito Humming Toadfish Festival during the spawning season. The sound is produced by rapid contractions of specialized muscles attached to the swim bladder and can last up to 50 minutes without a pause. Anyone who catches a midshipman

Embryos and juveniles in a nest. Danny Kent

Embryos and juveniles in a midshipman nest attached to the underside of a shoreline rock. Danny Kent

may feel a vibration and hear the occasional pig-like grunt. Bioluminescence and a drumming noise are assumed to be important during spawning, and in breeding and nesting behaviour.

Mature plainfin midshipmen migrate in the spring in May and June into the nearshore intertidal areas to spawn; June is the peak spawning time. There are two types of males. The larger males dig nests, attract females and protect the fertilized eggs. These "guarding males" can go for several months without feeding. Much smaller males, called "sneaker males," participate in spawning but not in the building of nests or protection of eggs. Spawning males will burrow under rocks and other objects to create spawning sites. Smaller males can sneak into a nesting site of a larger male and potentially fertilize eggs. Females will deposit a layer of eggs on the underside of a rock. The production of a small number of eggs (fewer than 200 eggs per female) is compensated for by the large size of the egg and the protection provided by the male. It is not known how long midshipmen live, how fast they grow, what their average age is at maturity or how they fit into the Strait of Georgia ecosystem. It is known that shallow, rocky intertidal areas are critically important for reproduction. The nests may contain eggs from several females. Males remain to guard the nests and fan the eggs with their large pectoral fins and tail. The large eggs resemble salmon eggs. Crescent Beach is a well-known spawning area where eagles feed heavily on the midshipmen. As many as 150 eagles have been observed taking the spawning midshipmen back to young eagles in their nests. Blue herons, common crows, glaucous-winged gulls and harbour seals have also been observed feeding on the spawning fish.

Midshipman eggs take several weeks to hatch into a well-developed embryo about 7 to 8 mm in length. The hatched juvenile remains attached to the underside of the rock by its yolk sac for about a month. The juvenile relies on its yolk, but it may also feed on plankton that is circulated through the nest by the male. At about 3.2 to 3.4 cm and well into the summer, the juvenile midshipmen look very much like their parents and leave the nest. Adults may grow to sizes between 30 and 38 cm. Unfortunately, there are no studies of the life history of this remarkable species.

TUBESNOUT

Aulorhynchus flavidus COMMON ●

IN THE STRAIT OF GEORGIA

Tubesnouts are common in the Strait of Georgia and particularly in Burrard Inlet in December. They are frequently seen by scuba divers throughout the strait among eelgrass in the nearshore area, often in large schools. They have also been observed in rocky crevices, kelp beds, over sandy substrate and around wharf pilings. This is a shallow-water pelagic species that is rarely caught on bottom.

DESCRIPTION

The tubesnout has a highly elongate, rigid body, almost cylindrical or pencil-shaped throughout its length. The head is elongate, shallow and narrow. The snout is drawn out, with the lower jaw forming a long tube, hence the name "tubesnout." At the end of the long tube is a very small mouth directed forward with small teeth. The eyes are large, almost the same width as the head. This species does not have scales but does have a series of bony plates covering the body. It is distinguished from pipefishes by the rounded sides and the spines on the back. The caudal peduncle, an area located just before the tail, is keeled to help in locomotion. There is a bright-silver patch between the operculum and pectoral fins, with a dark band running through the eye to the snout. The body colour varies from pale brown to green or yellow on the dorsal side of the fish and creamy white on the bottom.

A male tubesnout (with orange pelvic fins, a black cap over the eyes and a luminous blue snout) and a female (with a swollen belly, at top) face egg masses in seaweed on the right. Danny Kent

Danny Kent

LIFE HISTORY

The first record of a tubesnout in British Columbia was in 1866. Their range extends from central Baja California to the Alaska Peninsula, where they are found from the surface to 30 m deep. Juveniles can be found most commonly at about 5 m, but as deep as 12 m. This species is quite small, with a maximum length of approximately 18.8 cm. Swimming is mostly accomplished by the movement of the pectoral fins, which gives the fish a stiff appearance and is not conducive to fast, graceful movements. They swim slowly with many beats of their fins, constantly snapping at prey. They move in the water with rigid, jerky lunging movements.

This species reaches maturity within a year. The tubesnout exhibits sexual dimorphism; that is, breeding males have bright-red pelvic fins and a phosphorescent snout, while the females do not. Tubesnouts are highly territorial in spawning season. The male and female will occupy a nest site, with the male defending their breeding territory from other males until the female is ready to deposit the eggs. The male tubesnout is the nest builder and takes advantage of seaweeds as attachment points for nest building. The territoriality and nest-building traits are very similar to those of the tubesnout's close marine relative, the threespine stickleback.

Life history varies between central Baja California and the Strait of Georgia. Tubesnouts down south have been aged to nine years. They spawn in very deep water, perhaps year-round, and they occur in vast schools far offshore. In the Strait of Georgia tubesnouts spawn in late spring on bull kelp, sargassum and other shallow seaweeds, and individuals rarely survive beyond their second year. It is likely that tubesnouts can spawn throughout the year farther south, but spawning probably is concentrated in late winter and spring in the Strait of Georgia.

Tubesnout eggs surround the bases of bull kelp blades. Danny Kent

Eggs are approximately 2 mm in diameter and each mass of eggs may be as large as 3.3 cm in diameter. The colour of the eggs varies significantly from dark red to light yellow, but most are amber. Eggs are tended by the male until hatching, which takes place in two to three weeks; eggs from one spawning are the same colour, but colours vary among females from light brown to yellowish red. The young hatchlings grow very quickly, reaching 5.1 cm in two months. At this stage, they aggregate in small schools, taking shelter around structures. It might seem that tubesnouts are not abundant because they do not occur deep enough for most fishing gear, but scuba divers see huge schools along the shoreline. Tubesnouts are limited to shallow depths—to whatever the greatest depth of local seaweeds is—at all life stages, yet they occur near islands surrounded by deep water.

Windstorms tear loose bits of seaweed with tubesnout eggs attached, and the eggs thereby get dispersed in the wind-driven surface currents. Tubesnout larvae settle to the bottom and start schooling in accordance with bottom features within about 20 minutes of hatching, so they are truly non-planktonic, meaning they do not drift anywhere after hatching. If they mate with siblings, the offspring are just fine; this was found when reproduction at the Vancouver Aquarium was carried through ten generations without any signs of inbreeding depression (reduced survival and fertility), which was found in some other species, such as sailfin sculpin. The shape of the tubesnout's mouth gives an indication of their prey type: amphipods, mysids, crab zoea and other small crustacean plankton. They are also known to eat fish larvae, including their own.

THREESPINE STICKLEBACK

Gasterosteus aculeatus ABUNDANT ○

IN THE STRAIT OF GEORGIA

The threespine stickleback is an abundant species in the Strait of Georgia. It occurs in shallow water around wharves and piers and near the surface in open areas, but not in deeper waters. When not breeding, threespine sticklebacks are in schools that can be exceptionally large, perhaps containing several thousand individuals. Pelagic trawls fished at and near the surface can catch large numbers of sticklebacks just about anywhere in the Strait of Georgia. Threespine sticklebacks are commonly among the five most abundant species in catches in the top 75 m throughout the strait. They are common in most estuaries and are fun to watch as they hunt for prey.

DESCRIPTION

The threespine stickleback is elongate with a tough little body whose sides are covered in bony plates rather than scales. The mouth is small and directed slightly upward, ending with a pointy snout. The eyes are large. The threespine stickleback, aptly named, has two prominent dorsal spines followed by a third, smaller one just anterior to the fleshy dorsal fin. It has a long, slender caudal peduncle with a keel on each side just anterior to the caudal fin or tail. The caudal fin is blunt, not forked. The pectoral fins are large and fan-shaped; the pelvic "fin" is actually a spine. Anterior to the blunt anal fin is another small spine. The coloration can vary significantly depending on whether the animal is a freshwater or a marine inhabitant. Marine specimens can vary from silver-green to blue-black; freshwater specimens are muted in colour, often a mottled brown.

LIFE HISTORY

In British Columbia there is only one species of threespine stickleback. However, debating the possibility that the threespine stickleback is actually made up of multiple distinct species is an ongoing exercise for scientists, as the fish adapts to changing environments by changing its morphology. In the Pacific, its distribution extends from California to Alaska, into the Arctic in the Beaufort Sea and beyond as far as Japan and Korea. The same species is also found in the Atlantic from Chesapeake Bay north to Hudson Bay, Norway and Greenland and then south as far as Spain. It is also commonly found in fresh waters in Lake Ontario and throughout the Maritimes.

Danny Kent

The threespine stickleback is a small fish reaching a maximum length of approximately 10 cm. The fish can breed in both fresh water and marine waters. It thrives in all manner of habitats, from lakes to estuaries to open oceans. It is a nearshore species but has been captured in the middle of the Gulf of Alaska. In fresh water, male threespine sticklebacks have been shown to make nests and tend the young until they leave the nest. The species reaches sexual maturity at between one and three years of age. Consequently, some may spawn only once before dying. Their nests can be found in the ocean among seaweeds attached to floating docks. Males develop secondary sex characteristics, such as red chins and red bellies, believed to attract females and cause aggression in other males; the males also develop stunningly blue irises. There are some populations where the spawning males are black.

The stickleback has become a favourite of behavioural scientists as it is easy to handle, easy to procure and tolerant of captivity. Because these fish have been so popular to study, there is considerable information about their life histories, particularly regarding their courtship and mating habits. In the spring, the male leaves the school and defines his own territory in anticipation of spawning. He digs a nest in sandy substrate, removing the sand with his mouth. When he is satisfied with the size of the nest, he places pieces of aquatic plant and algae in a pile on top. The male stickleback then "glues" everything together with a substance produced by his kidneys and forms an acceptable nest shape with his snout. To allow himself access the nest, he wriggles his body through it, creating a tunnel slightly shorter than the length of his body. Once the nest is complete, the male stickleback changes into his breeding colours and commences courting females. The females will have also changed colour slightly and become very shiny; at this point they are obviously larger in pre-spawning condition as they have produced up to one hundred large eggs ready for a suitable male. As a female swims into a male's territory, he advances toward her in a zigzag pattern and darts at her. He continues to dart with intermittent pauses looking for acknowledgement from the female. If the female deems him a suitable mate, she swims toward him with her head up. The male then leads her to his nest. The male shows the female his nest with rapid thrusts of his snout in the direction of the entrance hole. He turns to his side and raises his dorsal spines toward her. This is the signal for the female to enter the nest. There she will rest with her head sticking out one side and her tail the other. The male prods her tail, initiating the release of her eggs. After depositing her eggs, the female swims out of the nest; then the male enters and fertilizes them. Once the male has chased off the last of his harem, he vigorously guards his brood and fans water over the eggs with his pectoral fins until they hatch. The male continues to defend the newly hatched offspring for several days, pursuing stragglers and returning them to the nest area in his mouth. The young then eventually leave the nest but remain in the company of other young sticklebacks. Large schools of these young sticklebacks are common around docks in July and August.

Prey for marine threespine sticklebacks includes copepods, amphipods, decapod larvae, barnacles, insects, young fishes and clam larvae. In the Strait of Georgia, hyperiid amphipods and calanoid copepods are a favourite prey early in the year. In fresh water, insects are an important food source. Threespine sticklebacks are prey for fish-eating birds, predatory fishes and seals. They are an important early host to several species of juvenile sea lice. With time, the sea lice leave the stickleback for a larger host.

BAY PIPEFISH

Syngnathus leptorhynchus ABUNDANT

IN THE STRAIT OF GEORGIA

Bay pipefish are generally abundant around the Strait of Georgia. They can be very common along the shoreline in estuaries and eelgrass beds. They are found around docks and even in floating seaweeds. One can count on finding them in the old tires and ropes hanging from the wharf at the Pacific Biological Station in Departure Bay. As might be expected, they are commonly seen by scuba divers around the strait but rarely occur in its open waters.

DESCRIPTION

Bay pipefish are long and skinny and perhaps look like a green pipe cleaner. They are mostly green but can have shades of brown and even purple. Pipefish have no pelvic fins. The shape, colour and vertical orientation of the fish provide effective camouflage when among eelgrass, which is necessary as gulls are an aggressive predator. Individuals are encased in bony rings that stiffen the body, giving it a segmented appearance.

A green-coloured bay pipefish is camouflaged among blades of eelgrass, a common plant habitat. Like many fish, the body colour will usually match the plant or other habitat the fish associates with. Danny Kent

LIFE HISTORY

Bay pipefish are part of the group of fishes that includes seahorses. This group has about 175 species worldwide, 25 of which are seahorses. Bay pipefish are not just straightened-out seahorses, however, as they have caudal fins and a different brood pouch than seahorses. They swim using a rapid motion of the small dorsal fin and steer by moving the head. They are found in shallow water from Baja California to Prince William Sound. A maximum length of 38.5 cm has been recorded, but most range from about 8 to 30 cm. Females are larger on average than males. Studies have used length and growth to estimate bay pipefish's age, but these studies are few. One study estimated there were at least three age groups in a population, with an average weight of 3.1 g for older fish. However, the maximum weight is about 20.3 g. If growth was constant during the summer and negligible in winter, the 20.3 g bay pipefish would be much older than three years.

Males can be distinguished from females by the lateral flaps of their brood pouch. Studies show that females generally are more numerous than males. Male bay pipefish use the brood pouch to hatch the fertilized eggs that

This male pipefish's belly pouch is filled with developing embryos. The species gives live birth to miniature adults, with no larval dispersal stage.
Danny Kent

they receive from the female. The brood pouch is a mid-ventral slit that extends about one-third of the length of the fish. The turned-in flaps of skin form a pouch that folds together once the eggs are inside. In the Strait of Georgia, males carry eggs as early as late May and as late as the end of August. Eggs are fertilized internally in an elaborate sex act in which males—as interpreted by humans—are choosy about their female partners, who compete to seduce a male. It is not clear exactly when the eggs received by the male are fertilized, but it is possible that they are fertilized as the female deposits them in the pouch. It is known that the eggs in a male's brood pouch can come from more than one female, but it is not clear if the eggs from a female can be fertilized by more than one male. A female depositing eggs in a male's pouch may do this over several days. The number of eggs in a pouch varies depending on the size of the male, with larger males carrying four to six times as many eggs as smaller males. Large males may have as many as 900 eggs in their pouch. Eggs hatch in the pouch in about 8 to 10 days, depending on the temperature. The juvenile bay pipefish are released at a length of about 1.9 cm. If most mating occurs about mid-June and the average daily growth is about 1 mm, it would appear that bay pipefish first mature in their second ocean year.

Bay pipefish can rotate their snouts and suck in prey from 2 to 3 cm away, and do so faster than a human eye can detect. The fish does not have a stomach to accumulate food for digestion so it feeds continuously during the day. Typical prey items are mysids, small shrimp and amphipods. Seahorses and pipefishes are used in Chinese medicine, and pipefishes are considered to be more potent than seahorses. There is international concern about the overexploitation of seahorses.

Rockfishes

Rockfishes in the Strait of Georgia are in the order Perciformes and family Scorpaenidae. This is a family of fishes that occur throughout the world's oceans from the intertidal to great depths. Its taxonomy depends to some degree on the reporting source, but one estimate is that there are 56 genera and about 420 species. Members of the family, which includes stonefishes and scorpionfishes, generally are spiny headed, have venomous spines and vary remarkably in appearance. There are two genera of rockfishes that are relevant to the Strait of Georgia. The genus *Sebastes* has about 102 species, mostly in the North Pacific, and these are the true rockfishes that are familiar to most people. There are three species in the genus *Sebastolobus* with the common name "thornyhead" that are generally grouped together with rockfishes. These three species occur in the northern and eastern Pacific. The common name "rock cod" was used when settlers arrived in British Columbia from the East Coast and added the name "cod" to many of the fishes that were new to them. The *Sebastes* fishes all have internal fertilization, with the larval rockfish born alive and some having received nutrients from the female. Fishes in the genus *Sebastes* also have swim bladders, whereas species in *Sebastolobus* do not. When the three fishes in the genus *Sebastolobus* reproduce, females release eggs that form a gelatinous mass.

There are 29 species of rockfishes in the Strait of Georgia, including one thornyhead. Importantly, prior to a report by Dick Beamish in 1979, all rockfish were believed to live for about 10 to 20 years. The new method of aging produced estimates such as 140 years for rougheye rockfish, 120 years for shortraker rockfish and others ranging from 36 to 90 years. The oldest, a rougheye rockfish off the west coast, has now been aged at 205 years old. In the Strait of Georgia, yelloweye rockfish have been aged at up to 98 years old. Interestingly, there is no reproductive senescence, as the older fish continue to reproduce annually. Fish probably live to be very old because their annual reproduction is influenced by ocean conditions that only infrequently are suitable for many juveniles to survive. The discovery in 1979 that rockfishes could actually be very old changed the science that regulated fishing. Unfortunately, this discovery was too late to stop the overfishing of many populations in the Strait of Georgia and elsewhere.

Milton Love and his co-authors indicate in *The Rockfishes of the Northeast Pacific* that there are 102 species of rockfish worldwide, and perhaps 96 are found on the west coast of North America—although several new species have been described since the book was published in 2002. It is difficult to be sure of the number of species in the Strait of Georgia, but excluding Juan de Fuca Strait and Johnstone Strait, there could be 29 rockfish species, including the shortspine thornyhead.

In 2002, there was a major change in the management of rockfish in British Columbia, brought about in part because of the concerns of environmental groups such as the Sierra Club. In the Strait of Georgia, the Department of Fisheries and Oceans reduced catches by 75 percent. The catch reductions were imposed on sport and commercial fishermen to limit the total catch of all rockfish species. The need to protect some areas from all fishing was also established. These rockfish conservation areas were identified through an extensive consultation process aimed to protect areas of rockfish habitat. As a result, at the time of this writing there were 129 rockfish conservation areas in the waters between Vancouver Island and the mainland and another 36 outside the strait and throughout the coastal waters of British Columbia. In the Strait of Georgia, it is estimated that about 30 percent of rockfish habitat is now excluded from all fishing. Monitoring occurs through a creel census that tabulates the number and species of rockfish caught in the recreational fishery. However, studies of the contribution these rockfish conservation areas to the rebuilding of the strait's rockfish populations are badly needed, even though it is expected it will take decades for the populations to rebuild. The recent establishment of areas closed to all hook-and-line fishing will help to protect inshore rockfish, but our poor understanding of the mechanisms that affect the production of rockfish means that the remaining individuals must be treated with great care. It is common sense that the constant removal of rockfish can only be balanced by an equally constant replacement with mature individuals. However, this constant replacement probably does not occur. If rockfish are to remain a colourful component of the Strait of Georgia ecosystem, there will need to be careful management and strong support from fishermen. If it is not accomplished voluntarily, concerned citizens should fill out the necessary forms to have the species protected under the Species at Risk Act.

ROUGHEYE ROCKFISH

Sebastes aleutianus

or BLACKSPOTTED ROCKFISH

Sebastes melanostictus UNCOMMON ◆

IN THE STRAIT OF GEORGIA

Recently, scientists using genetic techniques showed differences within the population of rougheye rockfish that are great enough to separate the aggregate of all rougheye rockfish into two species, rougheye rockfish (*Sebastes aleutianus*) and blackspotted rockfish (*Sebastes melanostictus*). Rougheye rockfish are found in deep water in the Strait of Georgia, but there are no estimates of either historical or recent abundances. Small numbers of rougheye rockfish were recorded by experimental trawl surveys in the mid-1970s. However, in recent surveys, none were found. Thus, it is probable that the species is now uncommon or even rare within the strait. The collarbones around the head of

Wirestock Creators / Adobe Stock

rougheye rockfish were used by some First Nations to make a broth that could treat illness. Chief Adam Dick recovered from a serious illness after consuming the broth.

DESCRIPTION

Rougheye rockfish are distinguishable by their red colour and the 2 to 10 spines on the ridge below the eye. These suborbital spines are the source of the common name "rougheye." The use of genetic identification to distinguish species that visually appear identical resulted in the term "cryptic species"; rougheye rockfish and blackspotted rockfish are thus cryptic species. There are some morphological differences relating to the number of gill rakers and the body proportions, but these differences are not always diagnostic and can be difficult to use routinely. To date, the only accurate way to distinguish the two species is through DNA. It is not known if both the rougheye and blackspotted rockfish species occur in the Strait of Georgia and we use the common name "rougheye rockfish."

LIFE HISTORY

Rougheye rockfish occur from Southern California around the Pacific Rim to northern Japan. Interestingly, they are not in the Sea of Okhotsk. They are found close to the bottom at depths of 25 to 900 m, but usually at about 200 to 600 m. They are associated with sloping areas of boulders and soft substrates that contain their preferred prey. Bowie Seamount, a submarine volcano west of Haida Gwaii, is an area that has relatively large abundances of either rougheye or blackspotted rockfish, and it is intriguing how these fish are recruited to this undersea mountaintop well off the coast of British Columbia. The maximum length of the rougheye rockfish is 97 cm and the maximum weight is 900 g. Lengths typically average between 40 and 60 cm. Fifty percent of males and females are mature by the age of 20 years and a length of 45 cm. Rougheye rockfish are the oldest of all rockfish and one of the oldest fish species on the planet; one individual was determined to be 205 years old.

A rougheye rockfish male (sometimes more than one) will inseminate a female during courtship. The female stores the sperm for up to a month before the eggs are fertilized. The eggs develop into embryos inside the female, and the larvae are released several months after insemination, usually from February to May. Rockfish are classified as ovoviviparous because they release live fish that were fertilized beforehand. There is some evidence that the female provides some nutrients to the larval fish just prior to release, as the larvae inside

Adisha Pramod / Alamy

of the female hatch several days before they are released; this type of birth is called parturition. Very little is known about the timing of birth or where the nursery areas are for most rockfish species. It is difficult to identify most of the species during this larval stage, though the use of DNA may help to solve this problem. After three to six months, it is believed the larvae settle close to the bottom near reefs and rocky areas. In general, the smaller fish remain in shallow water and move into deeper water as they age. Year class strength is mostly determined in this early period; thus the abundance of preferred prey items at this time is critical for their survival. Older fish tend not to school. They consume mostly pandalid shrimps that they find in the soft bottom, as well as amphipods, crabs, mysids and other fishes. Because these fish grow relatively slowly once they mature, they probably do not need to feed voraciously to survive.

Rougheye and blackspotted rockfishes are of concern to biologists because of the species' potential to be overfished. Overfishing is a threat mostly because there is uncertainty about the accuracy and amount of information that is needed to ensure fishing is sustainable. An added complexity is that existing information now needs to be related to a new species. When these issues are added to the uncertainty on how many fish of either species are in the Strait of Georgia, it becomes apparent how much work is needed to manage any fishery for these two species. In Canada in 2009, both were listed as species of special concern, meaning they may become threatened or endangered. The listing indicates that enough is known about the biology of the species and the current threats to their populations that their ability to survive at sustainable levels is worrisome. What is most worrisome is that this is a relatively large and desirable commercial species that is extremely long-lived and consequently may be unable to rebuild quickly from low abundances.

PACIFIC OCEAN PERCH

Sebastes alutus VISITOR ▼

IN THE STRAIT OF GEORGIA

Only three Pacific ocean perch have been caught in the thousands of pelagic research trawl sets in the Strait of Georgia since the mid-1990s, indicating that the occasional specimen can be found but that they are rare and likely not reproducing in the strait. Historical reports of Pacific ocean perch in the southern strait probably were related to the huge abundances off the west coast as well as to the species' long life of at least 103 years, resulting in some young individuals or adults finding their way into the strait.

DESCRIPTION

Pacific ocean perch generally are a bright-red colour but can be a paler red. There are dark-green to brown to black patches dorsally. The fish is elongate and has a protruding lower jaw with a distinctive knob in front, called a symphyseal knob.

LIFE HISTORY

Maximum sizes of Pacific ocean perch are up to 75 cm and 2.7 kg. The fish is distributed from Central California around the rim of the Pacific to off the coast of southern Japan. Males and females mature between the ages of 4 and 10. Males inseminate females in the summer and fall, and females then move into deeper water to give birth, mostly from about February to May. The

Vicky Okimura and Steve Axtell

discovery that all rockfishes can live to be very old was made by Dick Beamish in 1979 at the Pacific Biological Station in a study of Pacific ocean perch off the west coast of Vancouver Island. Females return to the shallower depths in the spring. The major diet item is euphausiids, but other items include fishes and squids. In the 1960s, before countries started declaring 200-mile limits to enable them to manage fish stocks for sustainable catches, there were large Russian and Japanese vessels fishing for Pacific ocean perch off the coasts in the Gulf of Alaska and south to Washington and Oregon. Catches off the west coast of British Columbia were as high as 60,000 t per year. By the mid-1970s the catches had declined precipitously and as of 2024 were about 5,200 t per year. The Pacific ocean perch was once called the longjaw rockfish.

YELLOWMOUTH ROCKFISH

Sebastes reedi VISITOR ▼

IN THE STRAIT OF GEORGIA

Records of yellowmouth rockfish throughout the Strait of Georgia come from the commercial fisheries using bottom and midwater trawls, as well as some hook-and-line fisheries. No specimens were retained to confirm the identification, but yellowmouth rockfish is an important commercial species that is closely monitored, indicating that at least some of the incidental catch records could be valid. There also were commercial fishing trips that had much larger incidental catches of yellowmouth rockfish in Queen Charlotte Strait. This greater abundance immediately north of the Strait of Georgia is evidence that an occasional individual or small school might find its way into the strait as we proposed for Pacific ocean perch. We list the yellowmouth rockfish as a visitor to the Strait of Georgia both because of the reports of incidental catches and as a reminder of the importance of retaining specimens caught in the strait; retention allows scientists to verify species identifications and can provide museums with specimens.

Kristina Castle

DESCRIPTION

The yellowmouth rockfish was recognized as a species in 1967. It was separated from similar-looking species such as Pacific ocean perch because of the yellow and dark blotches in the mouth and differences in the number of scales, spines, gill rakers and the symphyseal knob. Yellowmouth rockfish are red with a mixture of yellow-orange shades. There are diffuse dark blotches dorsally that do not extend below the lateral line. There is a distinct dark blotch on the dorsal operculum. The symphyseal knob is evident but moderate to small and not prominently pointing forward as in Pacific ocean perch. The lateral line is a thin pink.

LIFE HISTORY

Yellowmouth rockfish are distributed from the western and northern Gulf of Alaska to Northern California. This is an offshore slope species that occurs on bottom and in the midwater at depths from 75 m to 430 m, but mostly at 100 m to 400 m. The maximum size is 58 cm and 2.8 kg. The maximum age is 100 years, but in the British Columbia commercial fishery fewer than 1 percent are older than 60. The oldest yellowmouth rockfish caught in British Columbia were 99 years for a 32 cm female and 86 years for a 50 cm male. Length and age at 50 percent maturity are approximately 38 cm and 10 years. Larval fish are released from the female from February to June. The diet includes squid, shrimp and fishes. This is a slow-growing, long-lived species that has long periods of slow population growth. Occasionally there is a year of exceptional ocean survival of juveniles, such as in 1982 and 2006. In British Columbia, the commercial fishery is a bottom trawl fishery with about 16 percent of the catch in the midwater over rocky substrates. The population in British Columbia is currently considered to be healthy and even increasing. It is a well-managed fishery, but it is subject to surprises and requires constant monitoring.

BROWN ROCKFISH

Sebastes auriculatus UNCOMMON ◆

IN THE STRAIT OF GEORGIA

Brown rockfish are generally uncommon in the Strait of Georgia, but are more common in the southern strait than in the north. They occur in the Southern Gulf Islands in Stuart Channel and Saanich Inlet, west of Texada Island and in Baynes Sound, up Burrard Inlet in Indian Arm, and in Howe Sound, but generally are rare in these areas. They are very uncommon in northern or exposed areas. Brown rockfish tolerate silty, shallow water; this may explain their common occurrence in Puget Sound, which is shallower than the Strait of Georgia. Brown rockfish are curious and sometimes approach scuba divers, but usually have a hole in a crevice or rock pile into which they can safely retreat.

Danny Kent

DESCRIPTION

Brown rockfish are deep-bodied and live in shallow water, similar to the habit of their closest rockfish relatives, the copper, quillback and China rockfishes. Brown rockfish seen in the Strait of Georgia are easily identified by their distinctive brown and mottled base colour, together with a conspicuous dark-brown blotch on the rear of their gill cover, just above the attachment of the pectoral fin. In contrast to copper rockfish, whose white lateral line stripe is only on the hindquarters, the entire lateral line on brown rockfish is a white stripe. Brown rockfish do not have the large head of either a quillback or China rockfish, or the marked colour patterns of those species. They are closest in resemblance to a copper rockfish, except that the head is smaller and more smoothly contoured.

LIFE HISTORY

Brown rockfish grow to 56 cm in British Columbia, live to 34 years, and are 50 percent mature at 5 years old and 24 to 31 cm long. All are mature at 10 years and about 38 cm. They are included in a group of inshore and nearshore rockfish in British Columbia that prefer rocky habitats, from shallow areas to about 135 m. Brown rockfish are some of the shallowest-dwelling inshore rockfish and are more abundant in the southern part of their range, from Baja California to Prince William Sound. They are similar to copper and quillback rockfishes, with hybrids known to occur between male quillback and female brown rockfish in Puget Sound. DNA analysis has shown that these natural hybrids are fertile. Quillback rockfish also hybridize with copper rockfish in Puget Sound, making identifications difficult. Brown rockfish give birth to 52,000 to 339,000 larvae during late spring and may spawn several times in a year. They often eat crabs, shrimp, squid and fishes. This species has a large larva and is reportedly the easiest to rear in captivity, but there is no demand for it as the market is for live quillback and copper rockfishes.

REDBANDED ROCKFISH

Sebastes babcocki VISITOR ▼

IN THE STRAIT OF GEORGIA

Only one redbanded rockfish has been reported in the Strait of Georgia, from a research-related bottom trawl survey in the northern strait. No more have been found in any of the thousands of research trawl catches in the Strait of Georgia since the mid-1990s.

DESCRIPTION

This is a very distinctive rockfish with a pale-pink body and four red or orange vertical bands, one of them on the caudal peduncle in front of the tail. The first two bands bracket the spinous first dorsal fin, and the third goes through the soft dorsal and anal fins. The interorbital space is flat or slightly concave. The redbanded rockfish is a close relative of the similar-looking, vertically striped tiger rockfish. Fishermen call redbanded rockfish convicts or bandits because of their stripes.

Ashley / Adobe Stock

LIFE HISTORY

Redbanded rockfish are found off the west coast near hard bottom or rocky reefs and seamounts at depths of about 50 to 625 m, but mostly at depths from 150 to 350 m. The species ranges from the Aleutian Islands and Bering Sea to Southern California. The maximum size is 71 cm and 5 kg, with a maximum age of 106 years. In British Columbia waters maturity occurs at about 19 years and 42 cm, but can occur as early as three years in more southern areas. Females release live young from March to September. From 1915 through the early 1970s this species was considered synonymous with the more southern flag rockfish. This is an important commercially fished species and is caught using baited hooks on longlines set on the bottom. This colourful fish is popular in Asian markets.

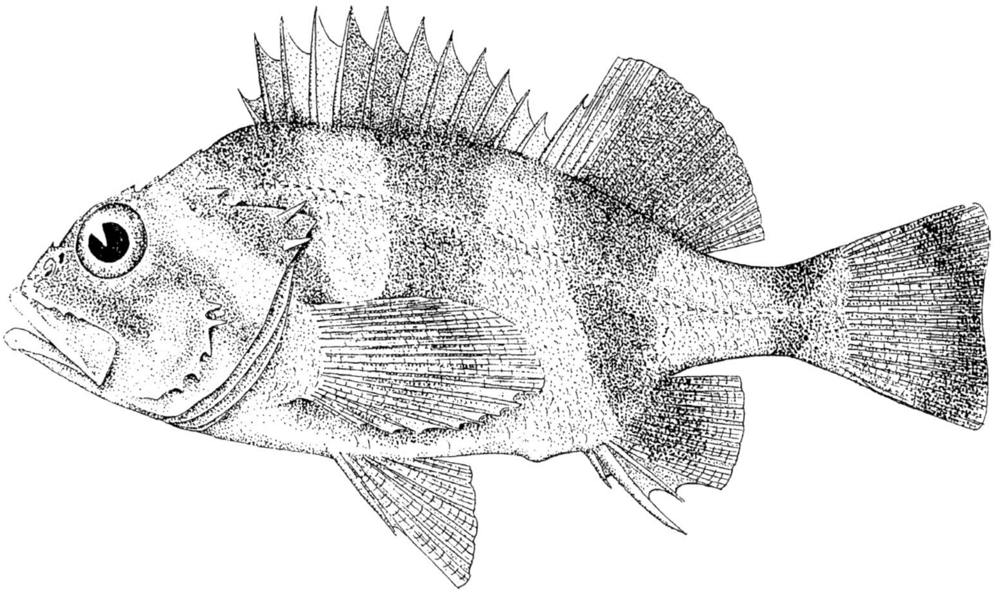

SHORTRAKER ROCKFISH

Sebastes borealis UNCOMMON ◆

IN THE STRAIT OF GEORGIA

The shortraker rockfish is uncommon in the Strait of Georgia. Small numbers have occasionally been caught in bottom trawl research surveys in the deeper areas of the central strait.

DESCRIPTION

The shortraker is another large rockfish that is a reddish pink to orange-pink, either with no spines or with one weak spine on the rim below the eye. It is readily identified by the large pores on the lower jaw, which are the largest among all rockfish. Otherwise, its appearance is similar to that of the rougheye rockfish. The gill rakers on the first arch are relatively short but their length is not a strong character.

Vicky Okimura

LIFE HISTORY

This species occurs from Southern California around the Pacific Rim to Japan. It occurs in depths of 25 to 1,200 m but is most common from 300 to 500 m. The larger and mostly older individuals may be solitary, living near ocean slopes with intermittent rocks and large boulders at the bottom. The biology of shortraker rockfish is poorly understood, particularly the reproductive biology through to the juvenile stages. For example, fish smaller than 35 cm are rarely found. Shortraker rockfish are difficult to age, resulting in changing estimates of maximum age. The current estimate is 157 years. It appears that 50 percent

NOAA

of males and females mature at about 45 cm. Mating probably occurs in the fall at depths of 300 to 500 m, but there are also reports that it occurs earlier, around June and July. Larval fish have been caught in the late winter and spring off Alaska, with maximum densities about April. However, it does appear that the reproductive period is protracted, with larvae being released even into the fall. Studies of other rockfish show that when the reproductive period is protracted, the surviving individuals may come mostly from a short time within it. Smaller shortraker rockfish feed on shrimp and amphipods, while larger fish in deeper water eat mostly squid and lanternfish.

SILVERGRAY ROCKFISH

Sebastes brevispinis RARE ★

IN THE STRAIT OF GEORGIA

Silvergray rockfish are rarely reported in the Strait of Georgia, but one was recently seen by marine naturalists Andy Lamb and Phil Edgell in Howe Sound and there have been scuba sightings in Active Pass. Only one has been caught in all of the thousands of pelagic research trawl sets in the Strait of Georgia since the mid-1990s. This individual was caught in March 2011 in the central strait. Another was caught in a research longline survey in 2004.

DESCRIPTION

The silvergray rockfish is a slim fish that can be dark green to dull silver-grey, as the name suggests. It has a white to pink belly and a protruding lower jaw with a strong symphyseal knob. The mouth is large with dark lips. The species name *brevispinis* ("short spine") reminds the observer to check for the supra-ocular spine at the top of the eye socket. If you grab the fish by the eye sockets without a glove and you get cut, it is a silvergray.

Mike Byerly

LIFE HISTORY

Silvergray rockfish occur from the western Gulf of Alaska to central Baja California. The maximum age is 82 years, and about half of the males and females mature after 10 years. The maximum size is 73 cm and approximately 5 kg for females and 70 cm and 3.4 kg for males. Mating occurs in the winter, and most larval fish are released from May to July. The largest females can produce almost two million eggs. The schools offshore may move into shallow water in the summer and back into deeper water in the fall. This is a long-lived pelagic rockfish that is found off the west coast over reefs and soft bottoms close to the edge of the continental shelf, from the surface to 375 m deep, but usually at 100 to 300 m. This species may loosely school or aggregate with other species like black, canary, dusky, Puget Sound or yellowtail rockfishes. Virtually all of the commercial catch is from between 80 and 400 m deep. In British Columbia, commercial catches occur together with those for canary and yellowtail rockfishes. There were exceptional large abundances offshore in British Columbia in the early 1980s and in 2000–01. The current fishery in British Columbia is well managed and considered sustainable.

COPPER ROCKFISH

Sebastes caurinus ABUNDANT

IN THE STRAIT OF GEORGIA

The copper rockfish may be the second most abundant rockfish in the Strait of Georgia, after the quillback rockfish. They are frequently caught in the recreational fishery and are the rockfish species most commonly seen by scuba divers throughout the strait. This is an inshore benthic species that is uncommon in the open areas of the strait or on the bottom deeper than about 90 m. They are more abundant along the western side of the strait, from Hornby and Denman Islands to the Gulf Islands area, where they are associated with rocky habitats at depths of 20 to 65 m and often found around structures.

DESCRIPTION

Copper rockfish are relatively deep-bodied and have a range of colours from dark olive and black to brown and copper to a yellowish red. Brown and yellow may be more common colours than copper. The colour may vary with latitude: fish in the north are a darker colour than the dull yellow to olive-pink of those in the south, although the darker fish have not been observed in the Strait of Georgia. All have patches of colour that are mostly pink or yellow and a white to pinkish belly. The typical pattern has several diagonal stripes of lighter pigment extending from the head to the belly. There is an unpigmented area around the posterior two-thirds of the lateral line; this appears as a white stripe in the mid-body extending to the tail. The symphyseal knob is weak.

LIFE HISTORY

Copper rockfish are found from the Gulf of Alaska to central Baja California. They are reported to occur from intertidal depths to 310 m, but most are in surface waters shallower than about 90 m. The maximum age for copper rockfish is 50 years, which indicates that recruitment into the population may be irregular, despite the fact that females produce 200,000 to 700,000 larvae each year, depending on the size of the female. About 50 percent of the females in the Strait of Georgia are mature at age six and a length of about 34 cm; the maximum length is 66 cm. Their rate of growth begins to slow at about age 15. Copper rockfish in Puget Sound average 35 cm and range in age from 3 to 15 years. Populations in Puget Sound are genetically distinct from those off the west coast. The larvae are released from the female about April and the young settle in the nearshore areas in July and August. Reef areas and rocky bottoms are the typical habitat for these bottom-dwelling fish, which are found

Andy Murch

both in small groups and as solitary individuals. Copper rockfish are thought to remain in the same general area that they settle in after their larval period. In the Strait of Georgia, the young fish feed on plankton but the larger fish prey on small fishes, shrimp and crab, or other invertebrates that live within the rocky bottoms of the reefs.

Copper rockfish have been actively fished for over 50 years and currently are highly esteemed as live fish in Vancouver fish markets. They support an important recreational and commercial fishery. The common size range in the commercial handline fishery in the Strait of Georgia is about 25 to 35 cm with weights between 700 g and 1.4 kg. Fish in this size range are about 6 to 12 years old. Copper rockfish are one of the most studied species of all rockfish because they are accessible in shallower water, and within the depths available to scuba divers. In Puget Sound, there is sufficient concern about declining abundances that protection may be needed through the Endangered Species Act.

DARKBLOTCHED ROCKFISH

Sebastes crameri VISITOR ▼

IN THE STRAIT OF GEORGIA

The only records of darkblotched rockfish in the Strait of Georgia are from Haro Strait and Saanich Inlet, making it a rare species in the strait. It is also very rare in Puget Sound.

DESCRIPTION

This is a deep-bodied, orange-pink rockfish with three to five saddle-like dark blotches on the back. The mouth is mostly black in large individuals and pink in small ones. Young darkblotched rockfish have five dark pigment patches along the back, sometimes extending down as broad bars, plus a black spot on the gill cover or operculum.

Vicky Okimura

LIFE HISTORY

Darkblotched rockfish occur at depths of 25 to 900 m from the Bering Sea to Southern California. Most adults in British Columbia occur near rocky- and soft-bottom mixed habitat at 145 to 435 m. Most individuals in the British Columbia fishery are 5 to 25 years old. Individuals begin to mature at approximately 28 cm and all are mature by the time they reach 34 cm and an age of 8 to 9 years. Mating occurs from August to December, and larval fish are released in single batches from November to June, with a peak in February. Females give birth to 20,000 to 610,000 young, depending on the age of the female; the young are 4 to 9 mm. Young will be perched on prominent bottom features, but move into deeper water as they mature.

The darkblotched rockfish is widespread off the west coast of British Columbia but is also a species of special concern in the province. It is classified as overfished off the coasts of Washington, Oregon and California, as the population was reduced by 84 percent as of 1999. Since this designation, the abundance has about doubled but is still very small. One reason for the overfishing was that this is a long-lived species that may produce large abundances of juveniles only every 15 to 29 years.

The species' maximum size is 58 cm and 2.8 kg. The maximum age has been estimated to be 105 years, with an average age of 23 years. It is caught in the trawl fishery for Pacific ocean perch from British Columbia to Northern California. Adults may rest on the bottom, but their major prey is plankton.

SPLITNOSE ROCKFISH

Sebastes diploproa COMMON ●

IN THE STRAIT OF GEORGIA

Splitnose rockfish occur near or on the bottom in the deeper areas of the Strait of Georgia and in the shallower water near the bottom between Halibut Bank and McCall Bank. They may be more common in the southern strait as evidenced by their consistent capture in recent shrimp trawls in the Fraser River plume area. There are no recent abundance estimates for splitnose rockfish in the Strait of Georgia, but they probably are dispersed in the deeper areas of the strait, where they might be one of the more common species of rockfish. Larval splitnose rockfish are commonly found in floating sargassum in the Gulf Islands in May and June, along with young-of-the-year tiger rockfish.

NOAA

DESCRIPTION

The splitnose rockfish has a small, uniformly rose-red rockfish body and a white belly. It is readily identified by the deeply notched upper jaw, which has prominent knobs on the anterior border of each side of the notch. The fins are red with black blotches.

LIFE HISTORY

Off the west coast, splitnose rockfish occur from Baja California to the northern Gulf of Alaska. The greatest numbers occur between Southern California and British Columbia at depths of 80 to 800 m. The species has a maximum length of 46 cm, but most fish are much smaller. A sample of 40 individuals collected in a research trawl catch in the Strait of Georgia in September 2011 had an average length of 25.6 cm in a range from 19 to 32.5 cm. They are reported to prefer softer bottoms that contain occasional rocks and debris. Splitnose rockfish are another long-lived species of rockfish. The maximum reported age is 86 years, though there is a report of one fish aged at 103 years. In recent years in Puget Sound, trawl surveys produced individuals that were up to 64 years old, with an average age of 27 and an average length of 25 cm. Recognizing that this is also a relatively small fish, the old ages indicate that the fish grows very little each year once it approaches the maximum size. This also means it probably eats relatively little and remains more or less in one place. Males reach their maximum length before females, but females are larger than males. Growth is related to large-scale climate indicators, which may indicate that survival is related to climate, but these relationships are difficult to study. Males and females mature at about 6 to 10 years old, depending on the location and their own development. Off the west coast of British Columbia, 50 percent of males and females are mature at about 27 cm. A 37 cm female produces 255,000 eggs, which are fertilized internally. The larval splitnose rockfish are found near the surface over a prolonged period, indicating that the reproductive period is also prolonged. Fertilization may occur in the summer, with larvae being released in the strait from the spring through to the fall. Juveniles that were one year old and about 4.6 cm were found near the surface in Puget Sound. These pelagic juveniles eventually migrate deeper in the water column. Depths are related to size and age; larger and older fish are found in the deeper water. The diets of the larval fish are mostly amphipods and calanoid copepods. Older fish mostly feed on euphausiids, but also consume copepods, shrimps and amphipods.

GREENSTRIPED ROCKFISH

Sebastes elongatus COMMON ●

IN THE STRAIT OF GEORGIA

Greenstriped rockfish have frequently been found in research trawl surveys throughout the Strait of Georgia from the mid-1990s until the present. It appears they are widely distributed on or near the bottom throughout the strait in rocky areas, rubble and sandy or even muddy bottoms. They have been found repeatedly in shrimp surveys off Comox, in Howe Sound and in the plume area of the Fraser River. In Puget Sound they are found at depths of about 40 to 213 m, but they are not common. The rare sightings by scuba divers suggest that the species is not common in depths of less than 40 m in the Strait of Georgia. It is possible greenstriped rockfish may become more abundant in the strait as it warms, because they are a species that prefers warmer waters.

DESCRIPTION

The four broken horizontal green stripes and the slender, elongate body distinguish this relatively small rockfish. There are also green streaks on the caudal fin, and the lateral line is within a pink-coloured area. Pectoral, pelvic and anal fins are pale red. The symphyseal knob is missing or very small.

The greenstriped rockfish's horizontal greenish stripes of pigment blotches always leave a clear space over the lateral line, with another horizontal space above the lateral line and a couple more below, giving the striped appearance. Danny Kent

LIFE HISTORY

Greenstriped rockfish occur from central Baja California to the western Gulf of Alaska. Off the west coast they are found at 12 to 495 m, but mostly at 100 to 250 m. In the Strait of Georgia, they are more common to the north and at depths of 50 to 200 m. In studies using submersibles, they have been observed resting individually or in small groups on soft bottom near aggregations of rock. They are most frequently observed dispersed individually on sand and mud bottom deeper than 40 m, but they can also occur in high-relief, rocky, "untrawlable" areas deeper than 80 m. They have a maximum length of 43 cm and a maximum age of 54 years. Most fish in commercial catches range between 25 and 35 cm. However, the average size in the Strait of Georgia is about 15 cm, and most of the fish that are kept are between 5 and 15 years old. Growth in older females and males slows at about age 20 and at about 33 cm for females and 28 cm for males. Females are larger than males, but both mature at about the same age, between 7 and 10 years, and at lengths of 18 to 24 cm. Off Oregon and Washington, 50 percent of males were mature at 24 cm and 50 percent of females at 22 cm, with all males mature at 31 cm and all females at 29 cm.

Depending on their size, females produce between 11,000 and 295,000 eggs. As with other rockfishes, female greenstriped rockfish release live young. Mating occurs from December to February, with the stored sperm fertilizing the eggs in early spring. Gestation lasts about a month and the larval fish are released in the late spring. Larvae are extruded at 5 mm and are pelagic for about 2 months, which allows them to be widely dispersed within the Strait of Georgia. The juveniles remain in the surface waters until the fall, when they will have reached lengths of about 3 cm; then they move to the bottom. Like many long-lived rockfish, the adults move into deeper water as they grow larger. As small fish, they eat euphausiids, copepods, small shrimp and amphipods. Larger individuals will eat squid and small fish. Adults are eaten by sharks, porpoises, salmon and seals, and probably other predators such as lingcod and bigger rockfish.

PUGET SOUND ROCKFISH

Sebastes emphaeus

COMMON ●

IN THE STRAIT OF GEORGIA

Puget Sound rockfish are found throughout the Strait of Georgia but are most abundant in the southern strait, particularly among the Gulf Islands and San Juan Islands. They are not commonly caught in bottom and off-bottom research trawls, indicating they are probably found closer to shore. They are the third most observed rockfish by scuba divers in crevices in nearshore rocky areas throughout the strait (copper and quillback rockfishes are the top two).

DESCRIPTION

This is a very small species, less than about 18 cm in length. It is the smallest of all rockfish. It is a distinctive copper red with dark-olive blotches, green bands radiating from the eye and a red mouth. The soft part of the anal fin is rounded, and there are seven rays, with the second spine longer than the third.

LIFE HISTORY

The Puget Sound rockfish occurs from Northern California to the northern Gulf of Alaska. They have greatly increased in abundance in Puget Sound in

Puget Sound rockfish are always small and have a pink to beige colour and a horizontal dark streak along the lower body. The body profile is slender and the dorsal fins are short. Danny Kent

the last 40 years, and it is likely they have also become more abundant in the Strait of Georgia. The reasons for these increases are not known. It is possible they are a consequence of the diminished numbers of large predators such as lingcod. In Puget Sound the species is most abundant in the San Juan Islands area. The fish are a schooling species and are found in stronger currents and steeper rocky areas. They occur from 3 to 152 m in Puget Sound and as deep as 366 m throughout their range. In Puget Sound most are found at less than 91 m.

Puget Sound rockfish have a life history strategy that differs from other rockfish. In addition to their small size, they reach sexual maturity early, are not long-lived, and release their larvae in the late summer rather than the winter and spring. The maximum age is reported to be 22 years, but average ages are much younger, in the range of 5 to 6 years. In several studies, the oldest fish were 12 to 13 years old. Females grow to larger sizes than males. Females mature at 11 cm and age 1 or 2. About 50 percent are mature at 12 cm and virtually all are mature by 13 cm. As with most rockfish, fertilization is internal, and sperm is stored in the ovary for up to several months. The developing eggs and larvae remain within the female for about 30 days. Larval fish are released at the end of August and early September in Puget Sound, and most likely at the same time in the Strait of Georgia. The short release period, or spawning season, must have evolved to match with the production of preferred planktonic prey, but this relationship remains to be studied. Young-of-the-year Puget Sound rockfish that are 2.5 to 3 cm are found in January and February in nearshore areas in Puget Sound. There is a six-month period in which the larval fish can disperse throughout the area. Adults mostly occupy steep rocky habitats with large boulders and strong currents. Adults are most active at dusk when they probably feed, but they are reported to be inactive at night. Puget Sound rockfish feed in the midwater on copepods, amphipods, arrow worms and gelatinous plankton. Because of the changing species composition of rockfishes, responsible stewardship is needed to ensure there are periodic surveys, including a comprehensive analysis of the biology and population dynamics of the Puget Sound rockfish and other rockfish species.

WIDOW ROCKFISH

Sebastes entomelas RARE ★

IN THE STRAIT OF GEORGIA

Widow rockfish are rarely found in the Strait of Georgia, but individuals have been observed by scuba divers in Active Pass and specimens have occasionally been caught in pelagic research trawl catches, including in Toba Inlet. There is a recent report of a widow rockfish in a school of yellowtail rockfish at the mouth of Howe Sound.

DESCRIPTION

This species is a slim rockfish that has a brown to brassy orange body and a lighter belly. It has no symphyseal knob on its jaw tip. The mouth is small and there are rough ridges above the eyes. The tip of the pectoral fin extends past the pelvic fin. Fins can be black, as is the inside of the body cavity. This black body cavity distinguishes the species from most other rockfishes and is the source of the species name *entomelas*, which is from Greek and relates to the black-walled gut cavity.

NOAA

LIFE HISTORY

The widow rockfish is common offshore with a centre of abundance from British Columbia to Northern California and a range from the western Gulf of Alaska to northern Baja California. The maximum length is approximately 59 cm, and the maximum weight is 3 kg; however, most fish in the commercial fishery are 35 to 45 cm. They occur pelagically at depths from 25 to 550 m, with most at about 140 to 210 m over rocky banks. Schools of thousands of fish form in the midwater at night over offshore pinnacles and tend to disperse during the day, which is not typical of most rockfishes. The maximum age is about 60 years, and both males and females mature between age 3 and 8. Most fish in the commercial fishery range in age from 5 to 30 years; even fish of similar size can differ in age by 20 years. As with most other rockfish, fertilization is internal, and after one to three months the young are born alive. The largest females can produce over one million eggs. The larval fish are released from the female from December to May. Off British Columbia, most young are released in May. Prey includes a variety of plankton species, small fishes and pelagic crustaceans that are commonly found in a deep layer of plankton. Widow rockfish support an important commercial fishery from British Columbia to Oregon, with populations in the United States from off Washington to California considered to have recovered from low abundances.

Andy Murch

YELLOWTAIL ROCKFISH

Sebastes flavidus COMMON ●

IN THE STRAIT OF GEORGIA

Yellowtail rockfish are found throughout the Strait of Georgia but have been more common in the southern strait, particularly in Howe Sound and off Victoria, where they are often seen by scuba divers. They have been observed from a submersible in Saanich Inlet and found occasionally in on- and off-bottom research trawl surveys in the Qualicum Beach area. It is reported they have been resident around Icarus Point for the past few decades. Juveniles could enter the Strait of Georgia through deepwater currents from offshore, but the age range indicates they are reproducing in the strait. In general, yellowtail rockfish probably can be considered common, but large schools have recently been reported around Entrance Island, Cape Lazo and the northern strait, possibly indicating an increasing abundance. There have been frequent recreational catches near Campbell River, perhaps signifying a recent large increase in abundance.

The fins of a yellowtail rockfish have a characteristic yellowish tint, which also appears on its head and body markings, but the back has a dark pigment with several clear spaces at the top. The head and body profile is a bit more slender than that of a black rockfish, its closest relative, which has the same dorsal pattern. Lee Newman

DESCRIPTION

Yellowtail rockfish are called "greenies" by fishermen because of their green to brownish-green colour. There are other distinguishing characters, such as a vertical posterior edge to the anal fin, but the green and yellow appearance of the body and the fins, with a bright-yellow tail, is a dependable way of identifying this fish.

LIFE HISTORY

Yellowtail rockfish are found from near Point Conception, California, to the eastern Aleutian Islands. They are most abundant from Oregon to British Columbia. They are mostly a midwater schooling species with adults found off the bottom or near to offshore reefs. Offshore, they may be on bottom at depths of up to 549 m, but most are found in water 110 to 200 m deep. Offshore, adults tagged and released in areas away from the capture location were able to find their way home from a distance of 22.5 km. A study in the Strait of Georgia from 2003 to 2008 found that ages ranged from 5 to 73 years, with most between 15 and 35 years old. In this study, few fish older than 30 years exceeded 60 cm. A few of these older fish weighed between 5 and 6 kg. The maximum age of at least 73 years indicates this is a moderately long-lived rockfish species. The average length of fish in the Canadian fishery off the west coast of Vancouver Island is 47 cm, in a range of 35 to 51 cm. Age at first recruitment to the fishery is between four and five years and the number of age-four fish in the fishery off the United States is used as an index of year class strength. Females grow to larger sizes than males and may be older. The largest reported fish was 70 cm. Growth rates of most males and females begin to slow between the ages of 15 and 20 years. Males begin to mature at about four years old and 30 to 35 cm. Half are mature at 32 to 44 cm and 5 to 9 years old, with all being mature at 37 to 54 cm. Females first mature at 4 to 6 years old and lengths of 27 to 36 cm. Fifty percent are mature at 6 to 11 years and 36 to 45 cm, with all being mature at 38 to 50 cm. The older and larger fish are in deeper water and spawn first. Mating occurs about mid-fall in waters off the west coast. The sperm is stored inside the female until the eggs are fertilized in the winter. Gestation takes about 30 days and larvae are released from late winter to early spring in British Columbia. Females can produce 56,900 to 1,993,000 eggs, depending on their size. Spawning occurs once a year and it appears that mature females spawn every year. Juveniles and smaller fish feed on zooplankton, including copepods and euphausiids. Larger fish also consume euphausiids but will also eat fish, including Pacific herring and lanternfishes.

The major source of adult mortality in yellowtail rockfish is fishing. We all learned in school that plants and animals evolve to maximize survival in a changing environment. Prior to fishing, these rockfish adapted to their changing habitat by living to old ages. They most likely are adapting to the effects of fishing, but how they are adapting is not known and, importantly, how these changes will affect their survival in an environment that is also rapidly changing are questions that should keep biologists and managers up late into the evening. This is an important commercially fished rockfish that requires skill and persistence to manage.

ROSETHORN ROCKFISH

Sebastes helvomaculatus

VISITOR ▼

IN THE STRAIT OF GEORGIA

The rosethorn rockfish is considered to be very rare in the Strait of Georgia. The groundfish database at the Pacific Biological Station in Nanaimo has two records from research catches and 10 from commercial catches over the years 1976 to 2009. These records probably are valid, as the rosethorn rockfish is readily distinguished by its colour pattern and the records are from researchers and commercial rockfish fisheries. These records are from bottom trawl catches in the Gulf Islands in 1976 and between Texada and Vancouver Islands in 2009. There are four commercial hook-and-line catches from 2001 and 2003 off the mouth of the Fraser River and in the area west of Texada Island. One report is from a longline fishery for spiny dogfish in 2006 in the Gulf Islands. Two are from bottom trawl catches in the waters west of Texada Island in 1997 and 2003.

NOAA Fisheries / Southwest Fisheries Science Center

The remaining three commercial records are from hook-and-line catches in 1994 and 2004 in an area between Texada Island and the Northern Gulf Islands. This is a deepwater species that most likely enters the Strait of Georgia from offshore and is not reproducing in the strait.

DESCRIPTION

This is a small rockfish that is distinguished by the four to five distinct white dorsal patches with pink or orange borders. The second anal fin spine is longer than the third spine and there are typically 16 pectoral fin rays. The body can have dark pigment dorsally below the dorsal fins but ranges in colour from pink to orange or yellow and may have greenish mottling. Most rosethorn rockfish seem to be red-orange to yellow-orange. The ventral is lighter pink or whitish. Fins can be pink, yellow and green.

LIFE HISTORY

Rosethorn rockfish range from western Alaska to central Baja California. This is a solitary species that commonly occurs in deep water on or just off bottoms that are mixed rock and mud. All of the records from the Strait of Georgia were on or close to the bottom. Offshore, most are found at depths of 80 to 350 m, but there are records of them occurring at 25 to 549 m. Males and females mature at lengths of 27 cm and have a maximum age of 87 years. Females release larvae from February to September, most giving birth from April to June. The varied diet includes crabs, shrimps and fishes. The species is fished commercially in some areas.

QUILLBACK ROCKFISH

Sebastes maliger COMMON ●

IN THE STRAIT OF GEORGIA

The quillback rockfish is probably the most common rockfish species in the Strait of Georgia. They are the most common rockfish caught in the recreational fishery and the second most common seen throughout the strait by scuba divers. Research surveys show they are more abundant in the northern areas of the strait.

Danny Kent

DESCRIPTION

The quillback rockfish is brown to dark brown or black with lighter yellow-orange patches under the dorsal fin. The posterior of the head is a spotted light-yellow to orange colour with a dark patch dorsally over the eye. Fins are dark brown except for the light-yellow colour of the dorsal fin's spine area. The dorsal fin is large and the skin connecting the anterior spines forms notches. This high, deeply incised dorsal fin gives the species its Latin name, *maliger*, which translates to "I bear a mast." There is no symphyseal knob.

LIFE HISTORY

Quillback rockfish occur from Kodiak Island, Alaska, to Southern California from intertidal depths to 274 m, but most are from the intertidal to 150 m. Their

Juvenile quillbacks have the strongest contrast of light colours forward and dark aft, and a slightly lower profile of the spiny dorsal fin. Danny Kent

maximum age is 95 years, but most fish are about 10 to 15. Fish in this age range are about 25 to 30 cm in length. About one-half of the juveniles mature at age 11 or at about 29 cm in length. Females grow to slightly greater lengths than males, and the maximum age of males is 80 years, which is 10 years less than that of females. The oldest in the Strait of Georgia so far was a 67-year-old female. The species' growth rate slows at about 35 cm long and age 30 for both males and females. The maximum length is 64 cm, but the average size in the fishery in recent years is about 35 cm. The larger individuals occur in deeper water in rock or mud-rock habitats. The highest numbers occur at depths of about 60 m, with relatively few deeper than 80 m. They occur in the Strait of Georgia mostly from 26 to 100 m.

They are solitary, territorial fish, grouped within the inshore rockfish category, and they most likely remain close to the areas where they reached adulthood. Mating occurs from November to February, with most in December when the male gonads are running ripe. Females can mate with several males and store their sperm. Larvae are released in March to July, mostly in April and May, and they then drift shoreward. The young fish remain in these nearshore areas, close to plants such as large kelp or to free-floating plant material close to the bottom. Settlement occurs at 3 to 9 cm long or six to nine months old. As the fish age, they move into deeper rocky areas. In the Strait of Georgia,

The youngest juvenile quillback rockfish, with freckling prominent, are likely to be found taking refuge in cloud sponges or among piled cobblestones. Danny Kent

quillback rockfish live on the bottom, where they feed on pelagic fish, shrimp, squat lobsters and crustaceans. The average age of mature females and males in the strait is about 21 years old, and the average length is about 31 cm. Longline research surveys catch fish that average 32.3 cm in length, 696 g in weight and an age of 18 years. Ages in these surveys ranged from 4 to 55 years. The 50-year-old fish was 39.2 cm long and 1.0 kg in weight. Juveniles at one year old averaged 11 cm and 23 g; at two years, 13.6 cm and 32 g; at three years, 15.3 cm and 44 g; and at four years, 17.7 cm and 71 g. There is an estimate that quillback rockfish make up about 30 percent of all rockfish in the Strait of Georgia. They are eaten by seals in the strait. It is a preferred species in live fish markets in Vancouver. New research has found that the quillback rockfish and its close relative, the copper rockfish, produce sounds, likely using specialized sonic muscles that drum on the swim bladders. Scientists at the Pacific Biological Station and the University of Victoria are studying sound production in reef fishes near Nanaimo using a combination of cameras and SoundTrap hydrophones, along with surveys performed via scuba diving and the use of remotely operated vehicles.

DUSKY ROCKFISH

Sebastes ciliatus VISITOR ▼

IN THE STRAIT OF GEORGIA

The southern limit of the distribution of dusky rockfish is Johnstone Strait, which is the northern boundary of the Strait of Georgia. However, there are records of occasional catches of dusky rockfish throughout the strait in the commercial fishing records of the Department of Fisheries and Oceans from 1988 to 2022. No specimens from the commercial fishery bycatch were retained, so the identifications cannot be confirmed. There were larger catches of the species in commercial fishing trips in Queen Charlotte Strait, north of Johnstone Strait. This abundance in Queen Charlotte Strait, along with the recognized southern limit of dusky rockfish at the northern boundary of the Strait of Georgia, indicates there may be validity in at least some of the reports of commercial fisheries catching the occasional dusky rockfish in the Strait of Georgia.

Kristina Castle

DESCRIPTION

Dusky rockfish are uniformly black to dark blue with dark black-grey fins and a black lining of the body cavity (peritoneum). The symphyseal knob is moderate to small, and the anal fin margin is perpendicular. The head spines are weak with a convex interorbital space. In the Strait of Georgia, identifications could be confusing the species with the black rockfish, which is not common in the strait. Black rockfish have five or six lighter blotches at the base of the dorsal fin; these are absent in ducky rockfish. The body cavity of black rockfish is whitish, while in dusky rockfish it is black. The symphyseal knob in black rockfish is not distinct.

LIFE HISTORY

Dusky rockfish are a nearshore species that are found in shallow water and to depths of up to 160 m. The species occurs from the western Aleutian Islands in the Bering Sea through the Gulf of Alaska to Johnstone Strait. Staff at the Vancouver Aquarium collected juvenile dusky rockfish off the northwest coast of Vancouver Island, so the distribution extends at least that far. Also, our commercial catch records extend the distribution into the northern Strait of Georgia. A new species of dusky rockfish has been identified, so the species' biology remains to be sorted out, but for the combined two species the maximum age could range from 67 to 76 years. Females are slightly larger than males, with a maximum size between 47 and 59 cm. About 50 percent of females mature at 11 years old and 43 cm long and release live larval fish between May and July. It is unlikely that they are reproducing in the Strait of Georgia. They feed mostly on plankton, including crab larvae, amphipods and krill. These are long-lived, slow-growing fish that would not need to consume large quantities of prey.

BLACK ROCKFISH

Sebastes melanops

UNCOMMON ◆

IN THE STRAIT OF GEORGIA

Black rockfish aggregate in schools just off the bottom in about 180 to 275 m of water, and on or near the bottom near kelp. Movements have not been well studied in the Strait of Georgia. The species was common in the 1980s and moderately available in the 1990s but is now rarely reported, and some believe they have virtually disappeared from the central and northern strait. They are present in the inshore benthic areas of the southern strait but are not common. Black rockfish dwindled in abundance near Vancouver with the advent of sport downrigger gear for salmon trolling in the 1960s. Almost none could be found by the late 1980s at previously known hot spots like Starboat Cove at Point Atkinson and near Whytecliff Park. The Vancouver Aquarium collected young-of-the-year black rockfish at Ucluelet in 1996, reared them in display tanks and then released them with various tags, under permits, at Point Atkinson in 1997, 1998, 2004 and 2005. The mature rockfish released in 2004–05 had radio tags that indicated they remained at the reef where they were introduced. There was a big year class in the Juan de Fuca Strait in 2006, but at Point Atkinson young were observed for the 2004, 2006, 2008 and 2010 year classes, so it is presumed that the fish at Point Atkinson were reproducing.

The black rockfish has the same dark band with light clear spaces along its back as its close relative, the yellowtail, but the ground colour is grey and has no trace of yellow. The head is also heavier and the fins are a bit less pointed. Danny Kent

DESCRIPTION

Very young black rockfish may be orange-brown with a black spot on their dorsal fin. Adult black rockfish are not colourful; they are black to grey dorsally and a lighter white or grey ventrally. Some blue and black spotting may be present, or they may have a mid-lateral light-grey band. Fins are dark with black spots. They are more slender than the more colourful, stouter species. Because they are the traditional bass-like shape, some fishermen call them bass. As with all rockfish, the spines are venomous and can inflict pain or cause infection.

LIFE HISTORY

Black rockfish range from the southern Bering Sea to northern Baja California. They are grouped with rockfish in the inshore and nearshore shelf category in British Columbia. The maximum size is about 69 cm and 6 kg. The maximum age is 50 years. They range from the surface to 366 m but are mostly near the surface, above 75 m. There is an equal ratio of males and females up to about age 10, when the percentage of females declines. It is not known when the larvae are released in the Strait of Georgia, but if the timing is like that of other species—in January to May, with most released in February—fertilization would be about 37 days earlier in mid-January and copulation several weeks before fertilization. It has been observed that some males may be mature in the summer. Females may produce between 125,000 and 1,200,000 eggs, depending on their size. A 16-year-old produces almost one million eggs. Off the west coast, 50 percent of females mature at about 38 to 42 cm and 6 to 8 years old. However, black rockfish are first fished when they are 3 to 4 years old and 25 to 30 cm long. Their growth rate begins to slow at 10 to 15 years old and 42 to 45 cm long. Off Oregon, the oldest males and females in recent fishery samples were 29 and 26 years old, respectively. Larvae and small juveniles are pelagic for several months to a year, eventually settling in the nearshore areas. In recent years, handline fishermen report that off the west coast black rockfish are widely distributed and likely one continuous population. Tagging studies show that most fish remain in one area, where they occur in schools over rocky bottoms. Adults feed on a variety of prey, including euphausiids, shrimp and various zooplankton. They also feed on other fish such as sand lance and even juvenile rockfish that they find when the schools are near the surface.

VERMILION ROCKFISH

Sebastes miniatus UNCOMMON ◆

IN THE STRAIT OF GEORGIA

The vermilion rockfish is uncommon throughout the Strait of Georgia. In 2005 there was a year class of vermilion rockfish evident at various locations around southern Howe Sound, with between one and six individuals on any particular shallow reef and all of them about three years of age. As these fish grew larger over successive years, most individuals disappeared into deeper water. They had probably been born on the outer coast and drifted into the Strait of Georgia as larvae or early pelagic juveniles. There is an occasional report from the southern strait and in the Gulf Islands.

DESCRIPTION

Vermilion rockfish are a species that some people will confuse with the similarly coloured canary rockfish. The difference is that the canary is bright orange, like a goldfish, whereas the vermilion is actually the colour called

Andy Murch

vermilion, which is a shade of red. There are three orange bands radiating dorsally from the eye. The background variegation and facial war paint–like stripes on a canary rockfish are a silvery white (especially white on the belly), but the background colour of a vermilion rockfish is usually greyer. That said, colours can be misleading, so also look carefully at the rear tip of the anal fin, which is slanted anteriorly in the vermilion rockfish. To confuse matters more, both the vermilion and the canary rockfish tend to occur in exposed outer coast locations with rocky reef structures to the bottom. Juveniles of both occur at sandy margins of such reefs in shallower water, and both species occupy progressively deeper water as they mature and age.

LIFE HISTORY

Vermilion rockfish occur from central Baja California to the Gulf of Alaska. They grow to 76 cm in length and live a little over a half century, up to 60 years. Approximately 50 percent of females are mature at 37 cm and 5 years old, with all mature at 47 cm and 9 years. Females have been known to produce up to 2.6 million eggs in a year. On the outer coast, it is the smaller and younger vermilion rockfish that are most abundant. Recent DNA research has proven this is yet another fish that had been considered to be just one species but in fact is two "sibling species" that are very hard to tell apart. The second species is in deeper water and is distributed south of British Columbia.

Chad King

DEACON ROCKFISH

Sebastes diaconus VISITOR ▼

IN THE STRAIT OF GEORGIA

The deacon rockfish was described as a new species in 2015. It was separated from the blue rockfish (*Sebastes mystinus*) using molecular and morphological data. However, the two species are very difficult to distinguish based on appearance and are considered cryptic species, which means the morphological differences between the species are not clear. According to the reported distribution of the new species, any previous report of blue rockfish in the Strait of Georgia would actually be deacon rockfish. There are only recent reports of deacon rockfish in the strait. We know of reports near Haro Strait and Burrard Inlet, recent sightings in Howe Sound and a scuba observation from the San Juan Islands. Because they are more common elsewhere off the west coast, we assume the fish in the Strait of Georgia arrived from offshore.

The body colour of a deacon rockfish may tend to cause misidentification as a black rockfish, but note the lack of clear spaces along the back; the small, slender jaw; and the prominent, zebra-like streaks behind the eye. Andy Murch

DESCRIPTION

Deacon rockfish are difficult to distinguish from not only blue rockfish, but also another close relative, the dusky rockfish (which is itself one of a pair of sibling species), which occurs in a more northerly distribution centred off Alaska. Most people also have trouble distinguishing blue rockfish from another dark-coloured rockfish, the black rockfish, which has a bigger head and larger jaw. A paper describing the deacon rockfish provides some distinguishing characters; to summarize, the deacon rockfish is uniformly dark with small speckling, compared to the blue-green colour of the blue rockfish, which also has dark blotches. The deacon rockfish also has a protruding lower jaw; this feature can be compared to the jaws of the blue rockfish, which meet at the tips of the lower jaw (symphyseal knob) and upper jaw or only very slightly forward.

LIFE HISTORY

The deacon rockfish is reported to occur from Central California to British Columbia, while the blue rockfish occurs from northern Oregon to Baja California. However, these distributions should be further studied using molecular identifications, particularly for the northern distributions. We do not know if the deacon rockfish has similar behaviour to the blue rockfish, but the following is based on what has been reported for blue rockfish. The species tends to school in the water column and is less inclined to hide among boulders or in cliff crevices. It occurs over reefs and pinnacles from the surface to 550 m deep near surf-swept outer coast shores, but most commonly from the surface to 90 m. It grows to a maximum length of 53 cm and lives to a maximum age of 44 years. Because this species has a small mouth, it feeds on plankton less than about 5 mm long. Because they can occur near the sea surface, these rockfish are easy to catch on hook and line.

CHINA ROCKFISH

Sebastes nebulosus RARE ★

IN THE STRAIT OF GEORGIA

China rockfish are rarely seen in the Strait of Georgia. They have been reported in Jervis Inlet, off Gabriola Island, and around Active Pass and the Southern Gulf Islands to the Sechelt area, in Howe Sound and around the San Juan Islands. They are rare in the catches of any of the bottom or off-bottom research trawl studies. Scuba divers often have difficulty distinguishing China rockfish from quillback rockfish, which may result in erroneous reports. China rockfish are sometimes reported in recreational fishery catches throughout the Strait of Georgia.

DESCRIPTION

This species is one of the most spectacularly coloured rockfish. It has an almost tropical appearance because of its dark-blue or black body colour with a broad, patchy, yellow-to-white stripe from the front of the dorsal fin down and along the posterior of the lateral line, and numerous small yellow or white spots. There is a yellow third dorsal spine with the yellow line extending posteriorly. It is reported that the spines can inflict some pain if they penetrate one's flesh.

The dark freckles on a light background of a quillback are distinguishable from the light freckles on a dark background in a China rockfish, though the two species are often mistaken for one another. Note as well that the first dorsal fin is quite dark on the China, except for a yellow swoosh down through the dorsal fin and along the lateral line (in contrast to the dark lateral line on a quillback). Danny Kent

LIFE HISTORY

China rockfish occur from Kodiak Island, Alaska, to Southern California. They are sedentary and solitary, occurring at depths from 10 to 100 m close to rocky bottoms and crevices. China rockfish are one of the rockfishes British Columbia includes in the inshore and nearshore group. Their maximum size is 45 cm long and 1.67 kg, but fish found in the Strait of Georgia would be smaller. The size at 50 percent maturity is 28 cm. By 31 cm, most are mature. The maximum age reported is 79 years, and the age at 50 percent maturity is between 3 and 6 years. Fish with an average length of 8 to 44 cm are 1 to 17 years of age. Larvae are released early in the year, perhaps late January to February. Juveniles, about 6 cm in length, have been found in 9 to 18 m of water. China rockfish diets have not been studied in the Strait of Georgia, but these fish most likely feed on crustaceans, small fish and other benthic organisms. If larvae are released in late January or February, fertilization would have to occur in November to December. While this species is very rarely seen in the Strait of Georgia, it is commonly displayed in commercial aquariums.

TIGER ROCKFISH

Sebastes nigrocinctus COMMON ●

IN THE STRAIT OF GEORGIA

Tiger rockfish appear to be moderately abundant in the Strait of Georgia and are sighted as solitary individuals near or inside crevices by scuba divers throughout the strait during daylight hours. They are most numerous in the central strait off the east coast of Vancouver Island between Nanaimo and Denman Island, but are also occasionally seen around the Northern Gulf Islands, the Sechelt area and off the Sunshine Coast. This was one of the species that was killed in large numbers when Ripple Rock was exploded in the northern Strait of Georgia in 1958. In longline research surveys, tiger rockfish are caught in the middle regions of the strait between 41 and 70 m. Samples of fish in these catches averaged 35 cm in length.

Danny Kent

DESCRIPTION

Tiger rockfish are so brightly coloured that they are a common subject for artists. The body is both red to pink and black to brown with five black or dark-red stripes. Two dark bands radiate posteriorly from the eyes. The fins have white to pink tips. The fish is capable of losing colour quickly when captured.

LIFE HISTORY

This is a solitary species, living near the rocky bottom or in caves and crevices. Tiger rockfish are found from the Aleutian Islands in Alaska to Southern California. Scuba divers report that it is territorial and even a little shy. Scuba divers who return to the same areas think that they see the same fish. However, these fish are rarely observed during the day. The common depth appears to be 50 to 200 m, though they have been reported at depths from 18 to 275 m. The maximum age is an astonishing 116 years, with a maximum length of 61 cm and weight of 3.4 kg. Reproduction probably occurs about the same time as it does for other species, with most larvae being released in late January or February. Young-of-the-year larval tiger rockfish are often found in floating debris and sargassum in May and June in the Gulf Islands. Because tiger rockfish are sedentary and not known to move long distances, rockfish conservation areas may help to rebuild their populations. There is no population estimate, so it is difficult to evaluate the overall population effect of conservation areas and catch restrictions and adjust the policy where necessary.

BOCACCIO

Sebastes paucispinis

RARE ★

IN THE STRAIT OF GEORGIA

Bocaccios are rare at best in the Strait of Georgia and may no longer exist there except for an occasional visitor from offshore. This species was considered a choice sport fish in the Howe Sound region of the strait before the 1970s but is no longer reported. There were also records in Active Pass in the 1970s. There have been no bocaccio catches in thousands of pelagic research trawl sets since the mid-1990s.

DESCRIPTION

The bocaccio is a large, slab-sided rockfish with a long head and an elongate, heavy lower jaw clearly extending beyond the upper jaw. Colours of bocaccios range from pale grey-brown to pinkish grey, sometimes with an orange cast. This species is unusual for having almost no head spines. The species with the closest resemblance to a bocaccio is the silvergray rockfish, which can most rapidly be distinguished by just grabbing the top of the head between the eyes with one's thumb and forefinger. If spines slightly cut the skin, you have a silvergray, and if the head is entirely smooth, you have a bocaccio. This is the reason for the species names *brevispinis*, for "short spine" (silvergray), and

Juvenile bocaccio (pictured here) have the characteristic long jaw and slender fin height of the adult, with a bit more spotted body colour. Danny Kent

paucispinis, for "few spines" (bocaccio). Italian fishermen in California called them "big mouth," or *bocaccio* in Italian.

LIFE HISTORY

The bocaccio is a schooling, semi-pelagic species that occurs from central Baja California to the Alaska Panhandle in the western Gulf of Alaska. It is a fast-growing rockfish, reaching a large size in less than four years; its maximum size is up to 98 cm and 10.7 kg. They may live as long as half a century, which, as rockfish go, is not very long. In a sample of about 4,000 fish, one male was 70 years old but most older fish were 50 to 55. Bocaccios have been overfished in the southern part of their range, where they were commercially sought, particularly in California during the 1980s and 1990s. By the end of the 1990s the species' California population was only about 2 percent of the biomass estimated during the 1960s. They were never a commercial target off British Columbia. Bocaccios can reproduce more than once a year, with peak births occurring in winter and spring. Juveniles and adults tend to eat fish and squid. They typically occur over a rocky bottom near ridges or walls at depths from 122 to 478 m, with most at 50 to 250 m. They will form schools, often with other rockfish species like widow and yellowtail rockfish.

Bocaccios off the west coast of British Columbia were listed as threatened in 2002 and endangered in 2013. Abundances had declined for decades, with the last good recruitment in 1978. However, in 2016, there was a massive abundance of larval bocaccios all along the west coast that showed up in the diets of many species. This was the first especially strong year class in about 80 years, which has completely turned around the trajectory of the population. There was no evidence of the larval abundance in the Strait of Georgia in 2016, indicating that the exceptional survival was only offshore. If abundances increase in the strait over the next years, it will probably be from migrations into it. This occurrence of exceptional recruitment is consistent with a theory of longevity wherein fish are believed to live to be very old because the ocean conditions for successful survival occur only occasionally.

CANARY ROCKFISH

Sebastes pinniger UNCOMMON ◆

IN THE STRAIT OF GEORGIA

The canary rockfish is an uncommon species in the Strait of Georgia and is listed as threatened in Puget Sound. The fish have been seen by scuba divers in the past in the eastern and southern areas of the strait. More recent sightings have occurred in Burrard Inlet, Howe Sound and the Gulf Islands area. They were caught infrequently in longline bottom research surveys throughout the strait and were the fourth most common rockfish in recreational catches.

DESCRIPTION

Canary rockfish are attractive fish that are popular in aquariums. Adults appear bright orange but are actually a mottled orange over a grey background. There is a thick, clear grey-white band that extends along the lateral line from the head to the tail. Three orange stripes on the head angle posteriorly, and the fins are orange.

LIFE HISTORY

The canary rockfish occurs from northern Baja California to the western Gulf of Alaska, with a centre of abundance from Central California to British Columbia. It is believed this is one homogenous population, though the distribution can

Adult canary rockfish have a goldfish-orange colour. Their key characters include the V-shaped bands on the face, which slope away from the eye; a clear, white stripe along the lateral line; and the sharp, vertical rear margins of the pelvic and anal fins. Danny Kent

be patchy. This is a large rockfish with a maximum length of 76 cm, but most fish in the British Columbia fishery average about 48 cm and range from 26 to 54 cm. Females are slightly larger than males. Maximum sizes in the Strait of Georgia are not known, but the average size in the various fisheries in recent years is about 42 cm. The flesh is white and has a delicate flavour. The spines are mildly venomous if they penetrate flesh. The canary rockfish has been sought after by sports and commercial fishermen and most likely has now declined to very low abundance in the Strait of Georgia.

Canary rockfish may occur in large schools or loose aggregations near the bottom or near rock ledges and pinnacles where the current is strong. Scuba divers enjoy viewing these fish in schools of small numbers just off rocky ridges. The young occur in shallow water and adults usually at depths of 100 to 200 m, but the range is from the surface to 838 m. The maximum size is 76 cm and 6.7 kg and the maximum age is at least 84 years, but most are 60 to 75. Surprisingly, only males are commonly found over the age of 50, and females rarely are over 30. Most males and females in the fishery range from about 5 to 20 years old. Growth of both sexes slows at about age 15. Males first mature at age 3 to 4 and at lengths of 18 to 28 cm. Fifty percent of males mature at about age 7, and virtually all are mature at age 15 and a length of 45 cm. Females first mature at age 8 and lengths of about 27 cm. Fifty percent are mature at age 13, slightly older than males. All females are mature at age 20. Females generally grow more quickly and to larger sizes than males, but they also mature a few years later. In British Columbia 50 percent of males and females are mature at 43 and 51 cm in length, respectively. Like most rockfish, fertilization and embryo development is internal. Insemination occurs late in the year, usually about two months before the live larvae are released about February. Juveniles migrate into shallow water around reefs and rocky habitats, where they may remain for up to three years. Larger and older fish move into deeper water. Females produce 260,000 to 1,900,000 larvae, depending on the size of the female. The fish are mostly plankton eaters, feeding on euphausiids and small shrimp. However, they will also eat small fish and bottom invertebrates.

REDSTRIPE ROCKFISH

Sebastes proriger UNCOMMON ◆

IN THE STRAIT OF GEORGIA

Redstripe rockfish are uncommon in the Strait of Georgia, though they were found on the bottom in deeper water throughout the strait in one research study, and rare sightings have occurred in the Northern Gulf Islands area and outside of Howe Sound. Redstripe rockfish, like greenstriped and quillback rockfish, take up residence in deepwater glass sponge reefs in the Strait of Georgia along with juvenile yelloweye rockfish. They have not been found in the off-bottom research trawl surveys from the mid-1990s to the present.

Andy Murch

DESCRIPTION

Redstripe rockfish are distinguished by the light-red stripe along the lateral line inside green areas and by the blackened lips. There also is no spine on the lower rim of the eye, and the symphyseal knob at the front of the lower jaw is prominent.

LIFE HISTORY

Redstripe rockfish occur from southern Baja California to the southeastern Bering Sea along the Aleutian Islands. They are a schooling species, found near rocky bottoms at depths of 12 to 524 m, but mostly between 55 and 300 m. The maximum size is 52 cm. They are not common in Puget Sound but are sometimes found at depths from about 20 to 210 m. They are more common in Juan de Fuca Strait and around the San Juan Islands. The maximum age is 55 years, indicating that this is another moderately long-lived rockfish. The mean age of fish off the west coast of Vancouver Island is about 20 years. Females grow to larger sizes than males. Off the west coast of British Columbia, 50 percent of males are mature at about 28 cm, and 100 percent of males would be mature at only a few centimetres larger than 28 cm. Fifty percent of females are mature at 29 cm, and they also would all be mature at a few centimetres larger. Growth of females slows at lengths of 41 to 42 cm, while male growth slows at 33 to 34 cm. The age at 50 percent maturity probably is about seven years for both males and females. Mature female redstripe rockfish store sperm for several months until the eggs are fertilized internally in the early spring. Most larval fish are extruded about May to July. In Puget Sound, most larvae are released in July. Larvae are about 5.4 mm when released and remain in the surface waters for about two months. Larger fish are found in deeper water, and it is believed that the larger and older fish tend to remain in one area. The larval rockfish feed on copepod eggs and nauplii, as do the adult stages. Juveniles feed on zooplankton, including calanoid copepods and barnacle larvae. Adults consume crustaceans, fish and cephalopods. It is important to recognize that long-lived, slow-growing fish do not need to eat large volumes of food. It is probable that they remain in the areas that are relatively safe from predators and acquire food when it comes to them.

YELLOWEYE ROCKFISH

Sebastes ruberrimus COMMON ●

IN THE STRAIT OF GEORGIA

Commonly known as the "red snapper," the yelloweye rockfish is the best known and most esteemed of the rockfishes in the Strait of Georgia. They are found throughout the strait and are the third most reported rockfish in the recreational fishery. They are more abundant in the middle areas of the strait and to the north.

DESCRIPTION

Fishermen are familiar with the yelloweye rockfish's bright orange-yellow colour with a hint of pink. The belly is a lighter and more orange colour and the pectoral and pelvic fins may be tipped with black. The dorsal fins may also have a white edge. The top of the head is spiny. However, the easiest distinguishing feature is the yellow eye, which gives the fish its common name. Juveniles have a distinctly different colour pattern, with two or three lightly coloured stripes. One stripe is lateral and the other is more dorsal; another stripe is sometimes found on the edge of the dorsal fin. The juveniles are also more red than orange. Juvenile yelloweye displaying the prominent horizontal white stripes will not be subjected to aggression by territorial adults. In isolated rock piles, scuba divers will typically see the largest yelloweye in the mature adult livery of the uniform red body, but closely adjacent smaller adults will display the horizontal white stripes.

Bill Pennell

Young adult yelloweye rockfish are bright red, with considerable black on the fins and with horizontal white stripes that signal to mature adults that these younger fish do not want to compete for territory or for mates ("don't beat up on me"). Lee Newman

LIFE HISTORY

Yelloweye rockfish are found from the Aleutian Islands to northern Baja California. The species is one of the largest and oldest rockfish. The maximum size is 91 cm and 12.6 kg, but the average size in the Strait of Georgia is around 45 cm, with many fish being 15 to 25 years old. This fish occurs over rocky bottoms with broken rock around reefs, crevices, caves and rock overhangs. These are solitary fish that do not move far from their settlement area. The oldest individual found in the Strait of Georgia to date was aged at 101 years. The oldest found in all of British Columbia was 118 years old, and the oldest from outside of British Columbia was 121. In the Strait of Georgia, males have a mean age of 27 years old and length of 48 cm. Females have a mean age of 30 and length of 47 cm. The species starts becoming vulnerable to fishing at age 6 or 7 and is fully vulnerable by age 16 or at about 40 cm in length.

Smaller yelloweye rockfish are found at depths greater than about 40 m and often in the same areas as canary rockfish. Larger individuals are in deeper waters, often near reef walls and close to the bottom. Offshore, the habitat depths range from 15 to 549 m, with an average of 50 to 200 m. Fish in the Strait of Georgia are genetically distinct from those offshore. Fifty percent of the fish are mature at a length of about 40 cm and an age of 18 years. The average size of the fish caught in the sports fishery in the Strait of Georgia is 47 cm, and in the handline fishery, 43 cm. This means a substantial number of fish are caught before they can reproduce, as there is no size limit. Since 2018, recreational fishers have not been permitted to retain yelloweye rockfish in British Columbia and must release them at depth using a descending device. This is done to counteract the barotrauma symptoms—caused by expansion of the gas in their closed internal swim bladder—that rockfish are subjected to when brought to the surface. Some of these tagged and released yelloweye rockfish have been recaptured years after release in Alaska.

At settlement, yelloweye rockfish are black and white, with red pigment gradually replacing the black, especially on the horizontal body stripes. Remember that red is black at depth, since red light penetrates very little depth in the sea. NOAA

These large, slow-growing fish may produce several million eggs each year. They mate late in the year, and most of the pelagic larvae are released in May and June and extending to September. It is not known how long the larvae remain in the surface waters, but juveniles settle in deeper areas after about six to nine months and at lengths of 3 to 9 cm. Additions to the population are sporadic, which indicates that larval and juvenile survivals are highly dependent on ocean conditions. Despite the production of large numbers of young, few survive.

Fish species commonly eaten by yelloweye rockfish are Pacific herring, sand lance, shrimp, cod and other rockfish. This means they are resident in a particular area where they feed on prey such as fish and crustaceans that wander into their habitat. In the Strait of Georgia, they are listed as a species of special concern under the Species at Risk Act because their biology makes them susceptible to overfishing. Once overfished, these large, solitary fish can take 25 years or longer to rebuild the population in local areas. Unless there is careful monitoring of the impacts of fishing, their population may decline even further. It will be some time before biologists can assess the effectiveness of rockfish conservation areas, but the existing areas are a good start.

STRIPETAIL ROCKFISH

Sebastes saxicola RARE ★

IN THE STRAIT OF GEORGIA

The stripetail rockfish has been reported in the Strait of Georgia in the past, but it is rare. We have no records in any of the research surveys or scuba diver observations. We list this species because a few specimens have been reported from Porlier Pass and Indian Arm in *Pacific Fishes of Canada* and *Fishes of the Salish Sea*; however, although stripetail rockfish are not easily identified, it is likely they are not currently resident in the Strait of Georgia.

Jen Blaine

DESCRIPTION

The stripetail rockfish has large eyes and horizontal green stripes on its caudal fin. It is a silver-pink, yellow-pink or red fish with four to five saddles on its back. The caudal fin has green streaks. The tip of the lower jaw has a prominent symphyseal knob.

LIFE HISTORY

Stripetail rockfish are distributed from the eastern Gulf of Alaska to central Baja California at depths from 25 to 547 m, but mostly from 100 to 350 m. This is a smaller rockfish with a maximum length of 41 cm and weight of 700 g. The maximum age is 38 years, and females grow to larger sizes than males. The youngest juveniles are found shallow enough for scuba divers to sight them, but at a half year of age they are below diving depths at over 45 m deep. The stripetail rockfish is commonly found off the west coast of Vancouver Island, in Queen Charlotte Sound and southern Hecate Strait over soft bottom, usually at 100 to 200 m, with larger fish living deeper, at up to 350 m.

SHARPCHIN ROCKFISH

Sebastes zacentrus RARE ★

IN THE STRAIT OF GEORGIA

Sharpchin rockfish have only been reported a handful of times in the Strait of Georgia and are also rare in Puget Sound. There are old records from Burrard Inlet. A more recent record is from a longline survey of rockfish in the Strait of Georgia in the early 2000s; they also have been seen in recent remote camera footage in deep water in the southern strait. There have been no specimens in any of the numerous pelagic research trawl or bottom net catches from 1995 until the present.

DESCRIPTION

This is a small rockfish with a maximum size of 49 cm and 1.1 kg. It is light red or yellow-red. A horizontal dark V extends from the eye to the end of the gill cover, and faint brown irregular patches appear on the body. The knob on the front of the lower jaw is prominent and sharp, resulting in the common name "sharpchin."

Tom Laidig

LIFE HISTORY

Sharpchin rockfish occur from the eastern Bering Sea and the Aleutian Islands to San Diego, California, near hard bottoms at depths usually from 100 to 300 m, but depths can range from 25 to 660 m. The oldest age recently reported is 73 years. They mature at ages of 6 to 10 and lengths of 21 to 28 cm. Sharpchin rockfish have been found inside glass sponges or among aggregations of feather stars (crinoids).

SHORTSPINE THORNYHEAD

Sebastolobus alascanus UNCOMMON ◆

IN THE STRAIT OF GEORGIA

The shortspine thornyhead is rarely observed in the Strait of Georgia, though it is routinely caught in bottom research trawls near or on the bottom in the central and northern areas of the strait.

DESCRIPTION

This is a bright-red to orange fish with white patches on the body. The head is large and has spiny ridges and large eyes. Fishermen have traditionally called them "idiots" because of the large eyes and head, but they have become a very valuable species and could just as easily now be called "ingots."

LIFE HISTORY

Shortspine thornyheads are found on soft bottoms in deeper water from the western Pacific off Japan and Russia to the Bering Sea and south to central Baja California. They occur at depths from 17 to 1,600 m, but most are at 200 to 800 m, where they tend to be solitary. The fishery for them uses bottom trawls or hook and line. The fish are esteemed in Japan and elsewhere in Asia. Once it was realized that rockfish could be very old, biologists found out that shortspine thornyheads could be 80 to 100 years old. Milton Love reports they can live to at least 133 years old. Thus, this is a slow-growing, long-lived species that is adapted to live in deep, cold water that can also have low levels of oxygen. The maximum size in British Columbia is 89 cm and 6.6 kg, and most individuals mature at 8 to 10 years old. The average size is about 25 to 35 cm. Females produce a pelagic, gelatinous egg mass that floats to the surface between about March and May. The largest egg masses can contain about two million eggs. The hatched larval fish remain in the plankton for about a year before they settle on the bottom in shallower water and move into deeper water as they grow and age. Shortspine thornyheads are in the genus *Sebastolobus*

Mike Byerly

and their eggs are fertilized externally, as with most fishes. All other rockfish in the Strait of Georgia are in the genus *Sebastes*; their eggs are fertilized internally and hatch within the female, with the young larvae released or "born" as free-swimming individuals. Shortspine thornyheads feed on shellfish, amphipods and other fish, but the food requirements would be minimal for the slow-growing adults.

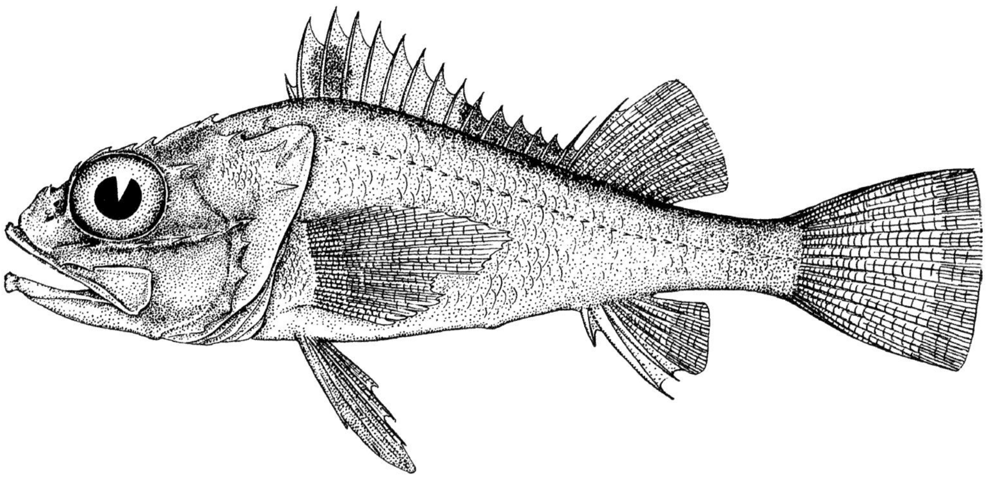

and Combfishes

The taxonomy of sablefish and greenlings continues to be an issue. Sablefish are in the family Anoplopomatidae, which has only two species, the sablefish and the skilfish. It is possible both species occur in the Strait of Georgia, but we only describe sablefish. Skilfish have been reported once in the Strait of Georgia. The report has some credibility, so there needs to be alertness for possible future reports, which we hope will result in a specimen retained and preserved. Greenlings are in the family Hexagrammidae, which consists of about a dozen species with genera that do not appear closely related. All are in the North Pacific and generally near or on the bottom in shallower water. Combfishes are in the family Zaniolepididae, which has three species. Sablefish, greenlings, lingcod and combfish are found in the Strait of Georgia.

SABLEFISH

Anoplopoma fimbria RARE ★

IN THE STRAIT OF GEORGIA

Sablefish found in the Strait of Georgia are virtually all juveniles in their first or second year of life. These fish are about 15 to 30 cm in length. They can be found throughout the strait, but are mostly found in its central and northern areas, where they are occasionally caught in bottom trawls or on longlines. There are no records of catches in pelagic trawls. They probably occasionally find their way into the strait in years when there has been a very successful spawning off the west coast. Any sablefish caught in the Strait of Georgia that is larger than 50 cm is noteworthy.

DESCRIPTION

Sablefish are not at all related to cod, but this sleek, dark-skinned fish will mostly likely have two common names forever. While "sablefish" is the official common name, if this fish is smoked in a fancy restaurant it will likely be called "Alaska black cod," and if a fisherman sells one to you, it will be called a "black cod." Sablefish are easily recognized by their rounded, elongate body that is mostly black, with grey, dark-green or blue shades and a lighter belly that is usually grey-white. There are two dorsal fins and one anal fin.

Danny Kent

LIFE HISTORY

This species occurs from Japan to off Russia, in the Bering Sea and along the coasts of Alaska and British Columbia to central Baja California. A maximum size of 117 cm and 25.5 kg has been recorded, though fish about 135 cm are reported from the Gulf of Alaska. Spawning occurs in deep water all along the west coast of British Columbia from January to March, and there is no apparent spawning migration. Eggs hatch at depths of about 300 to 400 m. The larval fish eventually find their way to the surface, where most migrate inshore in large schools and grow quickly. Between age two and five, the juveniles begin to mature and move to deeper water offshore. About 50 percent are mature by age five and at lengths between 53 and 58 cm, depending on their sex. Sablefish are known biologically for their long life span. In the late 1970s, Dick Beamish developed a method of determining the age of sablefish by tagging the fish and marking a bone in the head with an injected dye; when the tagged fish was recaptured, there were growth zones that had formed that marked the number of years after the tagging, confirming that the zones were annual. A few tagged fish were captured in fisheries for the subsequent 20 to 25 years. Most fish in the fishery are 4 to 35 years old, but the oldest known individual was aged at 113 years. Sablefish can now be cultured in fish farms. As the quotas diminish and the price keeps increasing, it is possible there may be sablefish farms in the Strait of Georgia. Sablefish are highly esteemed in British Columbia and in other countries such as Japan for their delicate and slightly oily texture, particularly when smoked.

KELP GREENLING

Hexagrammos decagrammus ABUNDANT ○

IN THE STRAIT OF GEORGIA

Kelp greenling are abundant throughout the Strait of Georgia, mainly in shallow water. In Puget Sound they are most abundant in subtidal zones between 7 and 12 m, and in the Strait of Georgia they are mostly in steeper, sloped habitats. The most common depth in the strait is likely 7 to 15 m. They are seen everywhere by scuba divers and are one of the most reported species caught from wharves or piers.

A male kelp greenling has a brownish-purple body and fins with relatively large, sparsely arrayed blue spots surrounded with dark, reddish rings. Lee Newman

DESCRIPTION

These are brightly coloured, attractive fish with an almost indescribable range of colours. The colour pattern is so different between the sexes that they were once believed to be different species. Body colour can range from grey or yellow to olive-brown to orange, blue and mauve. Males have large, irregular blue patches anteriorly with margins of black dots. Fins may have a bright-blue tinge, but they can also have yellow and even orange shades. Females tend to be more spotted, with red, orange or even blue spots. Body colours are reds, oranges, blues and browns. Fins are brightly coloured oranges, yellows and browns. This species is distinguished from other greenlings by two pairs of cirri on the head. There are five lateral lines, with the fourth lateral line extending to the end of the anal fin or beyond. There is a single small spine at the front of the anal fin. Both males and females have a spot (or ocellus) at the posterior of the rayed portion of the dorsal fin.

LIFE HISTORY

Kelp greenlings occur from Southern California to the Aleutian Islands in the Gulf of Alaska. They are found in kelp and rocky reefs to depths of 46 m. These are non-schooling fish that are territorial, particularly during the spawning season late in the year. The maximum length recorded is 53 cm, though most range from about 35 to 46 cm. There are only a few age determination studies, and apparently none used methods that could detect very old individuals. A maximum age is recorded at 18 years, but there are records of fish older than 20.

A female kelp greenling has yellowish fins and small golden-orange spots that can sometimes appear as horizontal lines.
Lee Newman

Females grow faster than males; consequently, mature females are larger than males of the same age. About one-third of males are mature at age two; 50 percent are mature at four and at lengths of about 30 cm. About 50 percent of females are mature at age five. Thus, in general, most kelp greenlings begin spawning at age five. Most spawning occurs from October to December. Females are batch spawners and produce egg masses that vary in size from about the size of a golf ball to that of a tennis ball over a 6- to 12-week period. Light-blue egg masses are demersal and adhesive at depths from about 5 to 17 m. A male guards the nest area, which may contain an average of 4 batches of eggs; up to 11 have been recorded. Tests show that in some cases the batches being guarded come from more than one female. Eggs hatch in 30 days at 10°C. Larvae are 7 to 8 mm and remain in pelagic waters for up to a year. Offshore, these larvae have been found to migrate as much as 800 to 965 km out to sea. In the Strait of Georgia, young fish eventually migrate to the bottom and become demersal at about 5 to 7 cm in length. On average males are found about 3 m deeper than females.

Lee Newman

Kelp greenlings are retained by commercial fishermen for the live fish market. They are also a popular sports fish, particularly among younger anglers. They can be caught from piers and wharves, from small boats in shallow water, and even from rocky shorelines, where they may be mistaken for juvenile lingcod. Kelp greenlings have small mouths but feed on a variety of organisms, including fish, crabs, snails, octopuses and algae. They are prey to seals, lingcod and even Pacific spiny dogfish.

ROCK GREENLING

Hexagrammos lagocephalus VISITOR ▼

IN THE STRAIT OF GEORGIA

Rock greenlings are rare in the Strait of Georgia as none have been reported in any studies or surveys in over 50 years. However, the species is reported to be in the strait in *Pacific Fishes of Canada* and *Fishes of the Salish Sea*. They are not easy to detect and it is likely they only very rarely visit the strait, possibly as juveniles that would leave it at maturity.

This juvenile was caught in Rupert Inlet, a north arm of Quatsino Sound, east of Coal Harbour (northern Vancouver Island). The conspicuous dark spot above the pectoral fin is diagnostic for the rock greenling. Lora Tryon

DESCRIPTION

During breeding season the male rock greenling is perhaps the gaudiest fish in British Columbia. When seaweed growth starts to accelerate in the spring, the male takes on an iridescent turquoise along the back, in marked contrast to patches of bright orange over the body and pale blue on the head, whereas the spawning female is a uniform dark green. The garish colours of the nest-guarding male actually disrupt his visibility among the bright-green, gold and red seaweeds that toss around in the surf, so he often remains hidden from scuba divers searching for discoveries. In sheltered areas near to open ocean exposure, a young juvenile with a conspicuous dark spot above the pectoral fin is a rock greenling.

LIFE HISTORY

The species is widely distributed in areas of heavy surf from Central California to Alaska, the Sea of Okhotsk and south to off Japan and the Yellow Sea. They occur on the wave-exposed outer coasts, where they are found on rocky outcrops among surfgrass or seaweed. Juveniles tend to occur in nearby sheltered areas. They mature in three to four years at lengths of 29 to 35 cm. From June to September, the male attracts a female into the holdfast of northern giant kelp or other seaweed species, and the eggs are laid so that the egg mass surrounds a stipe at the base of a blade or is embedded in the tangle of the holdfast. The bases of surfgrass blades are also used as a spawning substrate. The male guards successive egg masses laid by any females he can attract. This species is sought after for display in public aquariums because of its remarkable beauty, though people probably could not get away with wearing clothes with such clashing, bright shades of colour. Live specimens have been sold by professional collectors to public display aquariums for $1,000 each.

WHITESPOTTED GREENLING

Hexagrammos stelleri COMMON ●

IN THE STRAIT OF GEORGIA

The whitespotted greenling is widely distributed in the Strait of Georgia in nearshore areas with sandy or eelgrass bottoms, usually at depths slightly shallower than kelp greenlings, and often up into the intertidal area at high tide. Adults occur among rocks and algae, mostly in the subtidal area. Young-of-the-year are also seen in shallow seaweed areas. Unfortunately, there is surprisingly little known about this species.

DESCRIPTION

The colour of the whitespotted greenling is variable; the body colour ranges from green to orange to yellow to white. There are white spots on the body, but there also are spots of other colours. The fin colour is also variable, with a dark band or rows of dark spots. It is distinguished from other greenlings by the length of the various lateral lines. The first lateral line is short, extending to the midpoint of the spinous dorsal fin. The fourth lateral line is also short, extending past the base of the pelvic fin.

Danny Kent

Danny Kent

LIFE HISTORY

The whitespotted greenling is distributed from off the coast of Oregon to the Bering Sea and off the coast of Russia, including the Sea of Okhotsk and the Sea of Japan. Throughout its distribution, it occurs on the bottom to depths of 175 m, but mostly in shallow water. The maximum length is 48 cm. Spawning occurs from February to April in British Columbia. However, in the western Pacific there are reports of spawning in the summer in the north and in the fall in the south. Females are batch spawners and males guard the batches of eggs. Eggs are green, mauve or blue and may be deposited in rocks or in crevices. The development of eggs and the behaviour of larvae are probably similar to those of the kelp greenling. Russian researchers report a maximum age of 8 to 12 years, but age, growth and maturity studies need to be conducted for fish resident in the Strait of Georgia. Adults feed mainly on crustaceans and fish. Shrimp, crabs, molluscs and polychaetes are also consumed. Whitespotted greenlings may be retained by hook-and-line fishermen for the live fish market, so it is important to understand the impacts of fishing.

LINGCOD

Ophiodon elongatus ABUNDANT

IN THE STRAIT OF GEORGIA

Lingcod is perhaps the most charismatic fish in the Strait of Georgia and one of the most important species. It is found throughout the strait, with the juveniles occurring in the shallow water of such areas as the Gulf Islands, well away from adults. Adults prefer rocky bottom areas at intermediate depths in and around rock ridges and reefs; however, they occur pelagically and on soft bottom. Lingcod eggs were boiled and eaten by some First Nations.

DESCRIPTION

Lingcod is neither a cod nor a ling but the largest member of the greenling family. Greenlings are in a small family that is related to rockfishes and sculpins, but its species do not have spines on the head. There is no mistaking an adult lingcod because of its bucket-sized head, large mouth and prominent teeth. Dorsally, lingcod are shades of black, brown-blue to green. Ventrally they are a grey-green or cream. There is a mottling on the body. Males may have more tinges of pale blue and females more orange. They are a colourful group of fish that are an attraction in commercial aquariums.

Once eggs have been laid, the male lingcod remains on guard to chase away egg predators like greenlings or sea stars. The male fertilizes the eggs while they are being extruded by the female, and the creamy egg colour changes gradually to a grey as the embryos develop. Lee Newman

LIFE HISTORY

Lingcod are distributed from northern Baja California to the southwestern Gulf of Alaska. They are bottom fish ranging from the intertidal zone to 475 m; most are on the bottom at depths less than 300 m, but they occur on soft bottoms as well as in the midwater. The maximum size is 1.52 m and 59.1 kg. The larger fish are females, and few males exceed 75 cm. A method of determining age was developed by Dick Beamish in the late 1970s using sections of the dorsal fin rays. It was found that most female lingcod in the Strait of Georgia matured at age three to five and at lengths of 60 to 70 cm. Males mature at a smaller size, 45 to 55 cm, and at age 2 to 3. The maximum age is 25 years, but most fish in the commercial fishery offshore are 5 to 10. In the Strait of Georgia, the oldest female was 20 years old and the oldest male 14. Lingcod males in the strait may appear in the spawning area late in the year and begin spawning in the winter. Females arrive in January and February and select rocky, tide-swept passages with strong currents. Peak spawning is from the end of February to early March, but spawning ranges from the end of January to the end of March. Nests in the Strait of Georgia have been seen by scuba divers as late as June. The female usually deposits adhesive eggs in a rocky crevice. She layers eggs on the nest and the male then fertilizes them. The process is repeated until the female has released all her eggs and leaves the area. A 92 cm female can produce about half a million eggs. It has been shown, using DNA, that the sperm that fertilizes the eggs in a nest can belong to more than one male. In fact, as many as five males have been shown to contribute to a single nest. The male guards the nest containing the egg mass, which can weigh up to 14 kg. Depending on water temperature, the eggs take about seven weeks to hatch. If the male lingcod is removed from the nest, the eggs are quickly eaten by predators, including invertebrates. Males were once easily fished when guarding their nests, but nesting males are now protected. In the past, First Nations understood that the males must be left alone and thus did not

A lingcod holds still with its mouth agape as a juvenile spot prawn picks parasites off its lips and mouth tissue. This important hygiene service can be performed inside the mouth or on the body and fins by various small shrimps or fishes.
Lee Newman

allow fishing during spawning season, unless a father was teaching his son how to fish.

The male lingcod, during the fall and winter breeding season, has a very blackened colour pattern, and splits in fins or white spots where the skin has been bitten give evidence to the territorial aggression that occurs between males. Lee Newman

After the eggs hatch, the larval lingcod remain in the surface waters until about July or August, when they settle on the bottom in rocky areas. It is during this time that they disperse. The young lingcod grow quickly, reaching lengths of 17 to 35 cm by the end of their first ocean year. They are about 47 cm by the end of their second year. Once lingcod settle on the bottom, they tend to remain in one place as juveniles, though territoriality may vary depending on location. Early tagging studies at the Pacific Biological Station showed that adult lingcod released in another area will return to the original place of capture. Off the west coast and in Queen Charlotte Sound, large schools of females are occasionally caught in commercial trawls, presumably in some form of prespawning behaviour.

Lingcod were once fished by First Nations using spears, indicating they were once abundant in very shallow water in the Strait of Georgia. Decades ago, they would have been the dominant predator in the strait because of their size, abundance and feeding habits. It was lingcod that first provided commercial fishing opportunities for settlers arriving on the West Coast. In the 1920s,

The business end of a lingcod is the large mouth with its recurved teeth. Lingcod will pounce forward onto a prey item while expanding their jaws to engulf a large volume at the same time that the throat opens wide. A salmon over half the length of a lingcod can be swallowed back into the gut in an instant. Sometimes the lingcod will make a sideways grab and then must shift position for swallowing, typically head-first. Danny Kent

there were over 200 small boats fishing lingcod with handlines. By the 1980s, it was clear that lingcod in the Strait of Georgia were severely overfished. The declines were recognized, but lingcod were not viewed as being sufficiently important to restrict fishing to the levels needed to allow the population to rebuild. This attitude changed as management began to focus on ecosystems and as the Species at Risk Act became law. Abundances are increasing again with restricted recreational fishing allowed.

It is tempting to write that lingcod eat whatever they want. Adults feed mostly on fish such as Pacific herring, Pacific cod, walleye pollock, Pacific hake, Pacific spiny dogfish and other lingcod. Fishermen know that the best bait for catching lingcod is another lingcod, which is why attempts to culture the species are complicated by cannibalism. Lingcod also feed on octopus and squid. They have few natural predators except sea lions that prey on males protecting nests.

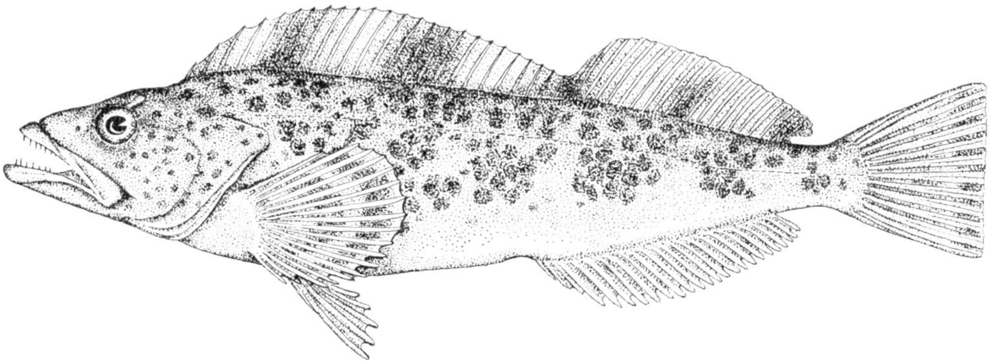

PAINTED GREENLING

Oxylebius pictus UNCOMMON ◆

IN THE STRAIT OF GEORGIA

Painted greenlings are not common in the trawlable areas of the Strait of Georgia, but they are observed in shallow, rocky reef areas by scuba divers. They are seen throughout the strait and are among the 10 most reported fishes by scuba divers, possibly because of their colour.

DESCRIPTION

The five to seven red bands on these relatively small fish are striking, resulting in the nickname "convict fish." The background colour can be creamy white to grey to golden brown. The red bands on the body extend onto the fins. Three bands radiate from the eye, one anteriorly and two posteriorly. Painted greenlings have two pairs of cirri on the elongate, pointed head and one lateral line. Large pectoral fins can have lines of red, white and maroon spots. There can be coarse speckling at the bottom of the head. Spawning males become a darkish black, and the red bars on spawning females turn brown.

LIFE HISTORY

Painted greenlings are found from Baja California to Prince William Sound around rocky reefs and rocky slopes. They occur at depths of up to 49 m, but most are found on shallow rocky bottoms in 5 to 20 m of water. Most range

Lee Newman

A close look at the pointed head of a painted greenling shows two pairs of fluffy cirri behind the eyes and on the nape.
Danny Kent

from about 14 to 17 cm in length and the maximum is 25 cm. They are rarely caught on a baited hook, and their small size makes them unattractive commercially. Painted greenlings are solitary, nonmigratory bottom fish and are virtually unstudied in the Strait of Georgia. Males tend to be separate from the females, which occupy rocky reef areas on the periphery of the male concentrations. Females may also be twice as abundant as males. Both sexes tend to become less active during the winter. A maximum age of eight years has been reported; all females are mature at age three and about 16 cm, and most males are mature at age two and 14 cm.

Spawning occurs during the spring and summer in Puget Sound, in the fall off California and most likely in the spring and summer in the Strait of Georgia. Females lay eggs on exposed reef substrates, where currents provide some circulation of water over the developing eggs. Spawning occurs over about a three-week period. Larger and older males accumulate egg masses from more than one female and then protect the golf ball–sized clusters of developing eggs until they hatch. The spawning cycle may be repeated with an approximately 10-day mating period and another 20 days of parental protection. On average, a male may protect 7 batches of fertilized eggs, but up to 10 have been observed. Females may produce about 3 to 4 batches per cycle. After hatching, the larval fish are planktonic for about three months, during which time they are dispersed by surface currents. Larval fish have been reported in April in the Race Rocks area off Victoria. Males and females remain within their territory for life. Males have a shorter life span, perhaps because of their vulnerability during their egg-guarding responsibilities. Painted greenlings feed on small crabs, polychaetes, amphipods and other crustaceans found within their territory. They are eaten by larger fish such as cabezon and by diving birds such as cormorants.

LONGSPINE COMBFISH

Zaniolepis latipinnis UNCOMMON ◆

IN THE STRAIT OF GEORGIA

Longspine combfish are probably uncommon throughout the Strait of Georgia, though they were commonly found in bottom research trawl surveys from 2003 to 2006 in the southern strait along the east coast of Vancouver Island. In these bottom trawl studies, catches of longspine combfish were about as numerous as those of whitespotted greenling and sturgeon poacher. They are also reported in Saanich Inlet and Nanoose Bay. The species is seen during night dives in Sechelt Inlet.

DESCRIPTION

The first three dorsal spines are elongated, with the second spine of the forward dorsal fin greatly elongated. This species, a member of the greenling family, is more slender and more elongate than other greenlings, and there is a single lateral line canal. The body has rows of orange to brown spots. The dorsal fin and the rest of the body become quite blackened in males during the winter breeding season.

The longspine combfish, a member of the greenling family, lives on the deep, soft bottom, where it props itself up on its pelvic fins. It has an extremely long second spine on the first dorsal fin. Danny Kent

LIFE HISTORY

Longspine combfish are found from Vancouver Island to southern Baja California. They can be locally abundant at deep depths on soft seabed with sand or mud rather than rocks. The maximum size is 30.5 cm, with individuals maturing at lengths of 12 to 14 cm. Most occur on or near the bottom at depths of 35 to 200 m. Perhaps because of its deep, soft-bottom habitat, few scuba divers get the opportunity to observe this species. Its spawning habits have only been observed in the displays of the Vancouver Aquarium, where it is known to spawn in batches, mainly in the fall and winter. The longspine combfish sits on the seabed with its head propped up above the bottom and the body and fins forming a triangle.

Sculpins

The 40 species of sculpins make this the most diverse group of fishes in the Strait of Georgia. There are four families around the world found mostly in the ocean, but some are in fresh water. The family Rhamphocottidae has one species called a grunt sculpin; it is found in the Strait of Georgia. The largest family, Cottidae, comprises freshwater and marine residents, and almost all of its approximately 300 species are in the northern hemisphere. These are mostly found on or near the bottom, but they range widely in shape and colour and are found at a wide range of depths. All species are reported to produce sounds. There are 32 species from this family in the Strait of Georgia. The family Hemitripteridae (sea ravens) has eight species with two sculpin species (silverspotted and sailfin) that occur in the strait. The first dorsal fin of these two species is distinctively long. The fourth family, Psychrolutidae (fathead sculpins), has about 30 species around the world. These are loose-skinned, bottom-dwelling species with a tadpole-like shape. There are five of these soft, blob-like species in the Strait of Georgia.

Red Irish lord, *Hemilepidotus hemilepidotus.* Gregory Jensen

GRUNT SCULPIN

Rhamphocottus richardsonii COMMON ●

IN THE STRAIT OF GEORGIA

This little fish with a unique shape is common in the Strait of Georgia, particularly in the Gulf Islands. It is seen by scuba divers throughout the strait and occasionally appears in on- and off-bottom research trawl catches. It could be the cutest fish in the Strait of Georgia.

DESCRIPTION

A frontal view of the grunt sculpin reveals two high cranial ridges running back from above the eyes and a tiny mouth. The mouth is used to snap sideways at small shrimp as the fish darts forward with a thrust of its tail and large pectoral fins.
Danny Kent

The appearance of the grunt sculpin is so distinctive that it cannot be confused with any other species. The fish appears hunchbacked and has a large head that is about one-half of the total length of the fish. The small mouth is at the end of a tapered, pig-like snout. The stocky body is curled and covered with embedded, multi-spined scales covered with skin, which makes the skin feel as if the fish were covered with prickles. The pectoral fins are large, and the lower pectoral fin rays have no connecting membranes. The grunt sculpin is quite colourful and many images of this photogenic little fish can be found online. The body is light tan to brown with mottled and oblique streaks of orange and cream colours. Light-brown streaks also radiate from the eye. The pectoral fin rays are orange and there is a blotch or bar of orange or red at the base of the tail. Small juveniles can be almost all white. Some think this is a cute little fish.

LIFE HISTORY

Grunt sculpin occur from Southern California to the Gulf of Alaska. Reports from the Bering Sea are not confirmed, but they are reported to be rare in Japan. The maximum length is just 8.9 cm. They are found from intertidal depths to 165 m and typically occur in shallow water near rocks or pilings. They are associated with debris on the bottom, where they may be observed inside of empty barnacle shells. In aquariums and on the ocean bottom, grunt sculpin swim

A side view of a juvenile grunt sculpin shows the species' typical colour pattern. The forward pectoral fin rays extend beyond the fin membrane and can be used like fingers for traction along the seabed. The side of the trunk is covered with tiny spines rather than scales. Danny Kent

with a hopping motion, almost walking, using the unconnected lower rays of their pectoral fins almost like legs. If the fish are handled, they can make a grunting or hissing sound that presumably is an adaptation to provide protection from predators. While on the bottom, they may move their eyes independently; this probably is an adaptation for finding prey and avoiding predators. The single long, sharp pre-opercular spine may also help to minimize predation. Spawning is reported to be in the winter, but the Port Townsend Marine Science Center reported that eggs were laid in the fall and hatched from December to the end of April. Eggs have been observed in a variety of structures and can be creamy white, yellow or orange. The egg masses are guarded by either the male or the female. In commercial aquariums, the female has been observed chasing a male into a spawning area, where some report that the male remains trapped until her eggs are laid. However, perhaps the male is not as trapped as he appears.

This is a popular species in commercial aquariums. The Vancouver Aquarium has successfully bred the species in captivity; fish ranging in age from five to nine years spawned throughout the year. Two fish have now survived for 22 years at the aquarium, which says something about the species and more about the care they get there. Grunt sculpin feed on copepods, isopods, amphipods and other crustaceans.

PADDED SCULPIN

Artedius fenestralis COMMON ●

IN THE STRAIT OF GEORGIA

Padded sculpin are common in the Strait of Georgia in intertidal areas to rocky nearshore areas at depths down to 55 m. They are seen occasionally by scuba divers throughout the strait and are sometimes caught by anglers or children fishing from a pier or dock.

DESCRIPTION

This species varies in colour from shades of grey to brown to yellowish, with darker brown bands forming wide saddles of pigment from the dorsal to the ventral. Like many sculpins, the padded sculpin has a large head and big, fan-like pectoral fins. There are no hair-like features (cirri) on the rather smooth head, except for a small projection at the corner of the mouth. The common name "padded sculpin" originated from the very prominent scales between the lateral line and the dorsal fin that join behind the second dorsal fin.

Padded sculpin have a relatively high head profile and short snout. The body has vertical dark bars and an embroidery of several rows of large scales below the first dorsal fin, extending past the second dorsal fin to the tail region. Lee Newman

LIFE HISTORY

The padded sculpin occurs from the Aleutian Islands to Northern California from intertidal depths to 55 m, but mostly at less than 20 m. They are found in rocky areas or on sand close to reefs, as well as in tide pools and eelgrass. The maximum size is about 14 cm. Padded sculpin spawn in the winter from February to May. The males provide parental care by defending the egg masses, which may be brightly coloured—from a maroon or purple to a grey-blue—on the underside of a rock. Females produce adhesive eggs and males may spawn with several females as part of the egg mass cluster that they defend. This sculpin feeds on shrimp, small fish, amphipods, clams and other organisms close to the bottom.

SCALYHEAD SCULPIN

Artedius harringtoni COMMON ●

IN THE STRAIT OF GEORGIA

The scalyhead sculpin is one of the abundant sculpins in the Strait of Georgia. It has been one of the most abundant fishes observed in dive surveys throughout the strait, and particularly in the Gulf Islands. However, it was not found in various research trawl surveys, which appears to indicate it is restricted to shallower waters that are frequented by divers. The scalyhead sculpin is reported to be territorial and to exhibit behaviour similar to that of a small dog who thinks it is a large dog.

Lee Newman

DESCRIPTION

The colour of the scalyhead sculpin varies, as with many sculpins. Most are brownish to olive, but reds, blues, purples, oranges and whites occur as well. There may be five to seven dark, saddle-like patches on the back. The gills of adults are a yellow-orange colour at the base. There are large fringed cirri on the forehead in front of the eye. Pectoral fins are large. Mature males have greatly enlarged, prominent cirri behind each eye. The first pre-opercular spine is bifid and covered in skin. There are scales on the head, which is proportional to the body, but there are none below the front of the eye or the forehead. Broad bands of scales on each side of the back join together on the caudal peduncle. The lateral line scales are enlarged and prominent.

LIFE HISTORY

Scalyhead sculpin are found from Southern California to the Aleutian Islands on rocky reefs around pilings at depths from the intertidal zone to 21 m. The body is relatively stout and has a maximum length of 11 cm. The spawning season is mostly in the winter, extending into the late spring, when courting males will be territorial while sequestering multiple females at the nest site. There are distinctive egg-like spots in the mouth and on the anal fin that may be associated with attracting females to a nest site. Male scalyhead sculpin have an enlarged genital papilla with a distinct tip that forms a structure analogous to a penis. Mature males differ morphologically from females: they have a bigger head, enlarged anterior teeth and large plumose (feather-like) cirri. Males may appear sufficiently different from females that in the past they have been misreported as a different species. Eggs are adhesive and laid in clusters that the male guards until they hatch in about two to three weeks, depending on water temperature. There is no information on the maximum age or age at maturity; further study would make a most useful student project.

The female scalyhead sculpin lacks the male's eye cirri and is usually smaller than a male. Females in a harem group stay in the protection of a territorial male through the winter/spring breeding season. The number of females in a harem is represented by the number of different colours of egg masses in the nest. Lee Newman

Colour variation is typical of many sculpin species, here exemplified by scalyheads in red, blue, orange, pink, white and green shades; there are combinations of colours of all varieties, usually enabling favourable camouflage. Lee Newman

SMOOTHHEAD SCULPIN

Artedius lateralis COMMON ●

IN THE STRAIT OF GEORGIA

The smoothhead sculpin is a common intertidal and subtidal species throughout the Strait of Georgia. Smoothhead sculpin are common in most estuaries and seen throughout the strait by scuba divers, but they are perhaps not as abundant as the padded or scalyhead sculpins.

Gregory Jensen

DESCRIPTION

No scales appear on the head, which may explain the common name "smoothhead." There is a band of scales on the upper body that do not join behind the second dorsal fin. The head is relatively long and pointed. Colours vary from grey to black to brown and green with occasional small dots of blue and pink. There are dark bands on the fins and bands of brown across the back.

LIFE HISTORY

The smoothhead sculpin is found from the western Gulf of Alaska south to northern Baja California. They are much larger than the tidepool sculpin, reaching lengths of up to 14 cm. Larger fish can be found around piers, in rocky tide pools, in shallow, sandy areas and in nearshore areas with mixed pebbles and rocks. They can be found under rocks at low tide. This is a shallow-water fish that occurs at depths of less than about 14 m. Mature fish spawn in winter, mostly in February, but extending into late spring in some areas. Eggs are adhesive and deposited along the underside of rocks in yellow, orange or red egg masses. Larvae of the smoothhead sculpin are found in the surface waters and settle in the intertidal areas on rocky shorelines. In contrast, the larvae of the padded sculpin are found in slightly deeper water and they settle subtidally at shallow depths. The ecological depth separation between these two species is more distinct early in life than during adult stages, when both species reproduce in the intertidal zone. Male smoothhead sculpin will guard the nest. After the eggs hatch in two to three weeks, the juveniles are planktonic for about two months before settling on bottom at a length of about 1.1 cm. This species is commonly caught by anglers fishing from shore or off docks. Like many sculpins, their diet consists of small invertebrates such as gammarid amphipods, crabs and shrimps, but also small fish such as other sculpins.

ROSYLIP SCULPIN

Ascelichthys rhodorus COMMON ●

IN THE STRAIT OF GEORGIA

The rosylip sculpin is considered to be common in the Strait of Georgia, but there are no reliable estimates of abundance. There are reports in Active Pass, Departure Bay and north to Comox. It is rarely seen by scuba divers, probably because of its nearshore residency on gravel substrate.

DESCRIPTION

The rosylip sculpin is the only sculpin in the Strait of Georgia without pelvic fins. The genus name *Ascelichthys* translates roughly to "fish with no legs," which relates to the absence of pelvic fins. It is without scales, giving the skin a smooth appearance. There is a branched cirrus above the eye and a single pre-opercular spine with an upward-curving single hook. It is olive to dark brown with mottling dorsally. The lip area may be red and there may be a rose edge to the top of the dorsal fin.

Scott Stevenson

LIFE HISTORY

This is a small sculpin, growing to 11 to 15 cm. It is found from Central California to Prince William Sound, Alaska. It is commonly found in tide pools under rocks and in the intertidal and subtidal areas. It is able to breathe air and may be found under rocks or seaweed, out of the water. Newly settled staghorn sculpin juveniles also lack pelvic fins and at this stage their pre-opercular spine is also simple like that of a rosylip sculpin, showing that these young juveniles cannot be distinguished using characters for adult stages. Spawning in the late winter to early spring is relatively short and is synchronized among groups of individuals. Large numbers of males migrate to the spawning areas before the females. Eggs are fertilized externally by a number of males as females deposit eggs in crevices between or under rocks. The blue or purple eggs are adhesive but are reported to stick to each other rather than to the rocks. Males do not guard the nests, but the presence of males in the spawning area provides some protection from predators. Larval rosylip sculpin are pelagic for several months, forming schools. After two to three months they settle on the bottom. The species has been poorly studied considering that it is abundant and readily available to students. A study of the age at maturity and rate of growth would be a useful contribution. It feeds on small invertebrates, and males will consume some of the eggs during the spawning period.

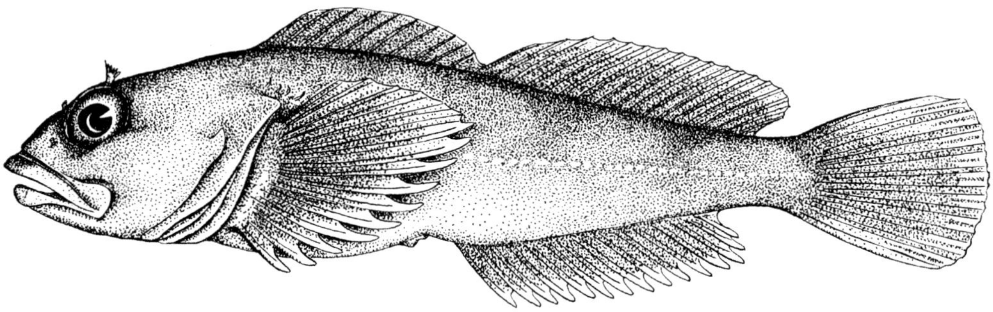

SPINYNOSE SCULPIN

Asemichthys taylori RARE ★

IN THE STRAIT OF GEORGIA

The spinynose sculpin is currently believed to be rare in the Strait of Georgia. Up to the early 1970s, it was frequently seen by observant scuba divers in all areas of the strait. It continues to be reported by scuba divers over shell fragment habitat at the base of rocky outcrops. The fish has not been found in Puget Sound.

A dark "five-o'clock shadow" along the lower jaw is usually a good identifier for the small spinynose sculpin. Lee Newman

DESCRIPTION

The spinynose sculpin has a trim head with a short snout and long pectoral fins. The face looks like it has a five o'clock shadow, as the top of the head is lighter in colour than the very dark lower head and jaw. The tail end of the body has disruptive bars of light and dark colour, which may tend to match the surroundings—for example, showing shades of pink if the rocks it lives on are covered with pink coralline algae.

LIFE HISTORY

This species occurs on sandy bottom at 5 to 18 m, and perhaps up to 50 m, in a restricted distribution from southeast Alaska to the Strait of Georgia. This is a small sculpin with a maximum length of 7.4 cm. The spinynose sculpin was originally named *Asemichthys taylori*, but the genus name was changed in 1961 to *Radulinus*. The recent book *Fishes of Alaska* uses the genus *Radulinus*, but the American Fisheries Society retains the original *Asemichthys*. The species was named after Reverend George Taylor, the first director of the Pacific Biological Station, which he and his son built themselves. Reverend Taylor collected the type specimen in 1910, one year after the station opened. The original common name was "Taylor's sculpin."

The nesting parasitism of this species on the eggs of the buffalo sculpin is described in the buffalo sculpin entry. Of interest is the apparent coincidence of population cycles in these two sculpin species. The nesting parasitism phenomenon was documented during scuba dives in the Strait of Georgia in the 1990s, when both species were relatively abundant. The spinynose sculpin had disappeared from collection records for an extended historical period, and this raised the concerns of the Committee on the Status of Endangered Wildlife in Canada.

Since the time of the documentation of interspecific nesting between buffalo and spinynose sculpins, the spinynose has become rare once again, and no records of this nesting were produced during the decade of the 2000s. (The buffalo sculpin was also becoming relatively rare during this period.) It is noteworthy that these two decades—the 1990s and the 2000s—were considered to have encompassed two different climate regimes. Climate regime shifts affect all marine life, and it is interesting to see documentation of fish species that receive no fishing pressure and probably no other human impacts going through these cycles of relative abundance versus rarity.

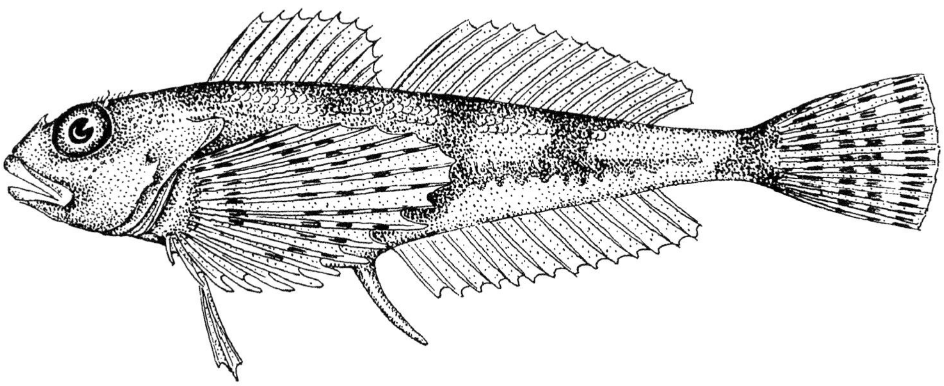

ROUGHBACK SCULPIN

Chitonotus pugetensis ABUNDANT ●

IN THE STRAIT OF GEORGIA

This species is an abundant sculpin that some would consider to be a pretty little fish. In the Strait of Georgia, it is one of the sculpins frequently found in experimental bottom trawl surveys and shrimp surveys, and it is seen by scuba divers just about everywhere, often during night dives.

DESCRIPTION

The scales on the upper head are rough, giving the fish its common name. The roughback sculpin is the only species in the genus and is distinguished by the size, number and arrangement of the dorsal body scales. Individuals are a slender shape with a large head, big pectoral fins and relatively large eyes. There is a deep notch in the first dorsal fin and a long first dorsal spine. The upper pre-opercular spine is antler-like. Individuals are grey-green to brown dorsally and white ventrally. There are brown-black, saddle-like bands dorsally. The body may also have small blotches of colour such as red, yellow or blue. Males in spawning condition may have bright-red or orange patches or eye caps above the eye. Dorsal fins may be tipped with black and some fins may appear transparent. Males have a large anal papilla, which is analogous to a penis and is used to fertilize females internally.

Lee Newman

Danny Kent

LIFE HISTORY

Roughback sculpin reach a maximum size of 23 cm and are found from Baja California through to northern British Columbia, just south of the Alaska–British Columbia border. They are common on soft bottoms at depths of less than 73 m, but have been reported as deep as about 140 m. Fertilization is not the typical internal fertilization that occurs in some fishes but an internal gamete association as the sperm does not penetrate the egg membrane and initiate fertilization until the egg is released into salt water. Spawning occurs in the winter and spring. Females have been reported to be batch spawners, producing up to three clutches of about 1,900 light-pink or red eggs per season. While there are conflicting reports about males guarding nests, it is unlikely they do so because these fish spawn on sand and bottom debris such as twigs. Females are reported to be mature at 6.6 cm, but a proper study of the roughback sculpin's reproductive biology would make a useful graduate student project. This is an ambush predator that patiently waits for prey. It appears to be mostly nocturnal, so likely feeds at night on shrimps and other crustaceans, with a preference for shrimps. In commercial aquariums such as the Vancouver Aquarium visitors may find it buried in the sand, which is normal behaviour for nocturnal fish.

Lee Newman

SHARPNOSE SCULPIN

Clinocottus acuticeps UNCOMMON ◆

IN THE STRAIT OF GEORGIA

The sharpnose sculpin probably is uncommon in the Strait of Georgia. A survey of fishes in the major estuaries in the strait in the 1970s reported that the sharpnose sculpin was present in the Squamish, Nanaimo, Cowichan and Campbell River estuaries, indicating it is dispersed throughout the strait. It is not reported by scuba divers, possibly because it is common in the intertidal areas, where it occurs with the similar-looking tidepool sculpin.

DESCRIPTION

The sharpnose sculpin is well named for its small, pointed snout on a relatively small head. The fleshy cirri on the head and along the lateral line are unbranched. The colour can be camouflage from green to light brown with darker blotches on the back and a cream to white belly. It is typical of all *Clinocottus* species in that its anus is placed one-third of the gut length forward from the anterior end of the anal fin (the end of the gut). This is distinct from the position of the anus at the very rear end of the gut cavity in all *Oligocottus* species. Since these fish usually do not roll over and hold still, it is important

danilately / iNaturalist

to take note of features on the head, especially the cheek spine, which is single in *Clinocottus* (with a single point) as opposed to a split, double-ended (bifid) single spine on the operculum or cheek of *Oligocottus* species.

LIFE HISTORY

This species occurs in the intertidal areas to 20 m from the Aleutian Islands to Central California. It is common in the rocky intertidal to subtidal zones but also on soft bottom and in eelgrass and seaweed, and even in fresh water. The maximum size is 6.3 cm, but most are about 4.5 cm. There are no age estimates. This fish spawns very high up the shoreline at the highest tide levels from January to April. The eggs are deposited in a single layer under rockweed, which cools the eggs and keeps them damp. It is probably the female alone that migrates up with the tides to deposit eggs, even though this species specializes in foraging right at the water line. It would make sense for the female alone to deposit eggs because the male is endowed with an enormous genital papilla, a copulatory organ. Most sculpins that spawn high in the intertidal zone have copulation and internal gametic association (i.e., the egg and sperm are united but fertilization remains incomplete until they meet with the calcium ions in seawater). Sharpnose sculpin prey on bottom invertebrates.

Gregory Jensen

CALICO SCULPIN

Clinocottus embryum RARE ★

IN THE STRAIT OF GEORGIA

Calico sculpin are rare in the Strait of Georgia. They do not appear to be common in estuaries around the strait, having been reported only in the Cowichan estuary. We have a scuba observation from Howe Sound, and there is an older record from Departure Bay.

DESCRIPTION

This species does not have scales, and the anus is well forward of the origin of the anal fin. The colour is variable but generally is darker and mottled green and maroon with darker saddles and a lighter belly. The name "calico" may represent the prominent dark bars on the fins, which contrast with the other body colours. Calico sculpin have clustered cirri on the head and lateral line but do not have cirri on the cheek (above the cheek spine) the way the mosshead sculpin does. For absolute certainty of identification, a single, unpaired

Goff Connor

Goff Connor

fleshy tubercle can be detected at the tip of the upper jaw in the groove of the upper lip; this feature occurs in only one other sculpin, the cabezon, a species so big and so different in habit and appearance that nobody would likely confuse the two.

LIFE HISTORY

The calico sculpin occurs from the Aleutian Islands to northern Baja California. It is an intertidal and subtidal species that has a maximum size of 7 cm. This is one of the smallest sculpins in the Strait of Georgia. It frequently occurs in tide pools but can be among rocks, algae or on sandy and gravel bottoms. This is another sculpin that we know very little about.

COASTRANGE SCULPIN

Cottus aleuticus UNCOMMON ◆

IN THE STRAIT OF GEORGIA

The coastrange sculpin is a common freshwater fish that is also found in estuaries and areas near larger rivers flowing into the Strait of Georgia. Technically, some have a catadromous lifestyle, which means they hatched in an estuary and migrated back into fresh water to feed. They probably are not common in the strait and certainly not commonly reported. In a study of estuaries around the strait, they were reported only in the Cowichan and Courtenay estuaries.

DESCRIPTION

This sculpin can be identified by its having no distinct gap between the two dorsal fins, one pore under the chin and no palatine teeth, which would appear as pads of small teeth on the roof of the mouth. It is dark brown, green or grey dorsally with darker blotches on the sides and a white belly.

Bob Armstrong

LIFE HISTORY

Coastrange sculpin are found from Southern California to Bristol Bay in the Bering Sea and the Aleutian Islands. There is an isolated population north of the Aleutian Islands along the shore of Alaska in the Bering Sea. This is a small sculpin that is usually about 6 to 8 cm but can be as large as 17 cm and can live to about eight years. Females mature at age two to three and about 5 cm. Individuals that spawn in estuaries migrate into the estuaries at night and spawn there from January to March. Yellow to orange eggs are deposited on the underside of a flat rock, and the male guards the nest until the eggs hatch. Coastrange sculpin are solitary, nocturnal carnivores.

PRICKLY SCULPIN

Cottus asper

UNCOMMON ◆

IN THE STRAIT OF GEORGIA

The prickly sculpin can be a catadromous species, meaning it may spawn in salt water and return to fresh water after spawning. The species is common in fresh water, where there are two forms: the coastal form and an inland form that remains resident in rivers and lakes. The other catadromous species in the Strait of Georgia is the coastrange sculpin, and both may occur in the same area. Prickly sculpin were found in surveys conducted in the 1970s in the Nanaimo, Cowichan, Courtenay and Campbell River estuaries, but it is likely they occur in most estuaries around the strait.

DESCRIPTION

The joined dorsal fins of the prickly sculpin distinguish it from the coastrange sculpin, which has separated dorsal fins. Prickly sculpin are a mottled dark brown to green with dark patches or bars on the body and dark bars on the fins. There are prickles on the body, which give the fish its name. Like most sculpins, they have a big head and a large mouth. Many fishermen know these fish as "bullheads."

Andy Murch

LIFE HISTORY

The species ranges from Seward, Alaska, to Southern California. The maximum size is about 20 cm, but few are larger than 15 cm. The maximum age is eight years, with individuals maturing at age two to three. Spawning is from February to June, mostly from April to June. Males will dig a nest under debris or under a flat rock, where females deposit adhesive eggs on the underside of the structure. Eggs can be yellow to orange, and more than one female can contribute to the nest, which is guarded by the male. After spawning, some adults and juveniles will move into upstream freshwater areas from July to September. This is a nocturnal fish, feeding at night on a wide range of items from invertebrates to small fish and even juvenile sculpins. In fresh water, the species can be a major predator of salmon eggs.

BUFFALO SCULPIN

Enophrys bison COMMON ●

IN THE STRAIT OF GEORGIA

Buffalo sculpin are seen by scuba divers throughout the Strait of Georgia from Campbell River to Burrard Inlet. They are common in rocky and nearshore habitats, including in major estuaries such as the Campbell River and Fraser River estuaries. Buffalo sculpin may be the most abundant of the larger sculpins. It is captured in bottom research trawls on sandy bottoms in the deeper water of the southern strait, but not in shrimp trawls.

DESCRIPTION

The head of the buffalo sculpin resembles the North American bison because of the horn-like spines on the pre-opercular bone (a bone at the front of the operculum or gill flap). These spines are sharp, so the fish must be handled carefully. The fish appears to be mostly head and large pectoral fins. The mottled appearance provides excellent camouflage and buffalo sculpin can change colour quickly, but divers report that it is easily approached when spotted. The predominantly green-brown colour has small blotches of black, white, yellow, orange and even pink, particularly on the head. There are three to four dark bands that extend across the body. The large pectoral fins also have dark bands. The eyes appear small in relation to the very large head.

Lee Newman

LIFE HISTORY

The buffalo sculpin is found from California to the Gulf of Alaska and west to Kodiak Island. In general this is a shallow-water species, occupying depths of less than 20 m, though the maximum depth reported is 227 m and they are on the bottom in the southern Strait of Georgia. Buffalo sculpins are recorded as reaching lengths of about 30 to 37 cm. Spawning occurs in late winter and early spring from February to April or even May. Females

Lee Newman

may produce 19,000 to 32,000 eggs, sometimes in two batches during a one- to two-month spawning season. Eggs can be a variety of colours—purple, red, pink, orange, yellow or tan; however, each female has ova of only one colour. Eggs are laid in layered egg masses in crevices and around rocks, pilings and barnacles, where they are exposed to currents. Egg masses can also be found in rocky areas close to low water depths. The number of females in a harem can be determined by counting the different colours of egg masses. The eggs all turn silver toward hatching time and no longer can be ascribed to a particular female. Males guard the nests until the eggs hatch in approximately six weeks, depending on temperature. In Howe Sound, the spinynose sculpin spawns only on top of the egg masses of the buffalo sculpin. This nest sharing has not been observed in other marine fishes. It appears that the benefit for the much smaller spinynose sculpin is the protection and pectoral fanning provided by the aggressive male buffalo sculpin. Scuba divers report that eggs in unguarded nests disappear rapidly.

It is fascinating that larval buffalo sculpin resemble the adults at the very small size of 9 mm. Anyone catching a buffalo sculpin may notice it can generate sound that is felt more than heard. The fish lacks a swim bladder but retains large swim bladder muscles that vibrate to deter predators. Some people consider this fish to be ugly, but the fish may have its own opinion about the appearance of the person who just yanked it out of the water. This is a species that is found in the Vancouver Aquarium, so readers can judge its attractiveness for themselves. Seaweeds and algae are reported to be major food items, which seems strange for a fish that is built to be an ambush predator. The fish also feeds on shrimps, crabs, amphipods, mussels and small fishes such as herring, sand lance and salmon. Buffalo sculpin are eaten by other fish when they are small and by large predators such as seals when they are bigger.

RED IRISH LORD

Hemilepidotus hemilepidotus COMMON ●

IN THE STRAIT OF GEORGIA

The red Irish lord is a larger sculpin that occurs throughout the Strait of Georgia but may not be abundant. It is caught on the bottom in research surveys in the southern strait. It is often photographed by divers throughout the strait because of its colour and its stationary behaviour when approached.

Lee Newman

DESCRIPTION

The fish is recognized by its single dorsal fin, which is notched to form three sections. The eye is large and there are four to five rows of scales immediately below the dorsal fin. There also are six to seven rows of scales in a ventral band below the lateral line. This is perhaps the best sculpin at camouflaging; colours vary to match the particular habitat. The mottled colours include red, brown, purple, orange and white to cream ventrally, with typical sculpin banding on the back and fins.

LIFE HISTORY

Red Irish lord are found from Central California to Alaska, and along the Aleutian Islands and Bering Sea to southeastern Kamchatka, Russia. This is a shallow-water sculpin found in rocky areas from the intertidal zone to 48 m.

The red Irish lord tends to be brightly coloured, including pigment cells over the eye, to increase camouflage. It also has embroidery scales extending around the front of the first dorsal fin. Danny Kent

Danny Kent

It has been found in tide pools in the Aleutian Islands, and specimens have been collected in the Fraser River estuary. Most occur at depths up to 50 m; it has been recorded at 275 m, but this is rare. They grow to lengths of 51 cm, but most are about 30 cm. The maximum age appears to be about six years. The fish mature at age four and lengths of 22 to 31 cm for males and 29 to 37 cm for females. The red Irish lord spawns mostly in the winter, but spawning may extend from October to March. Females deposit eggs in nests chosen in shallow rocky areas with fast-flowing currents. Eggs are reported to be pink, purple and blue, and the larger nests have over 100,000 eggs. Females and males both appear to guard the egg masses, but females are more common around the nests. It is most likely the guarding male is attended by a group of females intending to deposit eggs in his nest, so the females are staying near it for spawning purposes and the male is actually guarding the nest. Eggs in unguarded nests may be quickly eaten by predators. Red Irish lord feed on crabs, mussels and barnacles, and possibly small fish that pass by. The species is reported to be edible, but anyone deciding to handle it should be careful of its sharp pre-opercular spines.

BROWN IRISH LORD

Hemilepidotus spinosus RARE ★

IN THE STRAIT OF GEORGIA

The brown Irish lord is very rare in the Strait of Georgia. We caught three in a trawl in November 1998 near the surface in the northern strait. There was a shrimp research trawl catch in Howe Sound in 2003 and a bottom research trawl catch in the Gulf Islands off Pender Island in 2005. One more was caught at the surface off Cowichan Bay in June 2009.

DESCRIPTION

The species is brown and has dark saddles on the back, black or reddish fins, and is whitish ventrally. The dorsal fin is deeply notched between the third and fourth spines and between the spinous and soft rays. It is difficult to distinguish this species from the red Irish lord; for scuba-based identification it should be captured to make a clear distinction. The brown Irish lord has six to

Scott Stevenson

Scott Stevenson

eight rows of scales below the dorsal fins and above the lateral line, whereas the red Irish lord has a band of only four to five rows of scales. The numbers are virtually reversed below the lateral line, where the brown Irish lord has four to five scale rows and the red Irish lord six to eight rows. The red Irish lord tends to have a larger body and higher dorsal fins, as well as much more variable colours that include lots of red and pink hues, in contrast to the typical brown colour of the brown Irish lord.

LIFE HISTORY

The brown Irish lord occurs from Southern California through to the Bering Sea. It is a sedentary intertidal resident in rocky areas where there is seaweed cover. It is more typically seen on exposed shores of the outer coast. The species reaches only 33 cm, compared to 51 cm for the red Irish lord. The biology is unstudied.

NORTHERN SCULPIN

Icelinus borealis COMMON ●

IN THE STRAIT OF GEORGIA

Northern sculpin are commonly found throughout the Strait of Georgia. They are captured in bottom research trawls and occasionally in pelagic research trawls. They are also found in prawn traps. Northern sculpin are frequently seen by scuba divers throughout the strait, and often at night when they scoot along, disturbing a trail of sediment.

DESCRIPTION

The northern sculpin is distinguished by the anterior dorsal fin spines all being about the same length, and by the double row of scales below the two dorsal fins and extending back toward the tail (on the caudal peduncle). This is a brown or tan fish with black bars along the upper body and with light, filamentous cirri along the lateral line. There are four pre-opercular spines, and the lower, forward spine points directly downward, not backwards; the top rear spine is like an antler, with several upward secondary spines, typical of this genus.

Andy Murch

LIFE HISTORY

The range of the northern sculpin is from Puget Sound through all of British Columbia and north to the Aleutian Islands and eastern Bering Sea. It is fairly common on silty bedrock, soft bottom or gravel, and is reported to occur from shallow areas to 250 m or even deeper. It is one of the smaller *Icelinus* species, reaching only 11 cm in length. Its biology is poorly understood. It eats small shrimps and other crustaceans.

DUSKY SCULPIN

Icelinus burchami VISITOR ▼

IN THE STRAIT OF GEORGIA

There are historical reports of several specimens of the dusky sculpin from the southern Strait of Georgia in Howe Sound and off Nanaimo. More recently, there is one record in the groundfish research database at the Pacific Biological Station: from a bottom trawl catch in the northern strait in October 2003 at a depth of 115 m. However, no other catches in the past 50 years have been reported in any of the numerous recent pelagic and bottom research trawls or from any other potential sources (they would be too deep for sport scuba divers to see), which suggests this fish is a rare visitor to the Strait of Georgia.

DESCRIPTION

The dusky sculpin is similar to the northern sculpin, but its double row of scales under the dorsal fins does not extend back to the insertion of the second dorsal fin. The pelvic fins are very small and the pores on the head are notably large, possibly aiding them in detecting the motion of prey in deep, dark water, where visual cues are less helpful. The first and second dorsal fin spines are not longer than the third or fourth. The dorsal pre-opercular spine has antler-like projections.

LIFE HISTORY

The dusky sculpin occurs from Southern California to southeast Alaska in deep waters of 60 to 570 m. It can show up as bycatch of shrimp trawls and in prawn traps, usually near drop-offs offshore. It is found on gravel or rocky bottoms and reaches almost 13 cm in length. The biology remains to be studied.

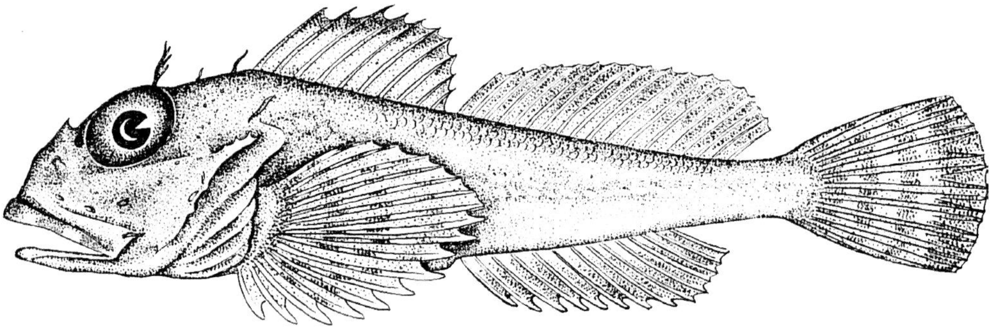

THREADFIN SCULPIN

Icelinus filamentosus UNCOMMON ◆

IN THE STRAIT OF GEORGIA

Threadfin sculpin generally occur at depths too great for sport scuba diving but have been sighted by divers in Jervis Inlet, off Gabriola Island and in Howe Sound. They are taken as a bycatch in shrimp trawls and prawn traps. They are caught occasionally in bottom research trawl surveys throughout the strait, but mostly in the southern strait. They do not appear to be abundant and probably can be considered uncommon.

DESCRIPTION

The threadfin sculpin is the biggest of the Strait of Georgia *Icelinus* species. This species reaches 27 cm in length and has two elongate dorsal fin spines and large head and mouth. It is important to distinguish this fish from the other *Icelinus* species with elongate first dorsal spines, the spotfin sculpin. The spotfin is smaller, occurs at shallower depths and has a conspicuous black spot near the back of the first dorsal fin, and its first dorsal spine is longer than the second, whereas the threadfin's first dorsal spine is shorter than the second. The threadfin has an upper row of body scales that extends to the back end of the second dorsal fin, whereas the same scale row on a spotfin sculpin extends only halfway.

Danny Kent

Gregory Jensen

LIFE HISTORY

This species occurs from the western Gulf of Alaska to Southern California at depths from 18 to 421 m, but there are reports of it occurring much deeper. It prefers a muddy bottom to one that mixes mud and rocky debris. Threadfin sculpin have spawned repeatedly at the Vancouver Aquarium, producing bright-green egg masses. This is another sculpin that needs essential biological information.

Danny Kent

FRINGED SCULPIN

Icelinus fimbriatus RARE ★

IN THE STRAIT OF GEORGIA

The fringed sculpin has recently been observed at the head of Sechelt Inlet in Porpoise Bay, where it was photographed by scuba divers. There was also a recent spawning observation in Saanich Inlet. Other observations have come from the Northern Gulf Islands area. Generally it is rare and has not been reported in Puget Sound. These fish, when seen by divers, were found adjacent to steep walls of rock, so the species may not occur in places where trawl nets could catch them, at least in the Strait of Georgia, where they have not been caught in any of the research trawl surveys over the past 25 years.

DESCRIPTION

This sculpin has a larger head than dusky or northern sculpins. It is a mottled brown or bluish grey. There can be purple on the head and in patches dorsally. The fins have dark bands that alternate with lighter brown-yellow bands. The lower body is covered with light spots that resemble eggs, a feature that some male sculpin (as well as unrelated fishes) use to attract females. The nasal spine on a fringed sculpin has cirri, and its jaw angle is fringed with them. The dorsal band of two rows of scales originates under the third dorsal spine and extends onto the caudal peduncle. There are two intersecting dark lines through the eye.

Fringed sculpin are hard to spot and difficult to identify, so they may be more common in the Strait of Georgia than has been thought. Lee Newman

LIFE HISTORY

Fringed sculpin range from the west coast of mid–Vancouver Island and from Pendrell Sound to Baja California. They grow to 19 cm in length. This is a deepwater sculpin that is reported to occur over soft muddy bottoms at depths of 30 to 265 m, though it also occurs on rock faces. Spawning appears to be in the early spring, with females producing about 200 to 250 eggs. There is one report of eggs being deposited in a tunicate. The larval planktonic stage lasts about two months.

SPOTFIN SCULPIN

Icelinus tenuis COMMON ●

IN THE STRAIT OF GEORGIA

Spotfin sculpin occur throughout the Strait of Georgia but are not abundant. The species has occasionally been recorded by scuba divers at night off the Sunshine Coast, in Howe Sound, off Powell River, in Jervis Inlet and Active Pass and off Gabriola Island. It also turns up occasionally in shrimp and bottom research trawls throughout the strait, where it could be considered to be somewhere between common and uncommon.

DESCRIPTION

This is a small, slender sculpin that is readily distinguished by the length of the first two dorsal spines. The first spine is longer than the second and about two times larger than the other dorsal spines. The top of the head is concave behind the eyes and there is a large, distinctive black spot on the first dorsal fin of males. There are two spines behind each eye and no nasal cirri. Two rows of scales extend dorsally to the middle of the second dorsal fin. Individuals are light brown with three to four dark brown saddles on the back. There are patches of bright colours such as orange, red, maroon and blue. Silvery areas occur between the gill openings and at the base of the large pectoral fin.

Gregory Jensen

LIFE HISTORY

The spotfin sculpin is found from Baja California to southeast Alaska. Its maximum length is 16 cm. It is found near reefs, or on sandy bottoms that may have shells and small rocks, at depths near the surface to 373 m, but most are at about 60 to 100 m. Spawning occurs in the winter and spring, with males guarding the nests. The diet is broad, including amphipods, polychaetes and small bottom-dwelling organisms. In general, the biology of this species is poorly understood.

LONGFIN SCULPIN

Jordania zonope COMMON ●

IN THE STRAIT OF GEORGIA

The longfin sculpin is abundant in the shallower waters of the Strait of Georgia, and particularly in the Gulf Islands. Until observed by scuba divers, it was thought to be rare; it was not caught by nets because of its preference for vertical rock surfaces. It is among the 10 fishes most commonly reported by divers throughout the strait. It was not reported in most research trawl surveys, confirming that it is restricted to areas and habitats that cannot be fished by a trawl net. It has been caught in bottom trawls in the southern Strait of Georgia on the United States side of the border.

DESCRIPTION

The longfin sculpin is noted for the red and blue stripes on its slender body. This fish always inhabits rocky bottoms, often on vertical faces. Danny Kent

Longfin sculpin are found from the intertidal zone to 38 m in a variety of habitats that include tide pools, kelp beds, rocky areas and rocky slopes. This distinctive fish is elongate and slender, growing to 15 cm. It is immediately identifiable by the three dark and light bars on the cheek below the eye. The dorsal and anal fins are similar in size. The lower pre-opercular spine is well developed, sharp and curved upwards. The upper spine is very small. There are prominent cirri on the head. Scales cover most of the body, and those below the lateral line are fused, forming plate-like ridges that give a washboard texture to the skin. The colours are variable and often match the surrounding habitat. Generally, the fish is light brown to olive green, but it may be red or even blue. There are six to eight dark saddles on the back and large dark blotches on the sides. The tail fin may be bright orange to yellow.

LIFE HISTORY

The species occurs from Central California to Prince William Sound in the Gulf of Alaska. It is not common in California and possibly not common in Puget Sound, but it is common in the Strait of Georgia. Divers report that this is a

Longfin sculpin males are darkened when guarding an egg mass. Danny Kent

skittish fish, but it is also territorial during the spawning season. Divers also report that it positions itself vertically on rock faces or under overhanging rocks. On the bottom, it is commonly observed resting on its large pectoral fins. Spawning has been reported as early as October, but the main season is in the winter. Males move into the shallow areas before the females do, and it is reported that a female will chase a male when she is ready to spawn. Males may be a blackish colour during spawning and guarding. After spawning, the male guards the egg mass, which is amber coloured but may include other colours. This is a sculpin that is commonly found in commercial aquariums, such as the Vancouver Aquarium, where it has spawned in captivity. Longfin sculpin feed on small crabs and shrimp. Polychaetes and harpacticoid copepods are also common food items.

PACIFIC STAGHORN SCULPIN

Leptocottus armatus ABUNDANT

IN THE STRAIT OF GEORGIA

The Pacific staghorn sculpin is one of the abundant sculpins in the Strait of Georgia. It is found just about everywhere from the intertidal areas to depths exceeding 100 m. It is among the 20 to 30 most abundant species in the strait; it is found in all major estuaries and seen occasionally by scuba divers everywhere except in Jervis Inlet. It occurs in most bottom trawl catches and even rarely in pelagic trawl catches down to 75 m.

DESCRIPTION

The Pacific staghorn sculpin gets its name from the upper portion of the preopercular spine, which is antler-like. It can be recognized by its scaleless body, the dark spot on the dorsal fin, and large pectoral fins that are golden yellow with dark crossbars. It is generally a mottled olive-grey colour with yellow and orange on the dorsal and pectoral fins.

LIFE HISTORY

This is a euryhaline fish, meaning it can live in a wide range of salinities. It occurs in bays, tide pools, estuaries and even the mouths of rivers from Southern California to the Bering Sea. Fish with this adaptability will

Staghorn sculpin have smooth skin (no scales), a black spot at the back of the first dorsal fin and a triple-pronged cheek spine above the pectoral fin. Lee Newman

Lee Newman

temporarily take water into their blood when they move into fresher water, then remove it when they move back into salt water. These changes in the blood are then compensated for physiologically by the fish. Juveniles are more tolerant of low salinity than adults and can be found several kilometres upstream in some coastal rivers.

This is one of the few sculpins that can be caught by anglers; be careful when recovering the hook as the spine can draw blood from a misguided finger. The species has been reported at depths from 90 to 100 m, but the depths of capture in the Strait of Georgia exceed 100 m. Small fish less than 7 cm long that are found in tide pools or around piers are about one year old. They grow to 48 cm and can reach an age of about 10 years. Spawning can occur from October to March, but most fish spawn in January and February. Females probably spawn only once a year, producing a batch of eggs that adhere to materials on the ocean bottom. Hatching success is higher at intermediate salinities, indicating a preference for spawning in subtidal areas. Juneau naturalist Bob Armstrong videotaped ravens on a tidal mud flat digging staghorn sculpin out of the mud and eating their egg masses from cockle or clam shells. It appears that the fish in the mud was guarding the eggs in the shells. It is not known if there is parental care, but this example appears to show there can be nest guarding.

The large mouth is an indication that the species is a voracious predator. It is an opportunistic one, feeding on fish such as sticklebacks, Pacific sand lance, flounders and other sculpins. It is also a predator of juvenile Pacific salmon, particularly when the juveniles are first adjusting to salt water, and a variety of other organisms such as amphipods, crabs, shrimp and other invertebrates. The Pacific staghorn sculpin is prey for a variety of birds and seals—and even Pacific salmon will eat small sculpin. The fish is often found in public aquariums, where it can be seen buried in the sand with only the eyes and part of the head exposed. Presumably this behaviour protects the fish from predators such as cormorants, herons and diving ducks. It is interesting that there are no reports of the fish being found in schools.

GREAT SCULPIN

Myoxocephalus polyacanthocephalus COMMON ●

IN THE STRAIT OF GEORGIA

Great sculpin are found in bottom research trawl surveys throughout the Strait of Georgia but are possibly more abundant in the southern strait. They are also seen by scuba divers off Campbell River, in Bute Inlet and Howe Sound, off Gabriola Island and in Active Pass, confirming that they occur throughout the strait. They should probably be considered common but not abundant. Scuba divers report that the strait's great sculpin may be readily approached. There are several reports both in Puget Sound and in the Strait of Georgia of the fish being captured in beach seines during the winter.

DESCRIPTION

The species is distinguished by its big head, large mouth and long, straight and smooth upper pre-opercular spine. The body is robust. There are few scales but some are embedded, forming fleshy, prickle-like structures on the head. The fish is a mottled green-brown with shades of pink, purple, yellow and orange,

Gregory Jensen

allowing it to blend into its surroundings. There are irregular dark saddles on the body and the belly tends to be white. It looks a bit like a big, elongated and flattened buffalo sculpin, but it has larger eyes and no plates on the lateral line.

Gregory Jensen

LIFE HISTORY

Great sculpin move into shallow water to spawn in tidal passes and rocky areas. They are distributed from Washington State to the Aleutian Islands into the Bering Sea, off the coast of Russia and south to the Sea of Japan. They can be found in shallow water down to 244 m and have been reported as deep as 775 m. In the Gulf of Alaska they are found from 200 to 720 m, mostly from 400 to 450 m. They can be in shallower water in the Strait of Georgia. This is one of the largest of the sculpins and the largest of the 25 species in its genus. This species is commonly caught incidentally in the bycatch of commercial fisheries for other species, usually on sand or mud bottoms. Studies in the eastern Bering Sea found that males ranged in size from 18 to 73 cm, with a maximum age of 15 years, and females from 13 to 80 cm with a maximum age of 17 years. Males were smaller than females. Few fish were older than 12 and most were between 4 and 6. Half of the fish were mature at 57 cm and about 7 years old. The youngest mature fish was 5 years old, and the smallest mature fish was 52 cm. The great sculpin spawns in the winter and is one of the most fecund of all sculpins. The species has been exclusively studied in the Sea of Okhotsk; Russian investigators reported a fecundity of 423,000 eggs.

This is another ambush predator. The dominant prey depends on the size of the fish, with smaller great sculpin feeding on crabs, shrimp and small fish. The large individuals feed on crab, octopus and fish such as salmon. Larger individuals are eaten by lingcod and harbour seals, and smaller ones by other sculpins, greenlings and skates.

TIDEPOOL SCULPIN

Oligocottus maculosus ABUNDANT

IN THE STRAIT OF GEORGIA

The tidepool sculpin is abundant in tide pools and exposed water at low tides throughout the Strait of Georgia. It is among the 30 most abundant species in the strait. In late spring and early summer it appears in tide pools that are only a few centimetres deep. Tidepool sculpin are the delight of parents and grandparents as they teach young children to enjoy and respect marine life.

DESCRIPTION

The colour of mature tidepool sculpin is drab, usually grey with accents of white and black forming vertical bands of disruptive light and dark. Juveniles are green and brown with reddish highlights on the fins. The colours match the tidepool sculpin's surroundings, providing excellent camouflage. Males have an elongated anal papilla and their first two or three anal rays are enlarged.

Gregory Jensen

The fish has a typical sculpin appearance: a large head and a long, thin body. It is distinguished by its size and location, but also by its lack of scales and having a forked, single, pre-opercular spine. It is separated from the fluffy sculpin by the relative scarcity of cirri, or hair-like projections, on the head and body, and from the sharpnose sculpin by the more posterior location of the anus, just ahead of the anal fin.

LIFE HISTORY

These fish are found from northern Baja California to the Gulf of Alaska. Reports of their presence in the Bering Sea and off the coast of Russia appear to be incorrect. Tidepool sculpin can be as small as 1.5 cm and are usually difficult to see until they dart for cover when alarmed. A study of their biology in Puget Sound showed that the smallest individuals in their first year of life were most abundant in the highest pools or areas closest to shore. Larger and older fish are less abundant overall and more common in pools farther away from shore and in deeper water. The maximum depth is a reported 102 m. The reduced numbers of larger fish may relate to predation on the smaller fish, particularly by the great blue heron. The maximum size is about 9 cm, and the maximum age is five years.

The tidepool sculpin tolerates a wide range of temperatures and salinities. In laboratory studies the fish were able to obtain oxygen through their gills and skin even when out of water for several days. Tidepool sculpin have also shown an ability to find their particular pool or be able to home: those removed from their pool were able to return to it from up to 300 m away, apparently using vision and smell. However, tagging studies indicate they do not usually move very far. Recent studies identified their ability to produce a chemical when alarmed that alerts other individuals nearby of the need to protect themselves.

Maturity occurs at age one, and spawning occurs from November to May. In Puget Sound, spawning occurs twice yearly, with a synchronous spawning throughout the sound over three to four weeks from about late December to January and a second asynchronous and extended spawning over a six-month period to late spring. Egg masses are green or deep red, and single eggs are extruded into spaces among barnacles or mussels, where the eggs adhere to each other to form a cluster. These are not adhesive to the substrate. Small tidepool sculpin feed on harpacticoid copepods and amphipods, barnacles, shrimp, isopods, polychaetes and insects as they grow. They are eaten by larger intertidal fish such as kelp greenling, and by the great blue heron.

SADDLEBACK SCULPIN

Oligocottus rimensis RARE ★

IN THE STRAIT OF GEORGIA

The saddleback sculpin is a tide pool–dwelling sculpin that is rare in the Strait of Georgia. There are records from the Nanaimo and Cowichan River estuaries and older records from off Gabriola and Saturna Islands. There are no scuba diver reports or catches in shrimp trawls that might encounter the species in shallower water.

DESCRIPTION

The saddleback sculpin is more elongate and smaller than the other *Oligocottus* species. It has the fewest cirri and has a pre-opercular spine that is short and not bifid. The first two anal fin rays are notably elongated on the male. Its colours are combinations of light olive green and pale blue, sometimes with red highlights. There are four to five brown saddles along the back. The highly variable colour probably is a way of matching surroundings. The body is covered with small prickly scales or papillae.

LIFE HISTORY

This species occurs from northern Baja California to southeastern Alaska. It reaches 6.5 cm in body length. It also occurs lower in the intertidal zone than the common tidepool sculpin, to depths of up to 8 m, often in subtidal rocky areas and kelp or eelgrass beds. Very little is known about the biology of this very small sculpin, and this is one more species that would benefit from the curiosity of some graduate students.

Gregory Jensen

FLUFFY SCULPIN

Oligocottus snyderi VISITOR ▼

IN THE STRAIT OF GEORGIA

The fluffy sculpin is rare in the Strait of Georgia but has been reported in a survey of the Cowichan estuary. There are no records from Puget Sound and we have no scuba observations from the strait. Because it is found in Juan de Fuca Strait and in the San Juan Islands, it is probable that any specimens in the Strait of Georgia arrived from outside the strait. It is also possible that the specimen from the Cowichan estuary was not a fluffy sculpin.

DESCRIPTION

Like other species of *Oligocottus*, fluffy sculpin can best be identified to the genus level by noting the bifid pre-opercular spine on the cheek and the posterior location of the anus. At the species level, the fluffy sculpin has the most developed cirri on the head, cheek, lateral line and dorsal fin base of all the *Oligocottus* species; some individual tidepool sculpin have relatively well-developed cirri,

The fluffy sculpin may be named for the hairlike cirri, but this species is very tricky to identify because its colours are mostly a reflection of the background seabed. The very pointed snout and the cirri are clues, as is its habit of following the waterline up the shore in shallow water. Gregory Jensen

Gregory Jensen

but none at the base of the nasal spine, which is located ahead of the eye. The fluffy sculpin has no scales or prickles. Its colours are highly variable, from uniform emerald green, when living among green seaweeds or seagrass, to mottled pink and black against a rocky background with pink coralline algae. As a result they are hard to detect in a tide pool.

LIFE HISTORY

Fluffy sculpin range from the western Gulf of Alaska to central Baja California but are most common to the west and southwest of Vancouver Island. The maximum age is reported to be two years, and the maximum length is 9.2 cm. Spawning occurs throughout the year, mostly in the late winter and spring. This is a tide pool and shallow intertidal species occurring to about 6 m over bottom with small rocks and algae.

THORNBACK SCULPIN

Paricelinus hopliticus RARE ★

IN THE STRAIT OF GEORGIA

The thornback sculpin is rare in the Strait of Georgia and uncommon throughout its range. We have records from shrimp research trawl surveys. In 2004, specimens were caught off Comox, in Howe Sound and off the mouth of the Fraser River. Catches off Comox were also made every year from 2005 to 2009. There is one record in the Pacific Biological Station's groundfish research database of a catch of five on the bottom at 172 m in Howe Sound in October 2003.

The thornback sculpin is named for the rows of sharply recurved spines along both dorsal fins along the back. The species has a slender body, and its anal fin and first dorsal fin are very long. Danny Kent

DESCRIPTION

The thornback sculpin is mottled in colour with yellow and blue spots. The entire body is very thin, but the first dorsal fin is tall compared to that of other sculpins; the first few dorsal spines are as long as the more posterior spines. It is distinguished by backward-hooked spines along the base of the two dorsal fins, plus sharp backward spines on the snout and head. As with similar sculpins, the lower pectoral fin rays are separate from the fin membrane and are used like fingers to prop the fish up off the bottom.

LIFE HISTORY

This species occurs from the Gulf of Alaska to Southern California. It is found close to the shoreline, usually from 20 to 183 m on rocky bottoms, sometimes on cliff faces. The thornback sculpin is one of many elongate sculpin species and has a maximum length of 19.4 cm. This is another sculpin in need of study.

SLIM SCULPIN

Radulinus asprellus COMMON ●

IN THE STRAIT OF GEORGIA

The slim sculpin is found at moderate depths over soft bottoms throughout the Strait of Georgia, but particularly between Texada and Vancouver Islands. It is common, but not abundant, in bottom research trawls, and rare in shrimp trawls. It has been sighted by scuba divers only in Howe Sound and Active Pass.

DESCRIPTION

The slim sculpin is distinguished by its long, thin and rounded body with a relatively small head and large eyes. In front of the eyes there is a distinctive single nasal spine that is long, sharp and thin. A single row of scales above the lateral line extends to about the middle of the second dorsal fin, and there are no scales between this row and the dorsal fin. Both the dorsal and anal fins are long. There is a dark spot at the back of the first dorsal fin of males. It is light brown to olive green with a creamy white underside. There may be mottled areas of orange-brown. The male genital papilla is long and conical, presumably indicating that sperm is deposited internally.

The slim sculpin can be identified by the single row of prominent scales between the base of the first dorsal fin and the lateral line, the rows on each side meeting on top of the head just behind the eyes. The large eyes are right on the top of the low-profile head. Janna Nichols

LIFE HISTORY

The slim sculpin is found from northern Baja California to the Aleutian Islands and the Bering Sea. It occurs on soft sandy or rubble bottom in 18 to 283 m of water, and is believed to occur in deeper water at the southern limit of its range because of its preference for cooler temperatures. The maximum length is 15 cm. Spawning occurs in the winter and spring, and males guard the egg mass. This species would be relatively easy to study because the bottom trawl catches provide some understanding of where and when to collect them.

DARTER SCULPIN

Radulinus boleoides RARE ★

IN THE STRAIT OF GEORGIA

The darter sculpin is very rare in the Strait of Georgia and generally uncommon throughout its range. We have one record from a confirmed scuba observation in Howe Sound and four reports from the groundfish research database at the Pacific Biological Station. There is one report from an on-bottom trawl catch in 90 m off the mouth of the Fraser River in 1975, and two in the same general area at about the same depth in February 2004. A research survey reported another specimen on bottom at 113 m in the northern strait in October 2003.

The darter sculpin is best differentiated from the slim sculpin by the longer spines on the first dorsal fin, together with a longer, flatter snout. Danny Kent

DESCRIPTION

The darter sculpin is very elongate and has a round body. The colour is olive-grey, with three to four dark saddles dorsally and a whitish silver belly. There are dark-brown bars on the dorsal, pectoral and caudal fins. It has a single row of scales immediately above the lateral line extending to about the middle of the second dorsal fin. The pelvic fins are relatively short, slightly longer than the base of the pectoral fin. There is one small cirrus on the posterior of the eyeball. The darter sculpin has a protruding jawline, giving the face a duck-like appearance; it is like the slim sculpin but with a longer snout relative to the eye diameter. Both of these species are very different in overall appearance from the spinynose sculpin. The larvae of the darter and spinynose sculpins look so similar that there is debate in the scientific literature about whether misidentification has occurred. This close similarity in larval anatomy may indicate the validity of including all these species under the genus *Radulinus*.

LIFE HISTORY

The darter sculpin is reported from Southern California to Langara Island in Haida Gwaii and north to Kodiak Island. The darter and slim sculpins both occur in deeper water than the spinynose sculpin, usually on sand or gravel, away from rocky reefs. The darter sculpin apparently reaches 15 cm and is found in water 15 to 146 m deep. The biology remains to be discovered.

PUGET SOUND SCULPIN

Ruscarius meanyi RARE ★

IN THE STRAIT OF GEORGIA

This sculpin has been reported in Howe Sound in the southern Strait of Georgia and in Malaspina Strait, but there were no reports in any surveys or from scuba divers until recently. We have a recent scuba record from June 2017 near Malaspina Channel at 10 to 15 m.

DESCRIPTION

The Puget Sound sculpin can be distinguished by its small size—its maximum length is about 6 cm—and by the filamentous appendages on the head and the location and pattern of the scales. This is the smallest sculpin in the Strait of Georgia, but it has a robust body. There are scales on the head and upper body and some on the upper eye. It has a single long cirrus in front of the eye and another over the eye that has three branches. It is greenish with some brown and dark bars below the lateral line. There are distinct bars on the pectoral and caudal fins.

LIFE HISTORY

Puget Sound sculpin range from southeast Alaska to Northern California. It is an intertidal to subtidal species that is found around large rocks or on rock faces, usually from the surface to 20 m, but sometimes as deep as 82 m. They will hide under rocks and inside sponges. It is rare throughout its range and there is little understanding of its biology.

Danny Kent

CABEZON

Scorpaenichthys marmoratus

COMMON ●

IN THE STRAIT OF GEORGIA

Cabezon are common throughout the Strait of Georgia. They are not common in estuaries but are seen everywhere by scuba divers. Recently hatched individuals are reported in May and June in sargassum and bull kelp along tide lines. Adults will often allow divers to come quite close, resulting in some amazing pictures that can be seen online. They are caught occasionally in bottom trawls throughout the strait, and surprisingly they are captured in pelagic research trawls in the top 75 m of the strait. The cabezon appears to be less common now in the Strait of Georgia than it was decades ago, possibly because of direct fishing pressure or indirect via other species, and possibly because of predation by marine mammals.

The cabezon, the largest of the sculpins, has a marbled pigment pattern and a very big mouth (the Spanish word *cabezona* means "big-headed"). Janna Nichols

DESCRIPTION

This is another scaleless sculpin with a very large head, flabby skin and big, round pectoral fins. It has branched cirri above each eye and an unpaired flap of skin at the tip of the snout. Young fish may be red with patches of white. Older fish are well camouflaged with marbled brown, green and white patches that may change to match the surroundings.

LIFE HISTORY

Cabezon are distributed from southeast Alaska to central Baja California. This demersal and solitary species is the largest sculpin in the world. It is reported to achieve a maximum length of 99 cm and maximum weights of between 7 and 14 kg, which is large, but it is the girth that is impressive. A 72 cm fish was found to be 13 years old, and 50 to 60 cm fish ranged in age from 4 to 9 years. The maximum age may be about 16 to 17 years. Cabezon are a sought-after sport fish and even of some commercial value. The flesh may be a bluish green, as is seen in some lingcod, but it turns white when cooked. Apparently, cabezon were once eaten by the coastal First Nations.

In January 1923, the renowned ichthyologist Carl Hubbs and his wife reportedly ate eggs from a cabezon that was apparently getting ready to spawn. Both Hubbses became violently ill, resulting in the general belief that eating cabezon eggs is not advisable. Spawning is reported to occur from November to September, mostly from January to April. Males mature first, at about 34 cm long and age two, while females mature at 45 cm and age three. Large greenish to purple egg masses are ferociously defended by males. Cabezon are batch spawners that spawn several times. It is most unsportsmanlike for spear fishermen to kill a nest-guarding male. Juveniles, which differ in colour and shape from adults, are found in shallower protected areas with dense algal cover and deep rocky crevices, or in tide pools. Adults may be fished at depths from 15 to 75 m. Cabezon feed on a variety of organisms, with juveniles eating amphipods and small shrimp. As cabezon grow, they are able to feed on larger and more heavily armed prey such as molluscs and crabs. They also consume fish and octopus.

MANACLED SCULPIN

Synchirus gilli

UNCOMMON ◆

IN THE STRAIT OF GEORGIA

The manacled sculpin is scattered throughout the Strait of Georgia, where it is seen everywhere by scuba divers, but generally is not abundant. It is not seen in estuaries or caught in trawls in the strait, most likely because of its preference to attach to objects in shallow intertidal areas.

DESCRIPTION

This skinny little fish with a pointed head uses the joined pectoral fins to cling to vegetation and other structures, often in currents. It is supposed to be easily identified by its two pectoral fins being fused together under the throat area, but this species is usually seen by scuba divers in shallow water, where it hovers or rests on kelp blades that match its body colour. A diver will have difficulty deciding if the pectoral fins are fused. The small body size and the habit of hovering on seaweed that it matches, together with a characteristic flat profile of the lower jaw and head, make this species recognizable once its appearance has been learned. Its colour can vary from a kelp brown to greenish or reddish, as with other fishes that match seaweed backgrounds. It has pearly white streaks along the cheek and below the lateral line on the side of the body, and a series of light blotches on the side.

The manacled sculpin is kelp-coloured, with pearly spots and a downward tilted tip of its small jaw. It usually rests on kelp.
Gregory Jensen

LIFE HISTORY

The manacled sculpin grows to only 7.5 cm in length. This species occurs in shallow waters to depths of about 14 m from Southern California to the Aleutian Islands. Spawning in the Strait of Georgia occurs in late winter and early spring with eggs laid on kelp holdfasts. Jeff Marliave collected larvae in Howe Sound, and all catches were less than 20 m from steep, rocky shorelines. This is typical of small intertidal fishes with low egg production and with larvae that do not drift with currents. Crustaceans are the major prey.

ROUGHSPINE SCULPIN

Triglops macellus UNCOMMON ◆

IN THE STRAIT OF GEORGIA

Roughspine sculpin occur throughout the Strait of Georgia, where they are occasionally caught on soft bottoms with research trawls. The only estuary where the species has been reported is the Courtenay estuary. Roughspine sculpin have occasionally been seen by scuba divers off Campbell River, Powell River and Gabriola Island, in Active Pass and in Howe Sound. They appear to be more abundant in the southern strait, in the Baynes Sound area and in the Gulf Islands at depths of 25 to 108 m, but generally would be uncommon.

DESCRIPTION

The roughspine sculpin has a spine ahead of its eye, but this thin, elongate fish is more readily identified by the long, protruding lower pectoral fin rays, which can act like fingers on the soft bottom that this fish occupies. The other key characters—at least in the Strait of Georgia, where only two species of this genus occur—are the oblique folds of skin on the lower sides of the body and a dorsal fin of 10 to 12 spines followed by a second dorsal of 28 to 31 soft rays.

Lee Newman

Lee Newman

It can be well camouflaged, with an olive green to brown body, dark saddles and a cream colour ventrally. There are dark bars on the rays of the dorsal and caudal fins and the upper pectoral fin.

LIFE HISTORY

The species is found from the eastern Bering Sea and the Aleutian Islands to Washington State. Individuals tend to occur deeper than 15 m on sandy bottoms away from any large rock features. The depth range is reported to be 18 to 275 m, most often between 30 and 166 m. The roughspine sculpin grows to just over 30 cm in length.

RIBBED SCULPIN

Triglops pingelii COMMON ●

IN THE STRAIT OF GEORGIA

The ribbed sculpin probably is common at intermediate depths on soft bottoms in the Strait of Georgia. It was the second most abundant sculpin in a series of research bottom trawl surveys from 2003 to 2006, with catches slightly larger than those of the Pacific staghorn sculpin but smaller than those of the roughback sculpin. There are scuba diver reports from off Gabriola Island and the Sunshine Coast. The abundance appears to be larger in the southern strait, particularly off Victoria at depths of 22 to 42 m.

DESCRIPTION

The ribbed sculpin, like the roughspine sculpin, has a thin, tapering body with oblique skin folds on the sides. The pectoral fins do not have elongate protruding lower rays like the ones found on the roughspine sculpin. In Alaska, where the ribbed sculpin is one of the most abundant sculpins, identification becomes more problematic because these two species occur with four very similar other species. The ribbed sculpin is olive-brown with dark saddles dorsally and a whitish belly. Black spots below the lateral line may be connected in mature males. The dorsal fin has 10 to 13 spines and 23 to 26 soft rays.

Andy Murch

LIFE HISTORY

This is a circumpolar species, as it also occurs in the North Atlantic and Arctic Oceans. In the Pacific, it ranges from the Sea of Japan to the Bering, Beaufort and Chukchi Seas, to off the west coast of British Columbia and in the Salish Sea. The maximum length is 24 cm and the maximum age is six years. It sits on soft sand or mud bottom, where scuba divers are less likely to be paying attention, and it is hard to see until it moves, which tends to occur in short, slow swimming bouts. Bottom depths range from 4 to 482 m, with most at depths of 20 to 150 m. The preferred prey apparently are zooplankton and not crustaceans, like with most other sculpins. However, Russian studies list mysids and shrimp as the preferred diet.

SILVERSPOTTED SCULPIN

Blepsias cirrhosus UNCOMMON ◆

IN THE STRAIT OF GEORGIA

The silverspotted sculpin has been taken in beach seine hauls right inside Vancouver Harbour, on the shoreline of Stanley Park in First Narrows in an area of high currents and abundant seaweeds. It is most common in the southern strait, from Burrard Inlet to the San Juan Islands, but is generally uncommon in the Strait of Georgia. There has been an observation from the Nanaimo estuary, but it is not reported from other estuaries and rarely seen by scuba divers anywhere in the strait. This is a shallow-water species and not found in any of the bottom research trawl studies.

Kelp-coloured silverspotted sculpin swim among marine plants, sculling with their fins to hover, and resemble leafy bits of seaweed. Danny Kent

DESCRIPTION

The golden-brown base colour of the body is disrupted by opalescent spots on the body and fins. The elaborate size of the deeply notched fins—and their sculling motion, which disrupts the body outline—are features that also appear on the sailfin sculpin. The body is compressed and there are three pairs of prominent cirri on the lower jaw.

LIFE HISTORY

Silverspotted sculpin occur from Northern California through to Alaska, the Bering Sea, and Sea of Okhotsk, in the Sea of Japan and to South Korea. They grow to a maximum of 20 cm, 87 g and about age six, and they occur from the intertidal zone to about 95 m, but mostly in shallow water to 60 m. This beautiful species is usually found in seaweed or eelgrass beds on the outer coast, where the fish remains in continuous motion, slowly sculling among plant blades. It is difficult to notice because of its own seaweed-like appearance. The silverspotted sculpin has been extensively studied in Japan, where it is known as the "little dragon sculpin." This research has shown it spawns its eggs inside sponges, where the eggs are incubated by seawater pumped by the sponge and kept free of infections by the sponge's antimicrobial secretions. Spawning off North America occurs in the winter, but off Russia it occurs in the summer. Males are absent from the spawning areas because the eggs are fertilized with stored sperm as they are released. Prey are mostly amphipods and isopods but also small shrimp and polychaetes. This species is sometimes called a "sea raven."

SAILFIN SCULPIN

Nautichthys oculofasciatus COMMON ●

IN THE STRAIT OF GEORGIA

Sailfin sculpin are found throughout the Strait of Georgia, where the species is moderately common. They are occasionally caught on bottom throughout the strait and frequently seen by scuba divers. There is even a rare capture in a pelagic trawl in the top 30 m.

DESCRIPTION

There is no mistaking this fish. The first three to five dorsal fin spines are very long, forming a sail-like fin that often projects forward as the fish swims. The second dorsal fin has more than 26 soft rays. There is a prominent dark band through the eye and downward across the cheek. The oversized first dorsal fin projections are more ornate on the juveniles than on the adults, and there is a large cirrus on the eye. The colour is mostly a light tan with darker dusky and cream blotches on the back. The ventral surface is lighter. However, a variety of colours, ranging from oranges to reds to pinks and blues, are displayed in dozens of photographs available online.

Gregory Jensen

In its pelagic juvenile stage, the sailfin sculpin's first dorsal fin grows relatively large, and the eye stripe and fleshy cirrus above the eye develop as well. The parachute glide enables the feeding strategy of a jellyfish, rising upward then drifting slowly downward while catching prey. Danny Kent

LIFE HISTORY

This is another very distinctive and pretty little sculpin that is commonly found in public aquariums such as the Vancouver Aquarium, where it is bred. It is found from the intertidal zone to 110 m in the western Gulf of Alaska and from Kodiak Island through to Southern California. Reports of findings from Russia may not be correct. This nocturnal species is nearly motionless during the day, even hanging upside down in crevices or on rock faces. At night it swims using undulations of the long dorsal and anal fins. These wave-like motions mimic the motion that currents create among seaweeds.

Spawning occurs late in the winter and into early spring. Males and females copulate, allowing sperm to be stored and the female to release eggs over an extended period, as the eggs are not fertilized until they contact salt water. Spawning occurs in shallow water with eggs wedged among clumps of mussels or in cracks between large barnacles; the egg masses do not adhere to the substrate. Like other sculpins that copulate, it appears that neither males nor females guard the adhesive, bright-orange eggs. Species that copulate typically deposit eggs at sites where guarding is not possible. After hatching, the young fish have enlarged pectoral fins that are used for reducing their sinking rate. They are readily recognized as sailfin sculpin after about one month. They mature after one year, but maximum ages and patterns of growth have not been reported. The maximum size is 20.3 cm. At the Vancouver Aquarium, sailfin sculpin have lived for five years. Their diet consists of small benthic organisms, particularly small shrimps, crabs and other crustaceans.

Lee Newman

SPINYHEAD SCULPIN

Dasycottus setiger UNCOMMON ◆

IN THE STRAIT OF GEORGIA

Spinyhead sculpin are found on soft bottoms throughout the Strait of Georgia from Campbell River to the southern strait, where they appear to be more abundant—particularly outside of Howe Sound and farther south. Bottom trawl catches occurred at depths from 75 to 380 m. These fish have been reported by scuba divers in Howe Sound and Burrard Inlet, and from only one estuary (Courtenay). They have been caught occasionally in research and commercial shrimp trawl catches. Aside from the southernmost area, they are generally uncommon in the Strait of Georgia.

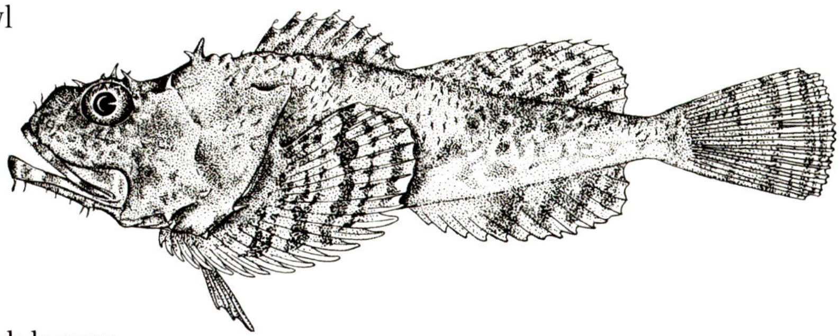

DESCRIPTION

This species looks attractive for a sculpin when it is pinkish grey. There are dark-brown areas on the body, with dark bars on the body and fins. The head is very large, appearing to make up much of the mass of the individual. There are large, prominent blunt spines on the head that give the fish its common name. The skin is loose, giving the fish a shaggy appearance. There are numerous fleshy projections, or cirri, on the body and head.

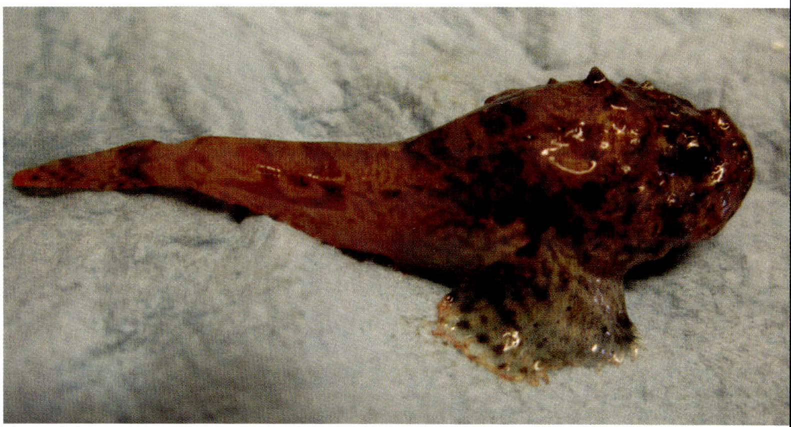

Dick Beamish

LIFE HISTORY

Spinyhead sculpin range from Washington State to the Bering Sea, off the coast of Russia in the Sea of Okhotsk, and south to the Sea of Japan. They are mostly caught as bycatch in commercial bottom trawl fisheries. Off the west coast of North America, they occur mostly in deeper waters, about 50 m to 300 m, but they have been reported as deep as 850 m. The maximum length is a reported 45 cm, but most fish range between about 10 and 25 cm. The maximum age is reported to be 11 years. Common diet items include shrimp, fish and benthic crustaceans, including amphipods and crabs. Information on the spawning ecology is lacking, but the fish most likely spawns in the winter, possibly in deeper water.

BLACKFIN SCULPIN

Malacocottus kincaidi COMMON ●

IN THE STRAIT OF GEORGIA

The blackfin sculpin may be common in the Strait of Georgia, as specimens are regularly caught in pelagic research trawls. Juvenile specimens collected in the mid- and deep waters of the strait appear to be this species, but a careful taxonomic study of this midwater sculpin may need to be conducted with DNA. There is some uncertainty about the similarity between this species and an Alaskan species called the darkfin sculpin (*Malacocottus zonurus*).

DESCRIPTION

The small fish with loose, gelatinous, scaleless skin that are common in the deeper trawl catches appear to be juveniles. Our specimens have distinct light margins outside of the dark bands on the fins. The caudal fin has distinct dark

Danny Kent

bands with a white margin. These globular fish are light grey or light brown with a diffuse mottled pattern of darker browns and blacks. They resemble tadpole sculpin with the colour pattern of blackfin sculpin. Recently, small juvenile fish were brought back to the Vancouver Aquarium, where they matured into large bottom-dwelling fish, typical of sculpins found on bottom in nearshore areas and prawn traps. This suggests that the pelagic form is a stage in their early life history.

LIFE HISTORY

The specimens from Alaska are not confirmed to be blackfin sculpin, but if confirmed, then the distribution would be from Alaska to Puget Sound. There are reports of a maximum length of 10.6 cm. More of this species' biology will be available once the pelagic specimens in the Strait of Georgia and the fish raised in the Vancouver Aquarium are identified using DNA.

The darkfin sculpin is an Alaskan species with a larger head and body than the British Columbia species, the blackfin sculpin. One of the difficulties with identifying these uncommon species is that few specimens are caught as mature adults in standard sampling nets. The other issue is that midwater trawling gear in the Strait of Georgia regularly catches the "globby" pelagic juveniles, which oddly resemble jellyfish, as the body is surrounded by soft gelatinous tissue. This tissue increases buoyancy by having lower salinity than surrounding seawater. Nobody can identify these globby individuals at the species level. The Vancouver Aquarium reared small juveniles from prawn traps, and within a couple of years they reached a length of 24 cm, a record size for a blackfin sculpin. The fish definitely has too small and rounded a head to be called a darkfin sculpin, which is listed as reaching 35 cm in length. The darkfin sculpin occurs in Alaska in deeper water than the blackfin sculpin does in southern British Columbia and Puget Sound, so the specimens that come up in prawn traps in British Columbia are probably blackfin sculpin. In Alaska, the darkfin sculpin is reported from the outer continental shelf and upper continental slope, making this species likely to be a rare occurrence in the Strait of Georgia. Experts note that misidentifications have occurred frequently between the darkfin and blackfin sculpins, so a DNA analysis is clearly in order.

TADPOLE SCULPIN

Psychrolutes paradoxus COMMON ●

IN THE STRAIT OF GEORGIA

The tadpole sculpin occurs in the midwater and deeper areas of the Strait of Georgia. Trawls fishing deeper than about 70 m usually catch anywhere from a few to several dozen specimens. This species is among the 10 most abundantly caught in sets deeper than 150 m, occurring as low as just off the bottom in the deepest areas of the strait. The numbers in a catch are not an indication of their abundance because they can easily fit through trawl meshes. Thus, our guess is that it is a common species in deeper waters of the Strait of Georgia.

DESCRIPTION

This is one of the smallest sculpins, growing to a maximum size of 6.5 cm. The fish has a distinct tadpole-like appearance, with a large head and a small tapering body. It has jelly-like, loose skin with distinct dark-brown bars on the fins and patches on the body, and a distinct dark bar on the pectoral fin. The spinous (anterior) portion of the dorsal fin is low and partially buried in the body.

Danny Kent

LIFE HISTORY

Tadpole sculpin occur on soft and rocky bottoms ranging from estuaries to 240 m. Marine naturalists Andy Lamb and Phil Edgell reported that they commonly occur in shallow, muddy bays, and we have a record from a shrimp trawl. However, J.L. Hart informed us in *Pacific Fishes of Canada* that they were caught at depths from 55 to 220 m, as well as in beach seines. A maximum depth of 1,100 m has been reported. Our numerous catches are from pelagic specimens in water as deep as 365 m. There is much more to be learned about this species and the specimens in the Strait of Georgia—for example, whether the pelagic specimens are the early phase of the life cycle.

This fish is found from Southern California to the Aleutian Islands, in the Bering Sea, and across to the Sea of Okhotsk in Russia and the Sea of Japan. It would be a most interesting study to compare specimens from all these areas, including an analysis of age and DNA. There is no information on spawning time, but it is most likely in the winter, as larvae off the west coast are found in the plankton from February to May and around the Fraser River in April and May. Tadpole sculpin eat barnacle larvae, crab larvae, amphipods and copepods. Scuba divers report that they may be unpalatable to predators.

GIANT BLOB SCULPIN

Psychrolutes phrictus VISITOR ▼

IN THE STRAIT OF GEORGIA

The giant blob sculpin has not been reported in the Strait of Georgia or other areas of the Salish Sea. However, there was a report from January 2019 in the Pacific Biological Station's groundfish research database of a larger specimen from a commercial bottom trawl catch off the mouth of the Fraser River at a depth of about 111 m. The giant blob sculpin is an offshore, deepwater, bottom-dwelling species, but we are accepting the report as valid as there is no mistaking the identity of larger specimens. There is no doubt that the fish was a visitor from offshore.

DESCRIPTION

This species is almost ovoid and has a very large head and thick skin. The blobby body results from a gelatinous layer between the skin and muscle. There are small cirri on the head and body. The large head is flattened, with

T. Mecklenburg

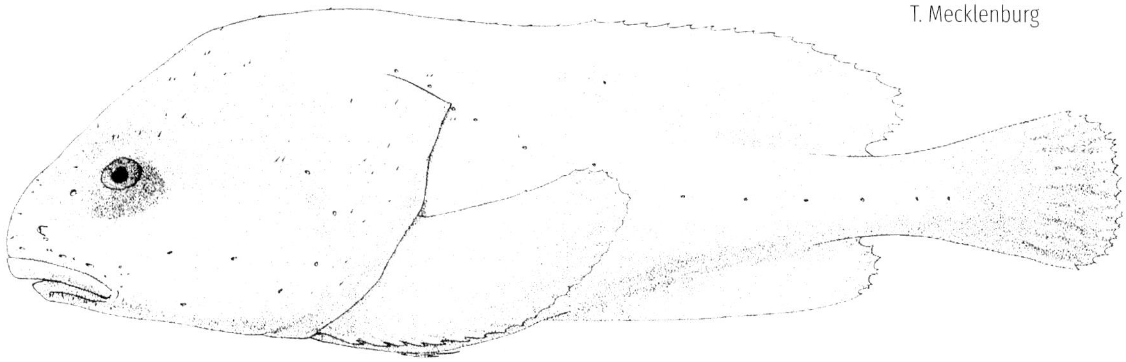

Archipelago

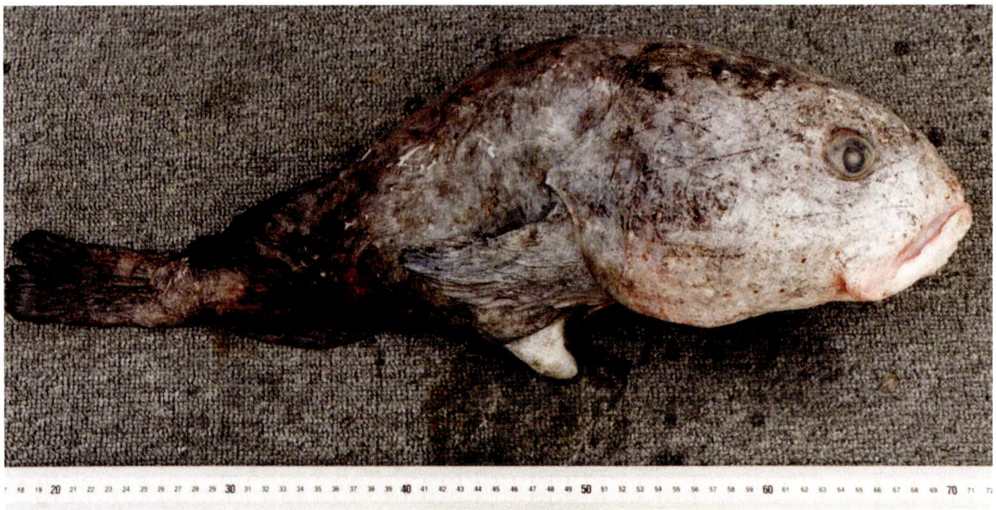

widely separated eyes and a curved mouth with fleshy lips. The anterior spinous dorsal rays are deeply buried in the loose skin, and the spinous and soft rays are continuous. The fish is mostly a uniform grey, black or brown. There may be some indistinct mottling, and the colour is lighter ventrally.

LIFE HISTORY

The giant blob sculpin is widespread, ranging from off the coast of Japan, in the Sea of Okhotsk, off the coast of Russia, in the western and eastern Bering Sea and south to off Mexico. It occurs on soft bottoms from 480 to 2,800 m. A maximum length of 70 cm has been recorded. There is a report of a spawning aggregation at 1,300 to 3,000 m off the coast of California, where numerous nests with large pink eggs were guarded by adults. Prey includes sea pens, sea cucumbers, molluscs, crustaceans and cephalopods.

SOFT SCULPIN

Psychrolutes sigalutes UNCOMMON ◆

IN THE STRAIT OF GEORGIA

This is a surface to deepwater pelagic species when it is immature and a crevice dweller as an adult in the Strait of Georgia. It is found in the surface 30 m in the summer and fall, and later occurs in rocky or silty bottoms nearshore and in the deeper waters. Thus, it appears this species has an extended period in the pelagic areas of the Strait of Georgia before it settles in the adult habitat nearshore around subtidal to deeper boulders and crevices on slopes and rock faces. Adults are reported by scuba divers only in Howe Sound and off Gabriola Island. Adults also occur occasionally in shrimp trawl catches off Comox, in Howe Sound and off the mouth of the Fraser River. They are not found in trawl catches in deeper water.

Jackson W.F. Chu

DESCRIPTION

The larval form of the soft sculpin is a red-orange and looks almost tropical. It has a large head, smooth skin and big, orange-red pectoral fins. The adult fish looks like a snailfish with its elongate but tadpole-like shape. Very large head canal pores occur above and below the jaws and posterior on the cheek. The fish is distinguished by soft flesh and loose skin; the anterior (spinous) and posterior (soft-rayed) portion of the dorsal fin are the same height and not divided into two fins. The dorsal and anal fins extend onto the base of the caudal fin. There is some uncertainty in the nomenclature, with the genera

Gilbertidia and *Psychrolutes* both occurring in publications. Drawings and pictures also differ in the way that the height of the anterior portion of the dorsal fin is shown. The anus is at the anterior end of the anal fin. It appears that different species or life stages may be described in various publications, making descriptions somewhat inconsistent. The uncertainty associated with identifying this species provides a good reminder of the importance of museums and the need to train ichthyologists. Reported colours vary, with most fish that are caught in the midwater trawls being greyish with some weak, darker mottling. However, browns and brown-orange colours and even pinks are reported.

LIFE HISTORY

The soft sculpin is reported to occur from the Aleutian Islands to Puget Sound, making the Strait of Georgia close to the southern limit of its distribution. There also are records from the Commander Islands off Russia. It is recorded to depths of 225 m. The maximum length is 8 cm, but the juveniles that are found in the midwater of the Strait of Georgia are much smaller. Studies on the west coast of Vancouver Island indicate the transition from the early pelagic habitat to the benthos occurs over a period in which the juvenile grows to about 75 percent of the adult size. Beginning in the summer, the brightly red-orange-coloured juveniles with large orange pectoral fins can be commonly found in the southern strait. In these larval and pelagic juvenile stages the fish would be expected to gradually leave the surface waters and enter a transitional stage in which they would migrate into deeper pelagic areas during the day and to the surface waters at night. Ultimately, they leave the open offshore areas of the strait and settle in the nearshore areas on bottom. Spawning is reported to occur in August but it might be later in the Strait of Georgia. A report from Barkley Sound indicated the female produces relatively large pink eggs that are laid in one layer. Females fan the eggs and a male guards the harem, rotating around and making stuttering jaw motions that keep other males away.

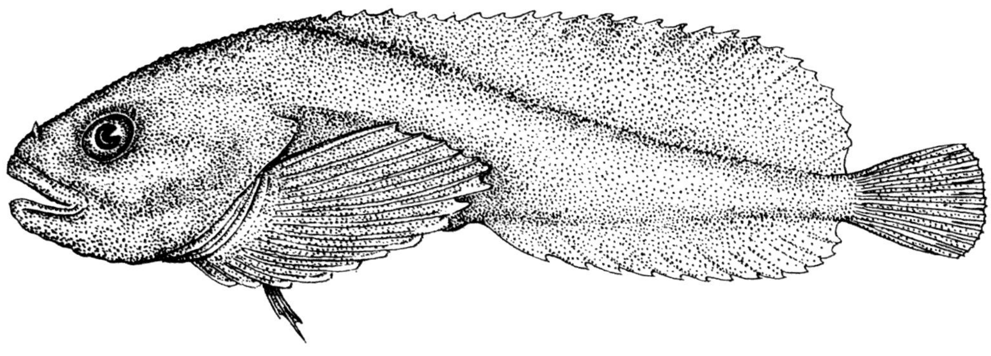

Lumpsuckers

The family Agonidae includes poachers, starsnouts, alligatorfishes and the rockhead. There are about 45 species in the family. They occur on the bottom from very shallow marine waters down to about 1,300 m. They are almost all in the oceans in the northern hemisphere. These are small fishes with long, thin bodies covered by bony plates. The large pectoral fins propel the fish with a sculling motion. There are 13 species of poachers in the Strait of Georgia. The family Cyclopteridae (lumpsuckers) contains 27 species around the northern hemisphere, one of them occurring in the strait.

NORTHERN SPEARNOSE POACHER

Agonopsis vulsa COMMON ●

IN THE STRAIT OF GEORGIA

This species is commonly distributed throughout the Strait of Georgia on mud bottoms at intermediate depths. It is seen by scuba divers virtually everywhere and recorded from the Nanaimo and Fraser River estuaries. Scuba divers report that it hides in crevices during the day but is active over soft bottoms at night. It is commonly found in bottom research trawl catches. Surprisingly, it also occurs in our pelagic trawl catches. It was the 10th most abundant species in catches in the upper 75 m, not including Pacific salmon and Pacific herring. It probably is more abundant toward the southern end of the Strait of Georgia.

DESCRIPTION

The northern spearnose poacher is consistent in form with other poachers, with an elongated body covered in bony plates. The snout is longer than that of other poacher species, contributing to the name "spearnose." In addition, the snout has two blunt, skin-covered spines that can be used to distinguish the species from other poachers. There are numerous barbels and cirri on the head. Coloration is brown on top and white to brown underneath. Six or more bands of darker colour form saddles over the body and fins. The northern spearnose poacher is known to some as the windowtail poacher because of the semi-transparent patch, or window, on the tail.

Scott Stevenson

LIFE HISTORY

In British Columbia the species has been found in waters as deep as 128 m, and from 5 to 393 m throughout its range from Southern California to the northern Gulf of Alaska. It is also occasionally observed in tide pools. The fish are known to spawn in finger sponges. The maximum size is approximately 20 cm. The biology remains to be discovered; this could occur in the Strait of Georgia as we know where they are.

SMOOTH ALLIGATORFISH

Anoplagonus inermis UNCOMMON ◆

IN THE STRAIT OF GEORGIA

The smooth alligatorfish is seen on cliff faces in the Strait of Georgia by scuba divers. There have been scuba observations from Campbell River, Active Pass, and across the strait to Howe Sound and Burrard Inlet. There also are reports from Malaspina Strait. Occasional catches have occurred in on- and off-bottom research trawl surveys, including in a lingcod bottom trawl survey in 18 to 27 m water in the northern strait. There are no records from estuaries.

DESCRIPTION

Perhaps the nicest feature of the smooth alligatorfish is that it is one of those poacher species that can be easily recognized and identified based on its overall appearance. This is another slender species with smooth body plates. Its slender body is concave along the surface of the back. There are no nasal

A close-up of the head of the smooth alligatorfish shows a concave depression as well as accumulated silt as it sits perfectly still, hoping to avoid being detected. Lee Newman

Smooth alligatorfish have a concave depression on the back of the body from the head to the single dorsal fin. The body plates are smooth and the head has no fleshy cirri projecting. Lee Newman

spines, just one dorsal fin on the rear half of the body, and the jaw has no cirri. It is grey or dark brown with a maximum size of 15 cm.

LIFE HISTORY

This species can occur from depths of about 114 m to just 2 m, on sand, gravel or rocky bottom, often near rock outcrops. It occurs from Northern California to the Aleutian Islands in Alaska. This is another species that is mostly unstudied.

GRAY STARSNOUT

Bathyagonus alascanus UNCOMMON ◆

IN THE STRAIT OF GEORGIA

The gray starsnout is found in the Strait of Georgia on sandy or muddy bottoms at depths more than 20 m. It occurs rarely in bottom trawls, and all records are from the southern strait, except for one off Campbell River. It is not reported by scuba divers or from estuaries. There are records from Howe Sound and Burrard Inlet.

Catherine Mecklenburg

DESCRIPTION

This species has the typical poacher form: a slender elongate body with scales modified into bony plates with spines. It has a larger eye than many poachers of comparable size. The beginning of the anal fin is obvious below the space between the two dorsal fins, and there is a single pair of plates ventrally in front of the pelvic fins. Undoubtedly it is called a "starsnout" because of the star-shaped pattern of five spines on the snout or rostral plate. The gray starsnout is indeed grey in colour, but may be anywhere between grey and light brown dorsally and light brown underneath. This species, like other poachers, has five or six darker pigment patches across the back and down the sides in a saddle pattern. The fins are pale and bars appear on the dorsal, caudal and pectoral fins. There are two spines on the infra-orbital ridge, but no spines on the suborbital bone.

LIFE HISTORY

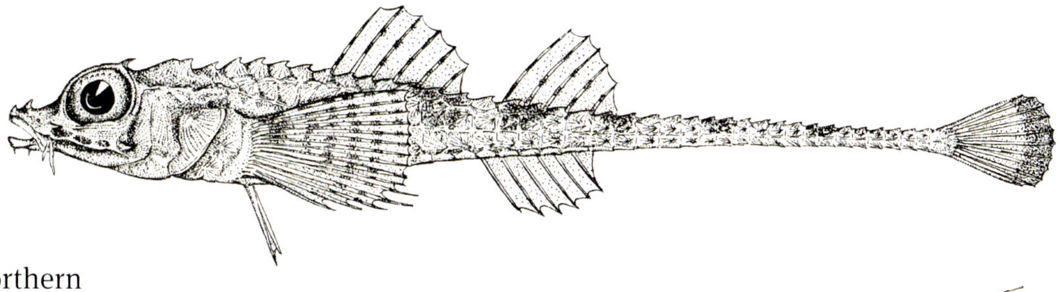

Gray starsnout are found throughout coastal British Columbia; their full range is from Northern California to the southeast Bering Sea, including the Aleutian Islands. The species is found on bottom, between 18 and 252 m, throughout its range. It reaches about 14.1 cm in length. Very little is known about the biology or ecology of this species. The gray starsnout has been found in the diets of nesting tufted and horned puffins in the Gulf of Alaska.

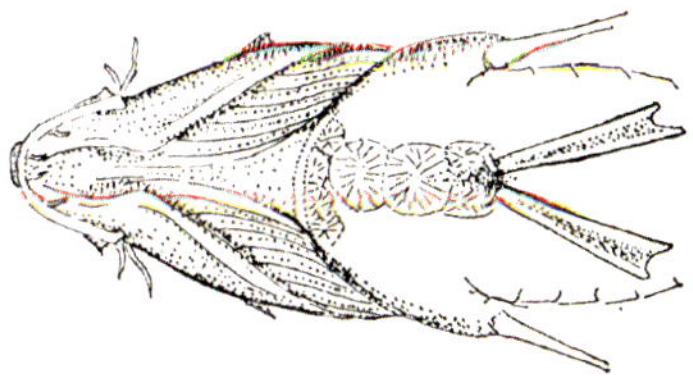

SPINYCHEEK STARSNOUT

Bathyagonus infraspinatus UNCOMMON ◆

IN THE STRAIT OF GEORGIA

Early collections of the spinycheek starsnout were made in the Strait of Georgia in Nanoose Bay, Porlier Pass and Burrard Inlet. There are other reports from Departure Bay and west of Texada Island. We have scuba observations from Howe Sound and Burrard Inlet but no trawl catches or estuary observations. Thus, we consider the species to be between uncommon and rare.

DESCRIPTION

This is an elongate, slender poacher, with a depressed (flattened) head and a small jaw at the tip of the snout. There are paired cirri or barbels at the back angle of the jaw. It has a brown or olive green body with darker blotches on a whitish belly. The anal fin starts just below the end of the first dorsal fin. The dorsal and anal fins are pale, but the tail and pectoral fins and the body may have dark bars. The body is covered with spiny plates and there are spines on the head that form the shape of a star. The spinycheek starsnout is very similar to another deepwater poacher of the same genus, the gray starsnout—so similar that a diver should not identify the species from only a visual sighting. Details of the fin placement, the pattern of plates under the throat, and the number and placement of barbels around the mouth are all key diagnostic characters, along with many other details that cannot be seen at a glance.

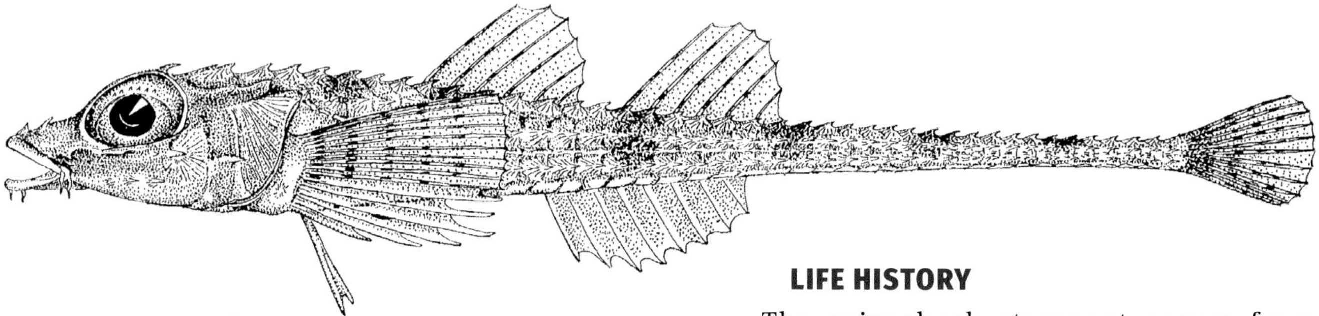

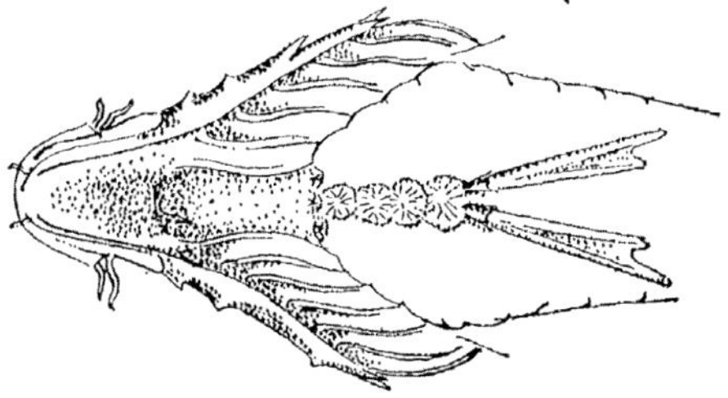

LIFE HISTORY

The spinycheek starsnout occurs from Northern California to the southeastern Bering Sea, on soft bottoms from 6 to 183 m deep. It grows to only 14.3 cm in length. While it is one of the deepwater poachers, its depth range is broad enough that scuba divers will see these fish in water as shallow as 20 m. Very little is known about the biology or ecology of this species.

BLACKFIN POACHER

Bathyagonus nigripinnis UNCOMMON ◆

IN THE STRAIT OF GEORGIA

The blackfin poacher was first caught in British Columbia in 1903 in Nanaimo Harbour at a depth of 98 m. The species is widely distributed around the Strait of Georgia at intermediate and deep depths, on or near soft bottoms. Most deeper trawl sets near or on the bottom will catch blackfin poacher throughout the strait, with the largest catches in the southern strait. In addition to the deepest areas, they also are occasionally caught in shallower water in shrimp trawl surveys. As might be expected, blackfin poacher are not seen by scuba divers. They are one of the more common species in the deep areas of the strait, but generally are uncommon.

DESCRIPTION

This fish has a highly elongate body, even for a poacher, and a longer head than that of the more common blacktip poacher. It is typically light brown dorsally and a slightly lighter brown ventrally. There may be a dark bluish-black colour on the pectoral fins. This species is called the "blackfin poacher" because of its uniformly dark-blue or black fins. Unlike the blacktip poacher, it has no maxillary barbel. The anal fin originates below the space between the two dorsal fins, and the lower jaw extends beyond the upper jaw.

Bill Pennell

LIFE HISTORY

Blackfin poacher are distributed throughout the North Pacific, from Northern California to the Bering Sea, off Kamchatka in Russia and south to the northern Kuril Islands. They occur at depths from 18 to 1,250 m and have a maximum length of 24.2 cm. They are found mostly on bottom at all depths, including some of the deepest areas. Blackfin poacher are reported to move into shallow waters in the summer months, which may explain why they are commonly captured in shrimp trawls. They are reported to live up to nine years. Like other poachers, they feed on amphipods, euphausiids and shrimps. The species is virtually unstudied throughout its range, including in the Strait of Georgia.

BIGEYE POACHER

Bathyagonus pentacanthus RARE ★

IN THE STRAIT OF GEORGIA

The bigeye poacher is rare in the Strait of Georgia compared to its cousin the blackfin poacher, which is more common in trawl bycatch. We have several records of catches of the bigeye poacher from bottom trawls in the waters west of Texada Island and off the mouth of the Fraser River. There is one record from a 10-year shrimp research trawl survey from off Comox, one report from the Courtenay estuary and one from the Gulf Islands area. There are no records from any of our deeper trawl sets throughout the strait.

DESCRIPTION

The bigeye poacher resembles the spinycheek starsnout except that the anal fin starts below the space between the two dorsal fins and that the plates below the head in front of the pelvic fins are paired instead of single. This species has dark pigment on all of its fins as well as bands of dark pigment on the body. The lower jaw does not extend beyond the upper jaw. As the common name implies, this fish has a big eye compared to similar species.

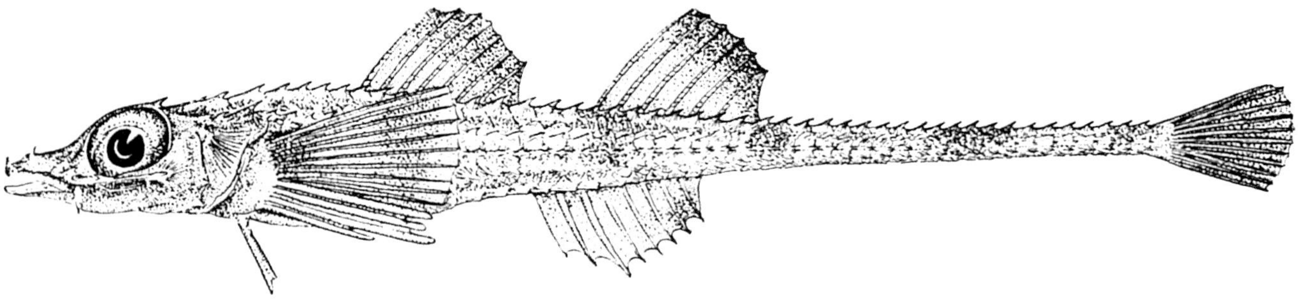

LIFE HISTORY

This fish occurs on soft bottoms in deep water and grows to 26 cm in length. The reported range is from the western Gulf of Alaska to near San Diego, California, at depths of 100 down to 910 m, but mostly at 100 to 375 m. The genus name *Bathyagonus* means "deep poacher." Although this species is known to live in deep water, it may go undetected if its habitat does not coincide with trawlable bottom. Exploration of deep ridges and other rough terrain could yield interesting results, but close inspection of a poacher's body is required for confirmed identity. It cannot reliably be identified from video or photo images.

ROCKHEAD

Bothragonus swanii

RARE ★

IN THE STRAIT OF GEORGIA

Although this species is well known from tide pools on fully exposed outer coast locations, it is rare in the Strait of Georgia. In the strait, a juvenile rockhead was collected at Foley Head at the entrance to Jervis Inlet, so it would appear the physical adaptations to living securely in wave surges work equally well for extreme tidal currents on rocky headlands in the Strait of Georgia.

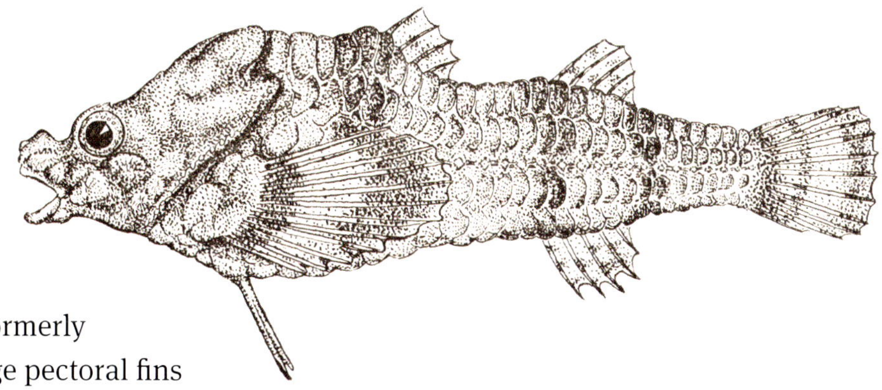

DESCRIPTION

The rockhead's appearance is dominated by its very large head, with two broad, heavy skull protrusions surrounding a pit on top of the head. It was formerly called the "pithead poacher." Large pectoral fins are positioned right below the large head bumps and pit. Juvenile specimens, as in various poacher species, have recurved body spines that become blunt as the fish grow.

Aaron Baldwin

LIFE HISTORY

Rockheads occur from Northern California to Kodiak Island, Alaska. In British Columbia, most records are from the outer coast. This fish attains a total length of almost 9 cm. It is usually found in the intertidal zone and sometimes to just 18 m deep, so this is a truly shallow-dwelling fish.

When this species hatches from eggs laid in kelp holdfasts, the larvae develop body spines before they settle. Once settled on the bottom, their body plates have strongly recurved, heavy spines that effectively anchor their bodies against currents. The spines hook onto the rock or gravel that the little 1.7 cm fish are resting on, and they easily resist being dislodged.

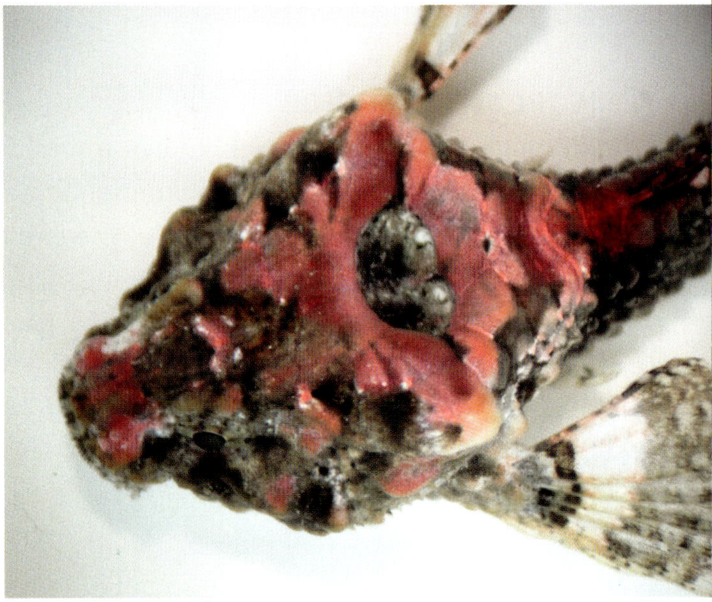

Danny Kent

Like many of the stouter-bodied poachers, the rockhead moves along the seabed by flexing its tail against the substrate and pulling forward or backward with its well-developed pectoral fins. When positioned against rough rock surfaces, the head is analogous to waffle-stomper boot soles. All the fish would have to do is push a bit upward against the overhead rock using its pectoral fins, and the pit and bumps would hold against the rough overhead surface. So far this is only speculation, but an enterprising student could record a video of a rockhead in a tank with Plexiglas "rocks" and strong water flows, simulating a wave surge or tidal rip, to test for this type of behaviour.

PYGMY POACHER

Odontopyxis trispinosa COMMON ●

IN THE STRAIT OF GEORGIA

Pygmy poacher are commonly seen by observant scuba divers throughout the Strait of Georgia, but they are not abundant. They are not reported in any other studies, including observations in estuaries and research shrimp trawls. They are reported at relatively shallow depths on sand or mud bottoms.

The pygmy poacher never exceeds about 8 cm in length. Its small size, the single barbel at the angle of the jaw and the moderate spines on the body plates are clues in the identification of this typical-looking poacher. Scott Stevenson

DESCRIPTION

Pygmy poacher are grey and brown with dark mottling that makes them hard to see. They are slender fish with moderate spines on their body plates, a vertical spine on the tip of the snout and a single barbel at the angle of the jaw.

LIFE HISTORY

This species occurs from Baja California to southeast Alaska at depths from 5 to 373 m, and perhaps at greater depths in the southern part of its range. The maximum size is reported to be between 9.5 and 13 cm and the fish mature at about 5.4 cm. They may spawn throughout the year. Biologists Danny Kent and John Fisher at the Vancouver Aquarium discovered that larval pygmy poacher were hatching from inside orange finger sponges. They dissected sponge tubes and found there tended to be more than one egg mass in a tube; the inner egg mass would hatch and the larvae would perish because the outer egg mass, which is less developed, would block the exit path. The last spawn to be deposited would thus be the winner in this competition for reproductive success. Sponges provide good water flow for respiration, their tissues tend to have antiseptic qualities and they hide the eggs from predators. Kent and Fisher recovered viable egg masses and reared the larvae in the laboratory, yielding juveniles of two different poachers, the pygmy poacher and the northern spearnose poacher, so this is an interspecific nesting competition over the use of the orange finger sponge as a spawn repository.

TUBENOSE POACHER

Pallasina barbata RARE ★

IN THE STRAIT OF GEORGIA

We have records for the tubenose poacher in the Strait of Georgia from scuba observations in the southern strait off Nanaimo, in Active Pass, off the Sunshine Coast and in Howe Sound and Burrard Inlet. Another report is from near the Cowichan River. There are no records from any of the research trawl surveys, suggesting the fish are generally rare to absent in water deeper than that visited by scuba divers. There also are no reports from estuaries, indicating this is an intertidal species.

DESCRIPTION

The tubenose poacher is perhaps the most slender of British Columbia's species of poachers, a family of fishes of relatively slender shape. The head is more elongate than that of other poachers, with no head spines at all. The tip of the lower jaw extends beyond the upper jaw with a long fleshy barbel of variable length. It is brown or grey with darker spots and can be yellow ventrally. In contrast to other poacher species, this fish swims up off the bottom amidst seaweeds or eelgrass, more like a tubesnout, which it closely resembles (except the tubenose poacher has a body covering of bony plates).

Tubenose poacher have typical poacher body plates, albeit smooth rather than sharp-pointed. Danny Kent

LIFE HISTORY

The tubenose poacher occurs from Bodega Bay, California, to the Bering Sea, off Russia in the Sea of Okhotsk, in the Sea of Japan and off South Korea. It is reported from the intertidal zone down to 128 m but is usually found much shallower. Tubenose poacher grow to 20.8 cm and mature at about 9 cm. They settle from the plankton in shallow depths near the shore, often in eelgrass. They have been found in brackish (low-salinity) lagoons. This species tolerates very cold waters and often swims in the water column, eating small crustaceans.

STURGEON POACHER

Podothecus accipenserinus COMMON ●

IN THE STRAIT OF GEORGIA

The sturgeon poacher is widely distributed in the Strait of Georgia on muddy bottoms at depths from 18 to 55 m. It has been common in bottom research trawl catches throughout the strait, particularly in the southern strait and in the area of the Fraser River plume. It is reported from the estuaries of the Fraser, Nanaimo and Courtenay Rivers. Scuba divers see sturgeon poacher in Active Pass, Howe Sound and Burrard Inlet. They are somewhere between common and abundant.

DESCRIPTION

Sturgeon poacher are grey to brown in colour on their dorsal side, with yellow to orange underneath. There is a very distinctive orange spot under each eye, as well as dark colouring in the form of saddles in patches over the back and

Andy Murch

sides. As the name suggests, the head has a striking resemblance to a sturgeon's. This is perhaps because of the pointy snout and fleshy bristles (cirri) at the corners and lower part of the snout. The sturgeon poacher's habits are also similar to those of sturgeons, as they reside on the bottom using their bristles to find food. They have very small teeth and make full use of their loose, fleshy lips to feed.

LIFE HISTORY

This species is distributed from Northern California to the Bering Sea, across the Pacific to the Sea of Okhotsk off Russia, and south to the northern Kuril Islands. It is found at depths from 2 to 300 m; juveniles have been reported to prefer shallow water where there is aquatic vegetation. The maximum age is about 11 years and the maximum length 33 cm. Three-year-old fish are about 11 cm and 10-year-old fish about 25 cm. Very little is recorded about their reproduction, but it appears they spawn in the spring and summer. Studies in the Bering Sea reported that sturgeon poachers spawn in June. The Vancouver Aquarium successfully cultured them in its laboratory in 1996. The sturgeon poacher has sexual dimorphism; that is, there are physical characteristics that differ between the sexes. In particular, large males have black pigmentation between the fin rays of their pelvic fins, whereas females and small males do not. Pelvic fins are always longer in males than in females. Like other poachers, it consumes amphipods, cumaceans, copepods, worms and shrimp. Sturgeon poacher are eaten by a range of predators, including sculpins, rockfish, birds and even sea lions.

BLACKTIP POACHER

Xeneretmus latifrons COMMON ●

IN THE STRAIT OF GEORGIA

This species was first reported in British Columbia off Bowen Island in 1903 at more than 42 m deep. It is found throughout the Strait of Georgia on soft bottoms at moderate depths. The blacktip poacher was commonly found in bottom research trawl surveys from 2003 to 2006 throughout the strait. These fish have occurred frequently in shrimp trawl survey catches off Comox, off the mouth of the Fraser River and in Howe Sound. They are commonly caught in bottom research trawls in the southern strait at depths ranging from 28 to 197 m. Some of the larger catches were in the waters west of Texada Island at depths of about 30 to 60 m. They are not common in the northern strait. They are rarely reported by scuba divers but are sometimes seen at night in Indian Arm in Burrard Inlet. The blacktip poacher has been recorded in the sandy sediment of Saanich Inlet by the VENUS monitoring camera.

NOAA

DESCRIPTION

The blacktip poacher is brown or tan with a pale belly. Its body shape is similar to the other poacher species found in the Strait of Georgia: elongate, with a flattened round head and thin body tapering to a fan-like tail. The blacktip poacher gets its name from the dark-black margin on the dorsal fin, which can be used in its identification. Another characteristic useful for identifying this species is the single barbel found below the bottom jaw. Its snout is shorter, its orbits longer and interorbital distance greater than those of other poachers. There is a single spine at the end of the snout, and there are three to six spinous processes on the dorsal margin of the eye.

LIFE HISTORY

Blacktip poacher are found from northern Haida Gwaii, British Columbia, to southern Baja California, from 18 to 400 m but are also reported from much deeper water. In Southern California, they are most abundant at about 120 m. They are fairly common offshore in southern British Columbia waters. They reach a maximum overall length of approximately 19 cm. In British Columbia, they have been documented as spawning in early spring. After the adhesive eggs hatch, the larvae are pelagic in the water column. In Southern California there is evidence that they may spawn all year. As with other poachers, there is very little known about many aspects of the life history or general biology. There is evidence that they live up to six years. One three-year-old specimen was 16.2 cm in length. Despite their hard body parts, blacktip poacher fall prey to flatfishes and Pacific hake. They eat amphipods, euphausiids, polychaetes and other small plankton.

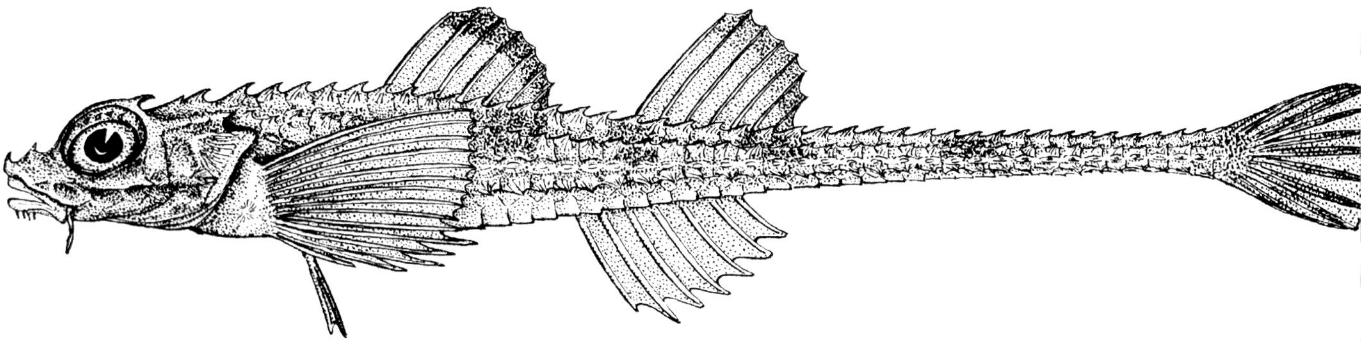

SMOOTHEYE POACHER

Xeneretmus leiops RARE ★

IN THE STRAIT OF GEORGIA

There were no records of the smootheye poacher from the Strait of Georgia until it was identified in a bottom research trawl catch at 70 m between Texada and Vancouver Islands in March 2012. Unfortunately, the specimen was not kept, but the identification was performed by a biologist who was familiar with the marine fishes in British Columbia and careful with their identification.

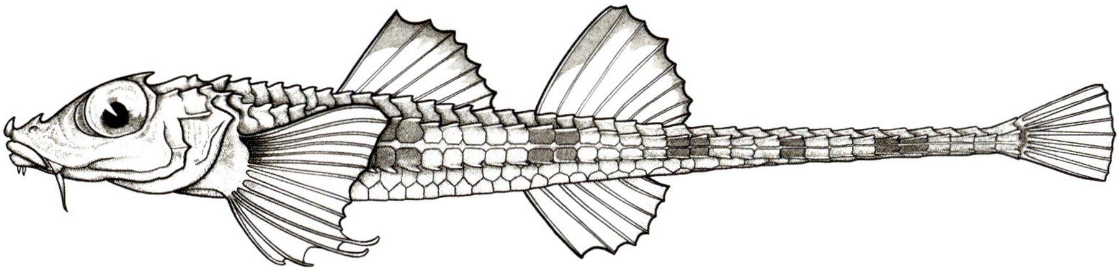

Meghan Walley

DESCRIPTION

The smootheye poacher can be recognized by the black markings on the margins of the first and second dorsal fins and the single barbel at the posterior edge of the jaw. The margin of the upper eyeball is smooth and without spinous processes.

LIFE HISTORY

Tom Laidig

This species is rare throughout its range from Southern California to the northern Gulf of Alaska. There is only one record from Alaska and no other records south to Rennell Sound off British Columbia. Very little is known about the species other than it is found on soft bottoms at depths of 37 to 400 m and may have a maximum total length of 27 cm. Our hope is that listing the species in this book will highlight the importance of keeping specimens and supporting museum collections.

BLUESPOTTED POACHER

Xeneretmus triacanthus

RARE ★

IN THE STRAIT OF GEORGIA

There is only one report of a bluespotted poacher in the Strait of Georgia, and it is from the Brentwood Bay area, near Round Island; the report was published in J.L. Hart's *Pacific Fishes of Canada*. There have been no records from the strait in all of the various trawl and longline research surveys or in the extensive coverage by scuba divers, indicating the species is rare in the strait. In contrast, it is not uncommon in Puget Sound, where it is found on muddy bottoms and in the diet of other bottom-dwelling fishes.

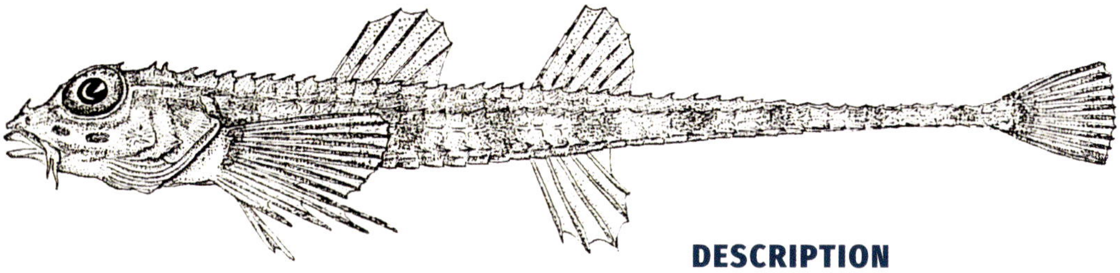

DESCRIPTION

The bluespotted poacher has a typical poacher appearance, with an olive-brown colour dorsally and blue spots. It is readily recognized by the two barbels at the tip of each maxilla. There are spinous processes on the dorsal margins of the eyeball and there is no pigment on the margins of the dorsal fins. The origin of the anal fin is below that of the second dorsal fin.

Tom Laidig

LIFE HISTORY

Bluespotted poacher are reported from British Columbia's central coast, including Hecate Strait, to northern Baja California. They live on soft bottoms at depths from 73 to 373 m. The Hecate Strait record was from a scientific bottom trawl survey in May 2005. The maximum size is about 17.8 cm and the maximum age has been estimated at between 5 and 15 years, but little else is known about the species. Because this fish is more common in Puget Sound, it is possible that it may be found in the Strait of Georgia in the future.

PACIFIC SPINY LUMPSUCKER

Eumicrotremus orbis COMMON ●

This bright-red juvenile of the Pacific spiny lumpsucker species displays the unmistakable rotund body form of this lumpfish. Lee Newman

IN THE STRAIT OF GEORGIA

This species occurs along shorelines in the Strait of Georgia, where it is often on stony bottoms in shallow water. It is reported by scuba divers from Campbell River to Burrard Inlet and in the Cowichan estuary and is occasionally caught in bottom research trawls and shrimp trawls in the strait.

DESCRIPTION

The ovoid (egg-shaped) body with protective conical tubercles makes the Pacific spiny lumpsucker unmistakable. Females tend to be greenish and their large eggs are often visible through the body wall during the fall to winter spawning season. Sometimes the male is more brightly coloured, including shades of red, and it is the male that guards the eggs.

Jon Bryan

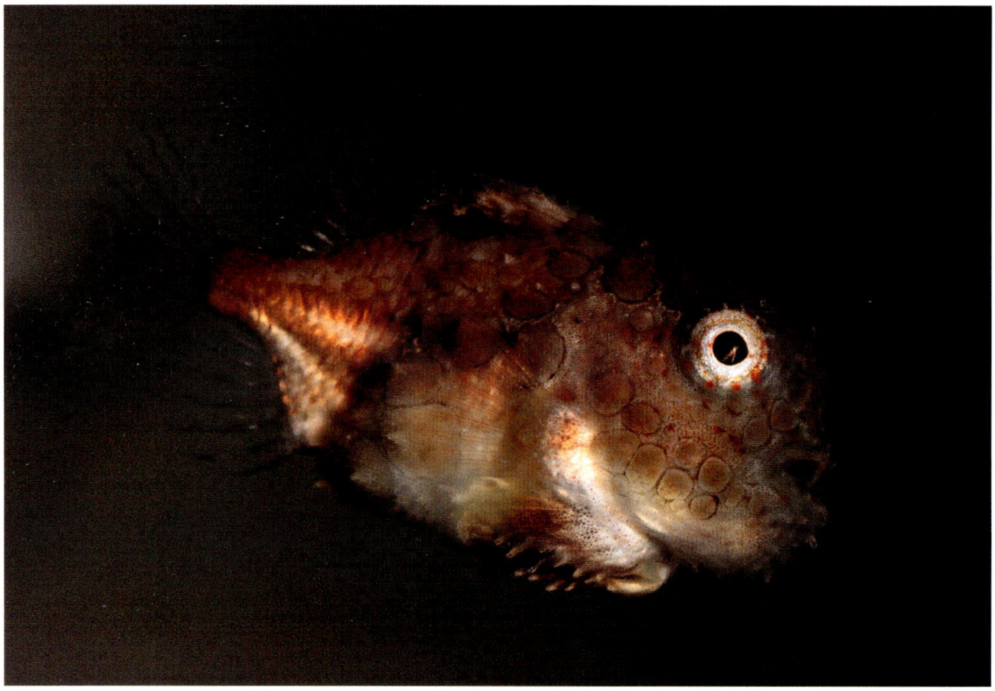

Individuals can be yellow, green, orange or grey, so body shape is the main characteristic to look for in Pacific spiny lumpsuckers.
Danny Kent

LIFE HISTORY

The range for this species is from Puget Sound north to around the Bering Sea, to the Sea of Okhotsk and to Hokkaido, Japan. This species grows to 12.7 cm in length. It is usually stationary, holding position on a rock with its adhesive pelvic disc, which has a thickened, fringed margin for aiding in gripping irregular surfaces. Unlike the larvae of the majority of marine fish species—which hatch with poor anatomical development and do not resemble juveniles or adults of their species—the Pacific spiny lumpsucker hatches as a well-formed juvenile with all its fin rays, including the functional suction disc. Right after hatching, the little juveniles will adhere to substrate, usually on new, smooth growths of seaweed, which provide the best surface for true suction. Later, when the pelvic disc has become more complex and muscular, the fish will be able to adhere to more irregular rocky surfaces. This lumpsucker is the only species of its genus occurring in British Columbia, which is lucky for us because the species are difficult to identify. In Alaska there are at least eight species that all look very similar—and that is only considering species in the genus *Eumicrotremus* and not a number of other genera of similar lumpsuckers.

Snailfishes

Snailfishes, in the family Liparidae, are found throughout the world's oceans. This is a remarkable group of fishes that are notoriously difficult to identify. The family is estimated to have as many as 400 species, but there is much to learn about the taxonomy of these fishes. These are scaleless, elongate, tadpole-shaped species with loose, gelatinous skin. Many have modified pelvic fins that form an adhesive disc on the bottom of the body. They range in size from a few centimetres up to 50 cm, and they occur from the intertidal zone to the deepest of any fish species, which is also the widest depth range of all fishes. The record now is 8,336 m, from a collection in 2022 in the Izu–Ogasawara Trench, south of Japan. We include 14 species that occur in the Strait of Georgia, and they are found in the deepest areas of the strait.

SMALLDISK SNAILFISH

Careproctus gilberti UNCOMMON ◆

IN THE STRAIT OF GEORGIA

Smalldisk snailfish occur near the bottom in deeper areas of the Strait of Georgia and are consistently found in our deep pelagic trawl catches, particularly in the southern strait. Their very small size would allow most to escape through the mesh, so there is little information about abundance, but they are not rare.

DESCRIPTION

This species has a tadpole-shaped body with a broad, depressed head ending in a blunt, rounded snout. The mouth is small and the upper jaw protrudes over the lower. The smalldisk snailfish has one dorsal fin extending from behind the head to the base of the small tail. There is one ventral fin extending from behind the anus to the tip of the tail. Not surprisingly, this species has a very small adhesive disc that is below a vertical line from the eye. It is distinguished by the gill opening that extends from the top of the pectoral fin to where the fin changes its shape, at about the 14th pectoral fin ray. The smalldisk snailfish varies in colour from creamy white to pale pink or light brown. It has blue eyes with silver coloration on the lower half. The stomach is black and has very loose, delicate skin.

Archipelago

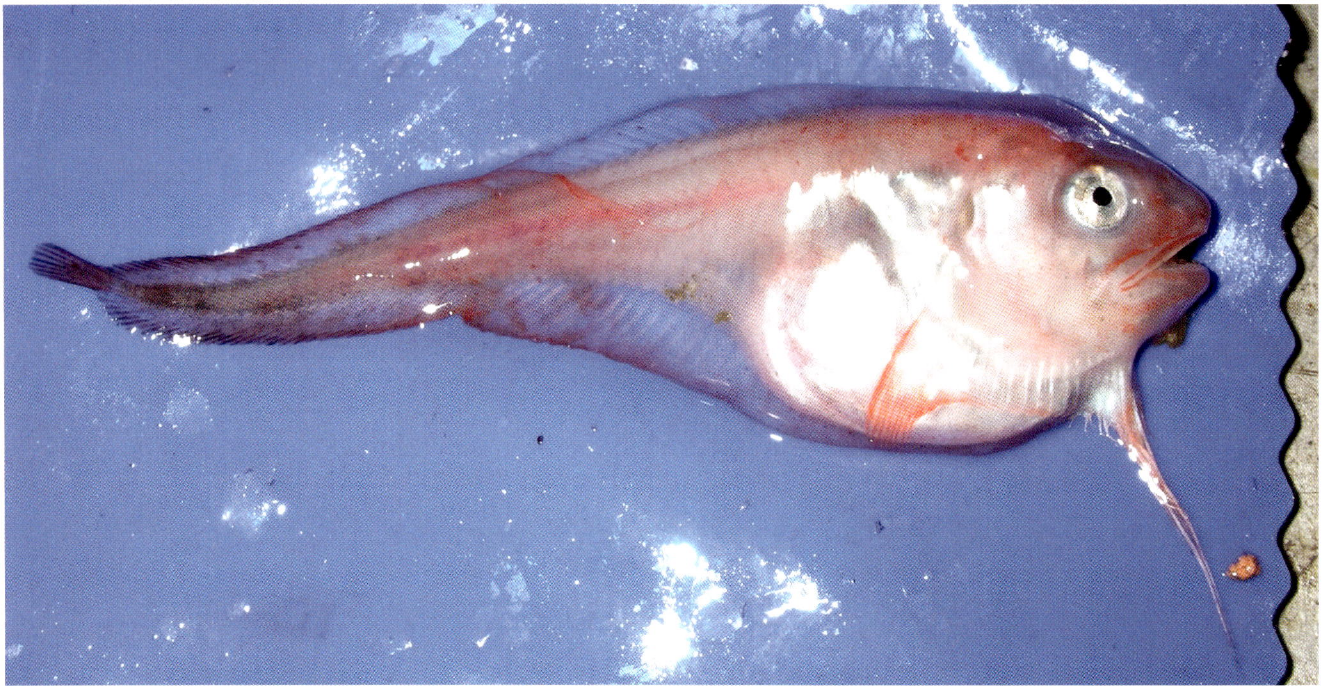

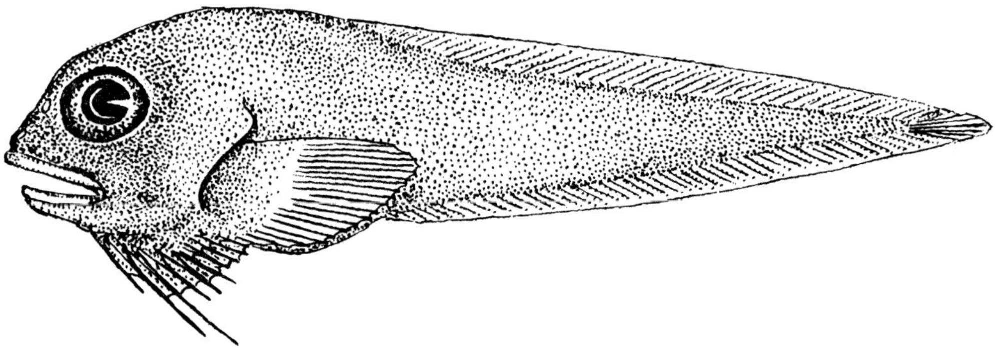

LIFE HISTORY

The smalldisk snailfish can be found from California to Alaska and the Aleutian Islands. The maximum length is approximately 12.7 cm. Despite not being rare in the Strait of Georgia, it is considered a rare species in British Columbia overall. It was first found in British Columbia waters in 1888 at a depth of 435 m. The fish occupy the deep water offshore and can even be found at a maximum depth of 886 m. No information is available on the life history of this species, including the reproductive or general biology. This is another species that needs to be studied to confirm the identification and to understand its life history.

BLACKTAIL SNAILFISH

Careproctus melanurus VISITOR ▼

IN THE STRAIT OF GEORGIA

We include the blacktail snailfish in the Strait of Georgia because of two reports over 55 years ago and because the species is readily identified by the pink colour and black caudal area. One report, in January 1953, was from Swanson Channel in the Gulf Islands and the second, in 1954, was from off the mouth of the Fraser River. There have not been any other observations from any of the numerous research activities or from commercial catches that are recorded in the groundfish database.

A recent study of the taxonomy of blacktail snailfish identified a new species within the distribution. The species that is found from British Columbia to

Tom Laidig

the north around the Pacific is now called the scorched snailfish (*Careproctus ambustus*), while blacktail snailfish's range extends farther south. The species that was found in the Strait of Georgia could have been either the southern or northern species, but until more specimens are found we retain the name "blacktail snailfish."

DESCRIPTION

The blacktail snailfish is a relatively large, soft-bodied and gelatinous snailfish. These fish are mostly pink or rose red, with black edges on the dorsal and anal fins and a black caudal fin. The disc is small, about the size of the eye. The gill opening is entirely above the pectoral fin and is relatively large. The snout is blunt and the dorsal and anal fins taper into the caudal fin. The lining of the mouth, gill cavity and tongue are black.

LIFE HISTORY

Blacktail snailfish occur from off the coast of Japan, off Kamchatka, into the Bering Sea and south of the Aleutian Islands to Baja California. This is a deepwater species found close to the bottom at depths of 61 to 2,286 m. The maximum size is 54 cm and 1.7 kg. Apparently, it is relatively common off the coast of California, where it is seen from submersibles. Research scientist Milton Love reports that spawning occurs from April to June, with some females depositing large eggs in the gill cavities of box crabs. Prey are mostly crustaceans.

SPOTTED SNAILFISH

Liparis callyodon RARE ★

IN THE STRAIT OF GEORGIA

The spotted snailfish has rarely been reported in the Strait of Georgia, though there are records from the southern part of the strait. There are also records from the Campbell River estuary, Discovery Passage, Bute Inlet and the Southern Gulf Islands. This is a shoreline species found in tide pools and other rocky shoreline areas. It has not been reported by scuba divers. Spotted snailfish are known to swim among eelgrass in the shallows off sandy beaches at night, indicating a preference for broad shallow-water habitats.

DESCRIPTION

This is another species with a deep notch in the dorsal fin, like the tidepool snailfish, but with fine black specks on its olive-brown, green or purple skin and bars of spots on the fins. The skin is loose and translucent with no prickles. The posterior nostril has a raised rim that forms a finger-like projection. As with the tidepool snailfish, the anus is equidistant from the anal fin and the pelvic disc or slightly close to the pelvic disc, but the spotted snailfish's disc is smaller than that of the tidepool snailfish.

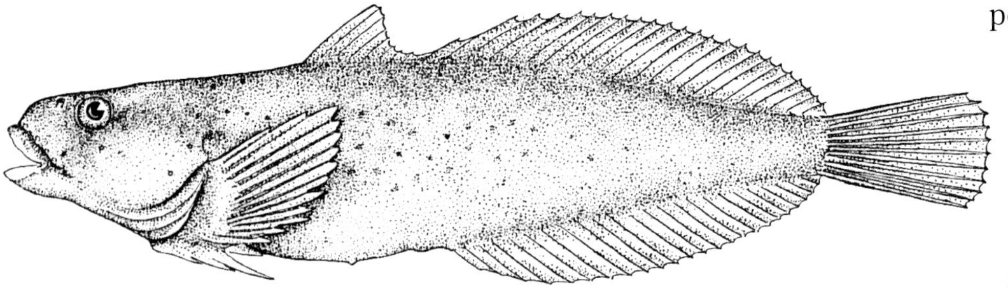

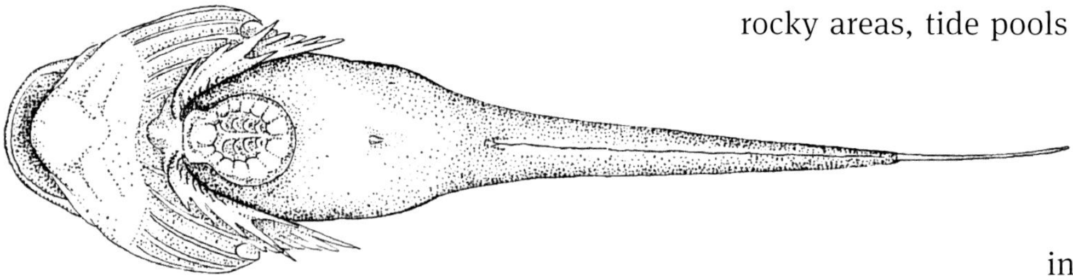

LIFE HISTORY

The spotted snailfish can be found in rocky areas, tide pools and intertidal areas from Oregon to along the Aleutian Islands and the Bering Sea, and across to Kamchatka in Russia. This fish spawns in empty barnacle shells on boulders in rocky intertidal areas. Egg masses containing fewer than about 400 maroon, orange and green eggs were collected in March in intertidal areas in Juan de Fuca Strait, but little else is known about the species. It might be found around the rocky shores of the Southern Gulf Islands. The maximum length is 12.7 cm.

RIBBON SNAILFISH

Liparis cyclopus UNCOMMON ◆

IN THE STRAIT OF GEORGIA

Ribbon snailfish are caught occasionally in shrimp trawls in the southern Strait of Georgia but are not common in the strait overall. The species has not been reported in estuaries or observed by scuba divers anywhere in the strait. In the late 1960s and early 1970s, ribbon snailfish were frequently caught by Vancouver Aquarium collectors using beach seines near Lumbermen's Arch, Stanley Park, and observed to spawn intertidally.

DESCRIPTION

This species has an elongate, compressed body with a broad, depressed head ending in a blunt, rounded snout. The mouth is small and the upper jaw protrudes over the lower jaw. Ribbon snailfish have one dorsal fin extending from behind the head almost to the base of the fan-shaped tail. There is one ventral fin extending from behind the anus, also to near the base of the tail. This species has an adhesive disc. Ribbon snailfish are olive-brown with darker spots and blotches on the body. As with other snailfish species, the ribbon snailfish has loose, smooth skin devoid of scales. It is distinguished by the gill opening extending to the 5th to 10th pectoral fin ray and by the dorsal fin barely reaching the caudal fin. The newly settled juvenile of any snailfish species will have a gill opening extending along the entire cheek, and only later during development does the opening become restricted and important to taxonomists for identification of adults.

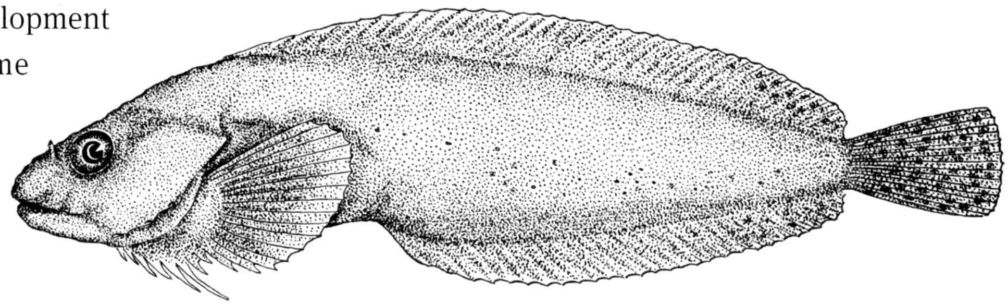

LIFE HISTORY

The ribbon snailfish is a benthic species. It is found from Oregon to Alaska and off the coast of Kamchatka, Russia. This is a rocky tide pool species that has been found as deep as 183 m. Its maximum length is 11.4 cm.

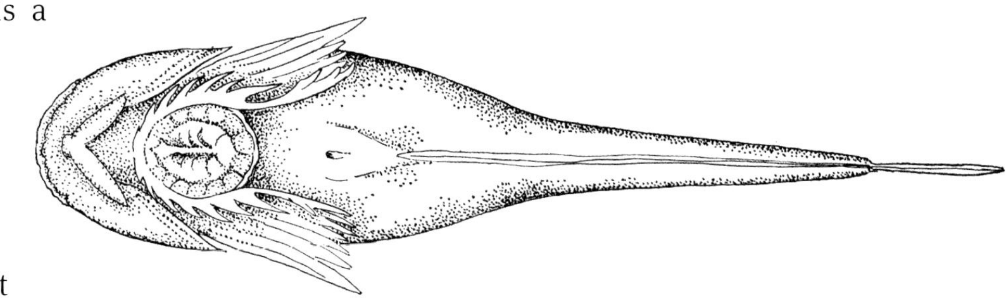

MARBLED SNAILFISH

Liparis dennyi UNCOMMON ◆

IN THE STRAIT OF GEORGIA

Marbled snailfish are not common in the Strait of Georgia. There is a record from the Cowichan River estuary and a scuba observation from Active Pass, but the species is rarely observed by scuba divers. There are five records in the groundfish research database at the Pacific Biological Station, and all are from bottom trawl catches in the Southern Gulf Islands area: one in February 1975, three in February 2004 and one in February 2018, all at depths from 90 to 226 m. It is also reported from off Denman and Saturna Islands and in Burrard Inlet, English Bay and Nanoose Bay in *Fishes of the Salish Sea*.

DESCRIPTION

Similar to other snailfish, this species has an elongate, compressed body with a broad, depressed head ending in a blunt, rounded snout. The mouth is very small and the upper jaw protrudes over the lower jaw. Marbled snailfish have

Scott Stevenson

one dorsal fin extending from behind the head to the base of the fan-shaped tail, and one anal fin extending from behind the anus to the base of the tail. The marbled snailfish is distinguished by the gill opening extending to the 10th to 15th pectoral fin ray, the 37 to 40 fin rays in the dorsal fin and 30 to 33 fin rays in the anal fin. This is another species with an adhesive disc. The marbled snailfish is olive to dark brown with wavy lines or spots of darker pigmentation. As with other snailfish species, it has loose, smooth skin, devoid of scales.

LIFE HISTORY

The marbled snailfish is a benthic species. The distribution ranges from Puget Sound to the Gulf of Alaska and the eastern Aleutian Islands. It is commonly found in British Columbia at depths from shallow water to 225 m. The maximum length for this species is approximately 30.5 cm. It is probable that in the Strait of Georgia marbled snailfish spawn in the winter, as females in spawning condition were found in mid-January. The diet of the marbled snailfish appears to be similar to those of other snailfishes in similar habitats and includes polychaetes and small crustaceans.

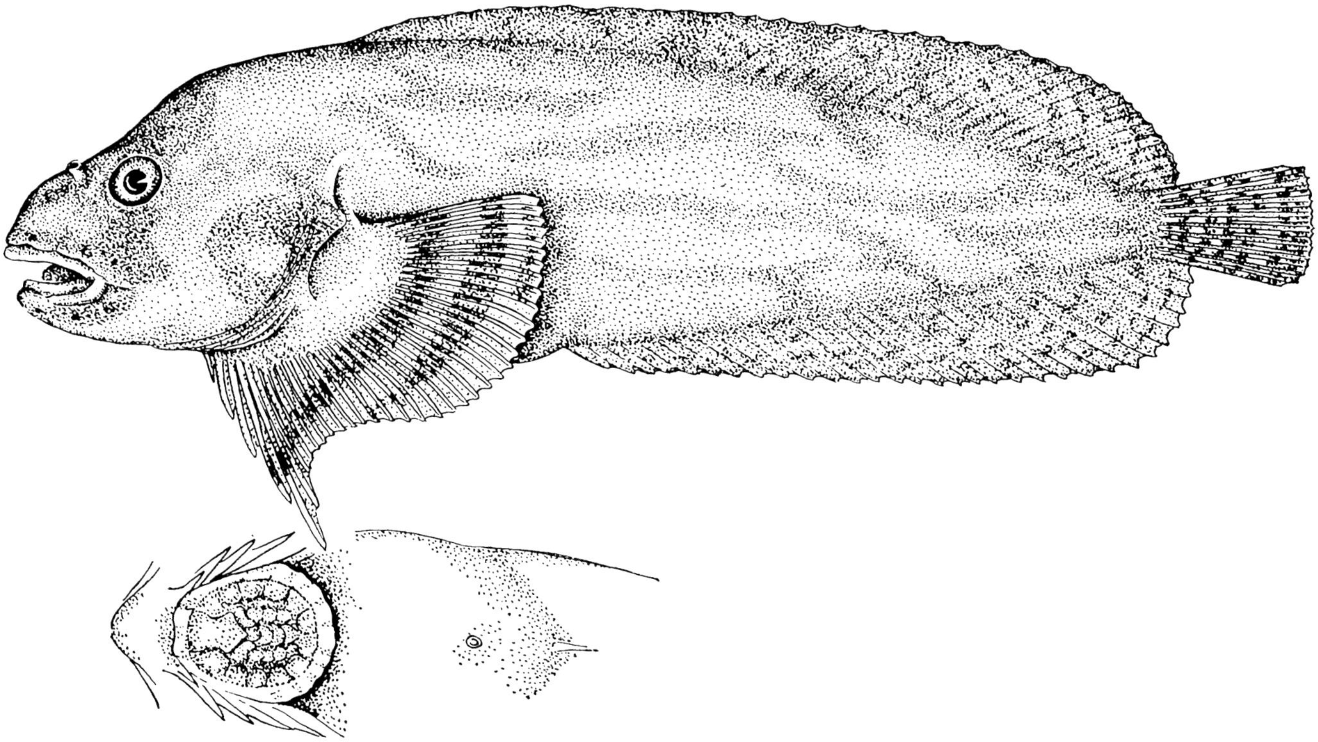

TIDEPOOL SNAILFISH

Liparis florae UNCOMMON ◆

Andy Murch

IN THE STRAIT OF GEORGIA

There are records of the tidepool snailfish from the Nanaimo River estuary near the Pacific Biological Station, Burrard Inlet, Saanich Inlet and off Texada Island. However, it is not commonly reported and not seen by scuba divers.

DESCRIPTION

The tidepool snailfish has an adhesive disc, like most snailfish. It has loose skin and would usually be a brown shade, but other possible colours include yellow, pink, red, green and purple. The tidepool snailfish can be very cryptic, but a careful observer may see it clinging to a kelp blade that it will match closely in colour as camouflage. The first lobe of the dorsal fin is high and often appears separate from the posterior of the dorsal fin, where the body depth is greater than in the gut region. The width of the suction disc is less than half the head length. The anus is closer to the suction disc than to the anal fin, or sometimes equidistant.

Jon Bryan

LIFE HISTORY

This fish is found in tide pools and in rocky shoreline areas to a depth of 15 m. The maximum length is approximately 18 cm, but most would easily fit in the palm of your hand. The tidepool snailfish is distributed from the northern Gulf of Alaska and Kodiak Island to Southern California. It is more common in the southern part of the range. It has been reported to be nocturnally active. In tide pools on exposed shores, it may outnumber all other species of snailfishes. Its diet includes polychaetes and crustaceans.

SLIPSKIN SNAILFISH

Liparis fucensis UNCOMMON ◆

IN THE STRAIT OF GEORGIA

The slipskin snailfish has been found in our pelagic trawl surveys in the deeper areas of the Strait of Georgia. It is rarely found in trawls in water shallower than 150 m. It is also found on the bottom in the southern strait and in Burrard Inlet, but possibly is uncommon. Because of the small size of the specimens in the catch, it is probable that many escape through the mesh, making it difficult to estimate abundances.

DESCRIPTION

Like many other snailfish, the slipskin has an elongate, compressed body with a broad, depressed head ending in a blunt, rounded snout. The mouth is very small and the upper jaw protrudes over the lower jaw. Slipskin snailfish have one notched dorsal fin that extends from behind the head to the base of the

Keith Clements

small, fan-shaped tail, and one anal fin extending from behind the anus to the base of the tail. This species has an adhesive disc. The slipskin snailfish is olive to dark brown with highlights of pink. Dark bands or blotches of pigmentation can often be found on the fins. It is distinguished by the gill opening, which extends down to the 12th to 16th pectoral fin rays, by its having more pectoral fin rays than dorsal fin rays, and by the tips of the rays in the dorsal and anal fins extending out of the fins. As can be surmised from its name, this fish has very loose, smooth skin.

LIFE HISTORY

The slipskin snailfish is a rocky-area benthic species. The distribution of this species ranges from the southeast Bering Sea to Southern California. It is a common species offshore and has been caught in British Columbia waters at depths of 388 m, but it also has been reported in rocky tide pools. It is believed this species spawns in late winter and early spring, perhaps even extending to September. Developing eggs have been taken from females in February in Burrard Inlet, and plankton tows in southern British Columbia waters from April to July have collected slipskin snailfish larvae at various stages of development. In addition, researchers have collected fertilized eggs from polychaete worm tubes from Barkley Sound in late May. A small (5.5 cm) slipskin snailfish was collected from Juan de Fuca Strait; it was found guarding a set of mussel shells in which four egg masses had been laid, each with more than 1,500 developing embryos. The egg masses' surface (where they were not in contact with the shell) had small circular depressions that were similar in shape and size to the adhesive disc of the guarding fish. The eggs were examined and it was determined that the stages of development were significantly different between the four egg masses, leading to the inference that they were from four separate spawning events. The tending fish was a male. Males of this species have also been reported to guard eggs that were deposited in polychaete worm tubes. Like certain sculpin species, this snailfish develops a "globby" larva, meaning that the body is filled with fluid. By having an abundance of low-salinity body fluid between the outer skin and the musculature and body organs, the fish is naturally buoyant and does not need a swim bladder. This buoyancy enables the species to grow to a relatively large size over a long period of time while swimming in the water column with little energy required to prevent sinking. The diet of the slipskin snailfish includes crustaceans and other shrimps. Its maximum length is 17.8 cm.

LOBEFIN SNAILFISH

Liparis greeni RARE ★

IN THE STRAIT OF GEORGIA

The lobefin snailfish is rare in the Strait of Georgia, with only one record, from Porlier Pass. We have no trawl catches or scuba records.

DESCRIPTION

This is a large snailfish with very small eyes. It has previously been described in a different genus as *Polypera greeni* and, not surprisingly, some scientists disagree with lumping the species into the genus *Liparis*. The dorsal fin in this species is strongly notched, forming an anterior lobe that resembles a separate fin. As with other *Liparis* species, positive identification depends on determining the size of the gill opening according to how far it extends down the pectoral fin (to pectoral rays numbers one to four) and the number of fingers (called pyloric caecae) that extend from the gut. This species has over 200. Neither of these features is of much use if one is looking at a photograph. Furthermore, all of the snailfish species have a gill opening extending all the way down the pectoral fin when they first settle from the water column. The young juveniles very likely do not have their full number of gut extensions formed either, so lobefin snailfish in these younger stages cannot really be given a positive identification using the characters described for full-sized adults.

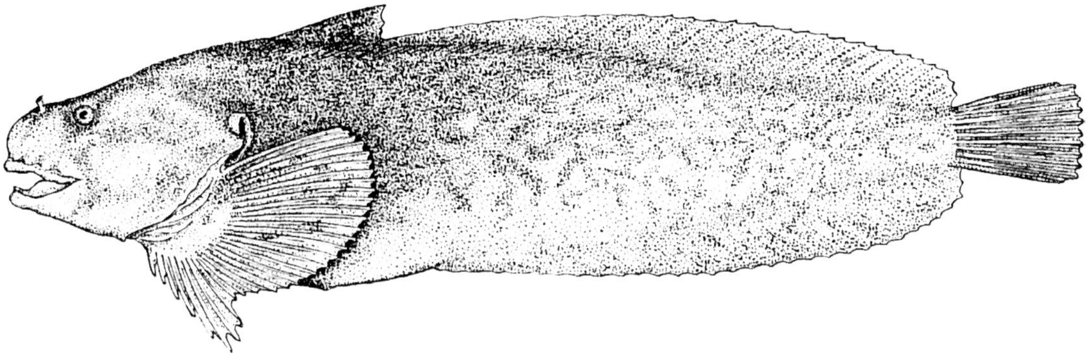

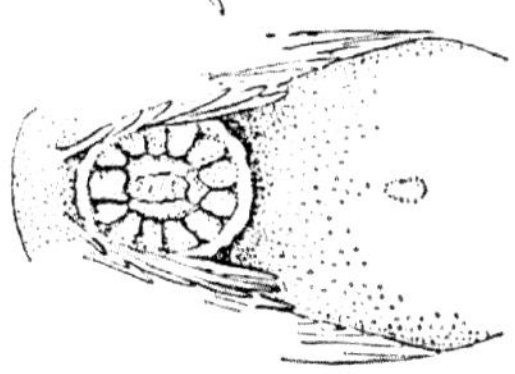

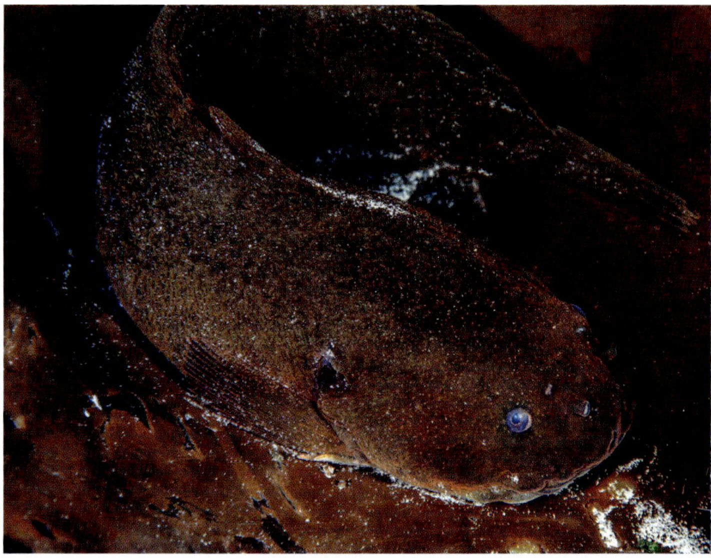

Scott Stevenson

LIFE HISTORY

The lobefin snailfish occurs from the Russian Commander Islands to the Bering Sea and from the Aleutian Islands to Washington State. It grows to 30 cm in length and can be found in tide pools, near the shore in shallow water or among seaweeds growing on dock floats. The body colour of golden brown or green matches the seaweed habitat. Very little is known about the biology of the species.

SLIMY SNAILFISH

Liparis mucosus RARE ★

IN THE STRAIT OF GEORGIA

Slimy snailfish occur rarely in the Strait of Georgia. We have one record from a surface trawl in the central strait in April 1994 and one from each of four areas around the strait during a bottom trawl survey in the summer of 2006. It is more abundant on the outer coast than in the Strait of Georgia.

Alex Hayman / iNaturalist

DESCRIPTION

The slimy snailfish, like the more common marbled snailfish, will have dramatic body striping in some individuals, but not all, so the colour pattern is a misleading character for identification. More reliable is that the round disc is more than half as long as the head and the dorsal fin has an anterior lobe that is broadly rounded but not always pronounced. The variable body colour can be plain or reticulate or striped horizontally, with shades from yellow to dark brown, red or lavender. The caudal fin base sometimes has a black bar, and the lips are dark.

LIFE HISTORY

This snailfish is found in tide pools off California and to depths of only 15 m. This fish grows to 12.7 cm in length. High-energy shorelines are preferred, where boulders and seaweeds are abundant. It occurs from southeast Alaska to Baja California. Spawning apparently occurs throughout the year. Prey is mostly crustaceans, as with many snailfish.

SHOWY SNAILFISH

Liparis pulchellus COMMON ●

IN THE STRAIT OF GEORGIA

Showy snailfish are found throughout the Strait of Georgia, particularly in the southern areas, but are most likely not abundant. The species is commonly found in shrimp catches and is also occasionally observed by scuba divers at night over soft substrates, frequently hiding in empty Lewis's moonsnail shells. It is one of the most common snailfishes in Puget Sound but is probably not abundant there either.

DESCRIPTION

As with other snailfishes, the showy snailfish has an elongate, compressed body with a broad, depressed head and a blunt, rounded snout. The mouth is small and the upper jaw protrudes over the lower jaw. Showy snailfishes have one dorsal fin extending from behind the head to near the end of the small,

Valerie Lyttle

fan-shaped tail. There is one anal fin extending from behind the anus onto the tail. This species has an adhesive disc. Showy snailfish are light to dark brown and may have wavy lines or spots of darker pigmentation. They have three pores on the underside of the jaw and loose, smooth and transparent skin devoid of scales. They have fewer pectoral fin rays than anal fin rays.

LIFE HISTORY

This is a benthic species found on soft bottoms at depths from near the shore to 183 m. Its maximum length is 25.4 cm. It has been caught in shrimp trawls at depths of 92 m and in beach seines in some estuaries throughout its range, which is from the southeast Bering Sea and the Aleutian Islands to Southern California. Sexual maturity is reached between 7 and 9 cm in length. As with many fish species, the number of maturing ova was found to be higher in females of greater weight and length. Studies off the coast of California showed the spawning period to be during the winter and spring. Gravid fish were found in abundance from late December until May, with only a few as late as July. A 1.2 cm larval showy snailfish was captured in the Strait of Georgia in May 2011, indicating that spawning probably was in the winter. The stomach of the young showy snailfish contained copepods, polychaete worms and amphipods. The stomachs of older showy snailfish have contained worms, decapods, crustaceans, isopods, mysids and some fishes.

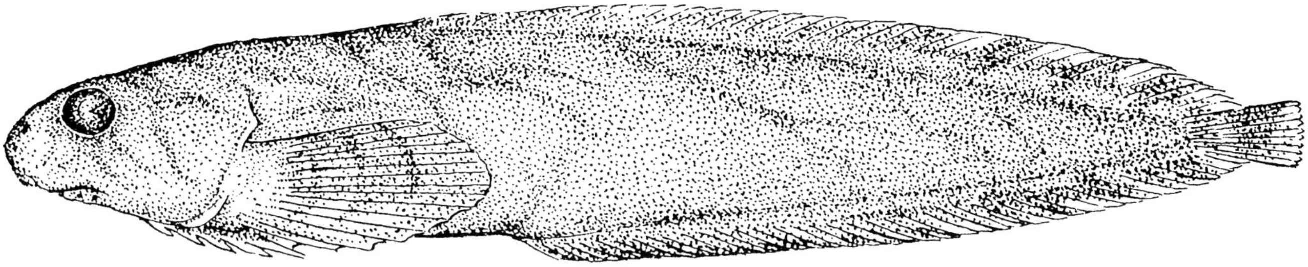

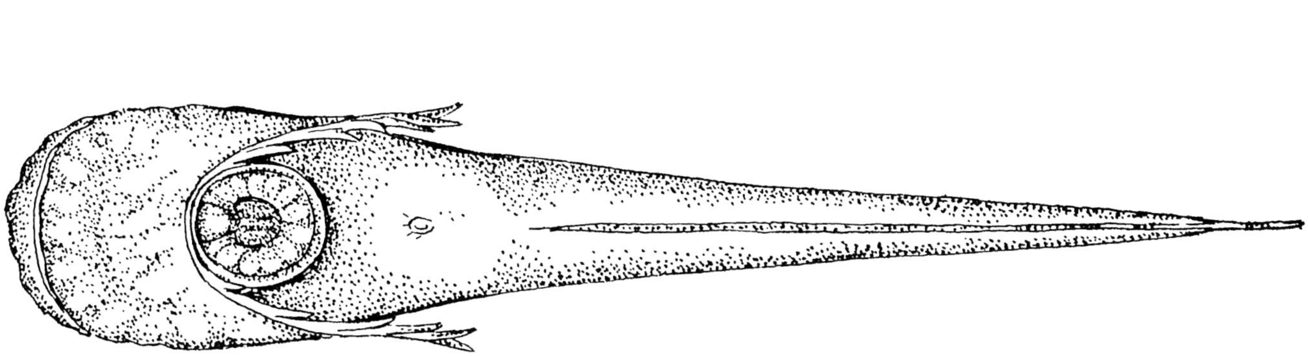

RINGTAIL SNAILFISH

Liparis rutteri RARE ★

IN THE STRAIT OF GEORGIA

The ringtail snailfish is rare in the Strait of Georgia. It has been recorded in the southern strait in Burrard Inlet, in English Bay and from the Cowichan River estuary. A photograph in the groundfish research database at the Pacific Biological Station shows a specimen that was caught in a bottom trawl in the Southern Gulf Islands area in February 2018. As diagnostic characteristics become more precisely defined for this species, it may become possible to obtain more precise habitat information.

DESCRIPTION

The species has loose skin, and the dorsal fin has an undulating profile. The base of the caudal fin has a vertical white band. The band may wrap around onto the posterior margins of the dorsal and anal fins, much as the black pigment does on the blackfin snailfish. The suction disc is relatively large, its width more than half the head length. There are discrepancies in the illustrations and descriptions in the scientific literature about the ringtail snailfish.

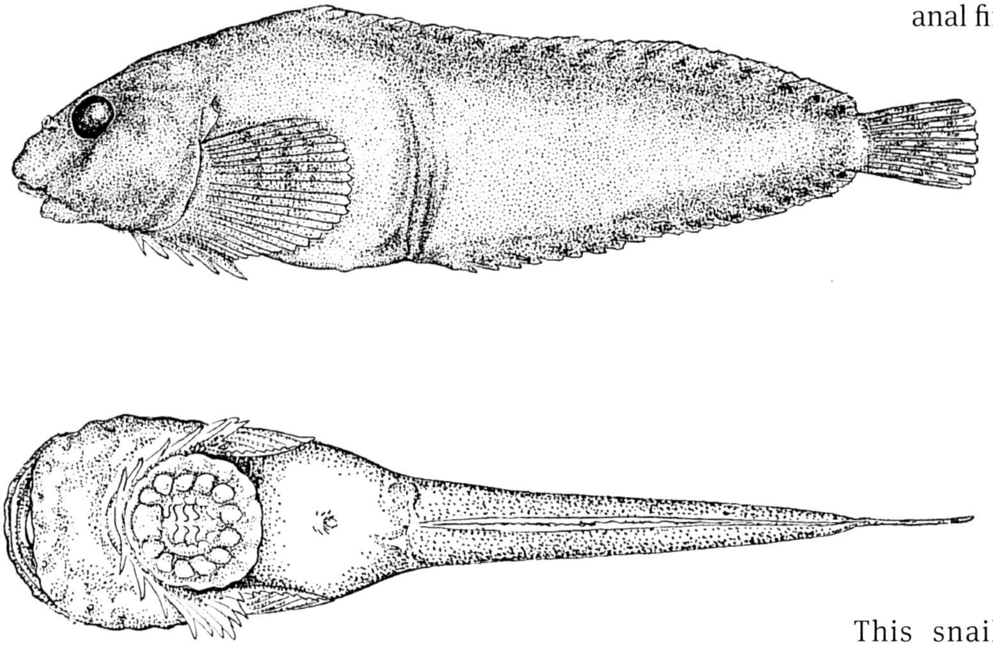

LIFE HISTORY

This snailfish ranges from the Aleutian Islands to Puget Sound but is rare in British Columbia. It may be rare throughout its range, and there may be misidentifications involving an unnamed species. Ringtail snailfish have been recorded from the west coast of Vancouver Island, where specimens have been collected from tide pools and soft bottoms to depths of 73 m. This fish grows to 16.8 cm in length. It is difficult to study the biology of these more rare species of snailfishes because of the uncertainties in identification at the species level. Clearly, this is one of a group of species that deserves expert attention.

PYGMY SNAILFISH

Lipariscus nanus COMMON ●

IN THE STRAIT OF GEORGIA

We routinely catch small numbers of pygmy snailfish in our deepwater pelagic sets. They probably get trapped among the other fish in the catch, as the species' small size would allow it to escape through the mesh of the net. The pygmy snailfish appears to be common in the deepest off-bottom areas of the Strait of Georgia.

DESCRIPTION

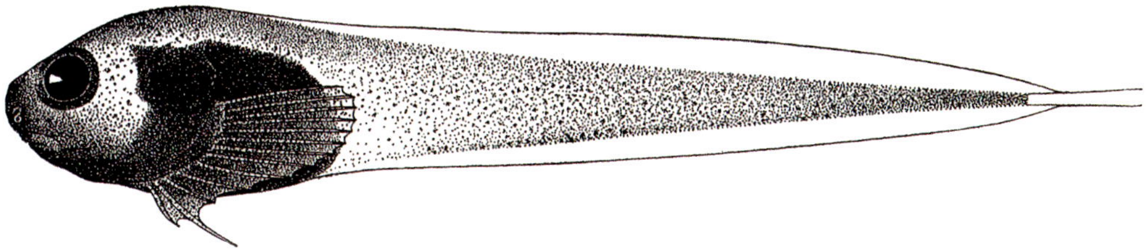

The pygmy snailfish has an elongate and compressed body with a broad, depressed head ending in a blunt, rounded snout. The body form in general could be referred to as tadpole-like. The mouth is small and the upper jaw protrudes over the lower. Pygmy snailfish have one dorsal fin extending from behind the head to the end of the body where it meets the tail. There is one anal fin extending in a similar manner from behind the anus, also to the end of the body where it meets the tail. The anus is just anterior to a line extending vertically from the gill opening. The species is distinguished by its loose, transparent skin, a black gill cavity and stomach area, and a gill opening that is entirely above the pectoral fin.

LIFE HISTORY

Bill Pennell

Pygmy snailfish are found from Central California to Alaska, the Bering Sea and the Sea of Okhotsk and off the coast of Hokkaido, Japan. It is a benthopelagic species that has been recorded in waters from 58 to 910 m deep. They are small fish, attaining lengths of only 7.1 cm. The species is not listed in *Pacific Fishes of Canada* as it was not well known until the late 1970s.

TADPOLE SNAILFISH

Nectoliparis pelagicus COMMON ●

IN THE STRAIT OF GEORGIA

The tadpole snailfish is another pelagic snailfish in the Strait of Georgia. These fish are commonly found in deep water throughout the strait. They appear to be most common in our pelagic trawl catches at depths below 150 m. One sample of 26 fish ranged in length from 3.4 to 5.3 cm, with an average length of 4.3 cm. These are small fish and not readily captured with larger mesh nets; hence, it is difficult to estimate the abundance. It is surprising that they are also found on bottom in the bycatch of experimental shrimp surveys in the Strait of Georgia.

DESCRIPTION

Similar to other snailfishes, this species has one dorsal fin that runs along the back from just behind the head to the tip of the tail. In addition, the anal fin extends from just behind the anus along the bottom of the fish to the tail tip. The tadpole snailfish has a highly elongated body with a short head, large eyes and a blunt snout. It has a large, rounded anterior body tapering to a point with a small tail, and resembles a tadpole. The pectoral fin has two lobes with a three-to-five-ray space between them. The short gill opening is entirely in front of the pectoral fin. The skin is loose and transparent. The pelvic disc is absent. In adults, the anus is on a forward-pointing papilla, directly below the eye. The anus is positioned more posteriorly in juveniles. This species reaches a maximum length of about 6.5 cm.

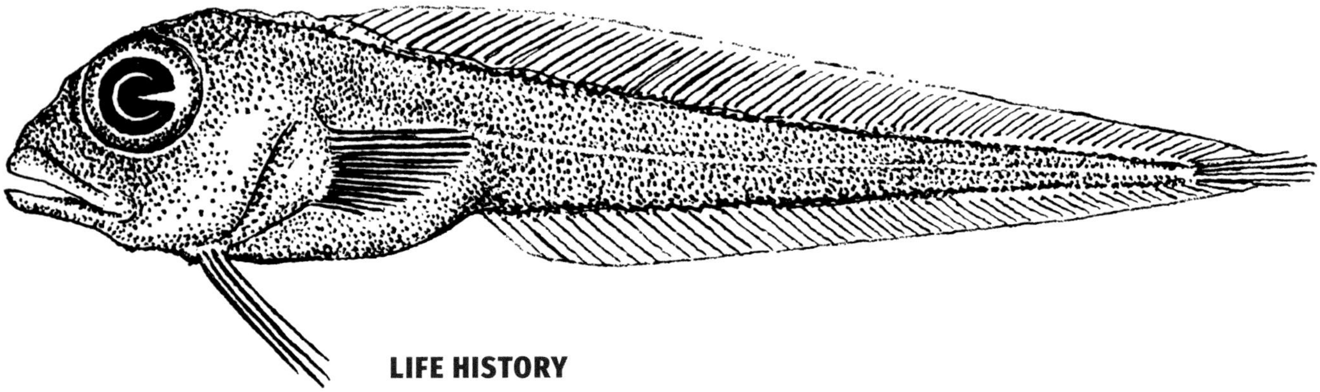

LIFE HISTORY

The natural distribution of the tadpole snailfish includes the northeastern Pacific from California to the Bering Sea, the Sea of Okhotsk off Russia and off northern Japan. This fish has been found at depths in excess of 3,383 m. Very little is known about the general biology and ecology of this species. It has been

Dick Beamish

recorded in areas that were stagnant and periodically devoid of oxygen. Bottom grabs from the same areas were mostly muddy and full of debris like leaves and twigs. In California, video recordings from remote vehicles captured two observations of this species at a minimum depth of 419 m and a maximum of 541 m. It curled its tail into its body in midwater and swam downward when disturbed.

Spawning occurs on the bottom, but upon hatching, larvae become pelagic and move upward to the top 100 m. Ichthyoplankton studies have shown that pelagic eggs and larvae were distributed in the top 100 m of the water column in the spring months. A study of the number of rays in the dorsal and anal fins found that the lowest number was in the Strait of Georgia, probably because the deep water is warmer and has lower salinities than deep water offshore. Beyond these small incidental studies, little is known about the tadpole snailfish. This is another species that would make an excellent project for a curious graduate student.

PRICKLY SNAILFISH

Paraliparis deani UNCOMMON ◆

IN THE STRAIT OF GEORGIA

The prickly snailfish has been found in the Strait of Georgia in deepwater trawl catches below about 70 m, but not commonly. Trawl nets with smaller mesh might capture more; however, it is likely the fish are uncommon. This species is considered to be benthic, which should indicate it is close to or on the bottom in deeper waters in the strait during the day, but the behaviour at night needs to be studied.

DESCRIPTION

The prickly snailfish has a thin, elongate, compressed body with a broad, depressed head and a blunt, rounded snout. The mouth is of moderate size and the upper jaw protrudes slightly over the lower. The head is small with very large eyes. Prickly snailfish have one dorsal fin extending from behind the head to near the almost indiscernible tail. There is one anal fin extending from behind the anus to just anterior of the tail. The species is distinguished by the large gill opening, which extends from the top of the pectoral fin to the front of the upper 10 to 13 pectoral fin rays. This species does not have an adhesive disc. It has loose, smooth, transparent skin devoid of scales.

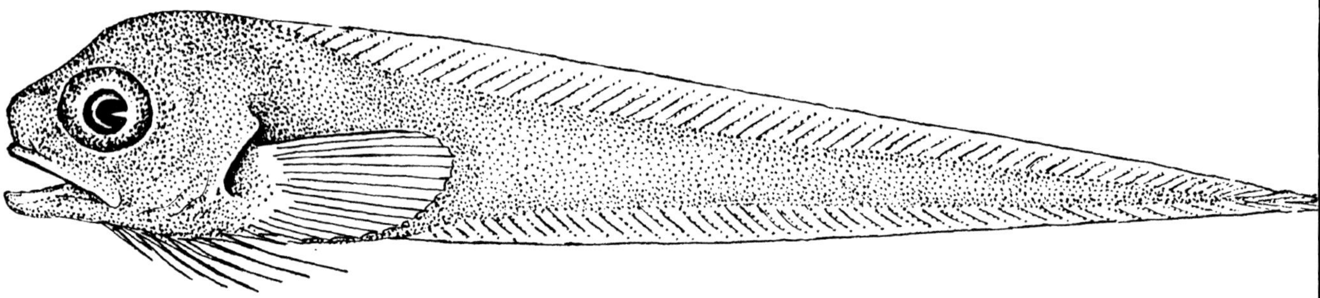

LIFE HISTORY

This is a deepwater pelagic species. It is found in the marine waters of Northern California, as well as those of British Columbia and Alaska to Shelikof Strait, but not off Oregon or Washington State or in Puget Sound. In British Columbia prickly snailfish have been found at depths of 203 m, but it is likely they are in much deeper water. Throughout its range this species has been caught at depths of between 18 and 1,008 m. The maximum length is about 10 cm. The presence of mature eggs in a female in February suggests this species spawns around that time of the year.

Jacks, Pomfrets and Manefishes

The jack mackerel, Pacific pomfret and bigmouth manefish are all in different families. Jack mackerel are in the family Carangidae, which comprises about 150 species. Most species in the family are found in all tropical and subtropical marine waters, with a few in cooler waters. These are streamlined, fast-swimming, voracious predators that may be seen in large schools. They vary in appearance, but most have a deeply forked, strong tail. Pacific pomfrets are in the family Bramidae, which comprises about 20 species that are found around the world, mostly in the open ocean near the surface. They are deep-bodied and flattened sideways with long pectoral fins and a blunt snout. The Pacific pomfret is an important commercial species that has a unique taste and texture that some consider gourmet. The bigmouth manefish belongs to the family Caristiidae, which includes about 19 species, all of which have an easily recognizable appearance. These species have a high, steep forehead, black body, large eyes, a high dorsal fin originating on the head, and long pelvic fins with dorsal, anal and pelvic fins fitting into sheaths at their base. Manefishes are found worldwide in deeper water, but they are rare and consequently poorly studied.

JACK MACKEREL

Trachurus symmetricus VISITOR ▼

IN THE STRAIT OF GEORGIA

There are three records of jack mackerel in the groundfish database at the Pacific Biological Station in Nanaimo. Two were caught in commercial bottom trawls: one in the southern strait in June 1998 and the other in the Northern Gulf Islands in November 2000. The third record is from a commercial mid-water trawl in the southern strait in March 2012. There are no other records from the Strait of Georgia.

DESCRIPTION

The jack mackerel is a streamlined fish that is built for fast swimming. It is a metallic blue to olive green dorsally and silvery ventrally. The caudal fin may be yellow to reddish. There is a deeply forked tail and a narrow caudal peduncle. The scales on the lateral line are modified into spiny scutes, and the lateral line declines abruptly below the origin of the second dorsal fin. The dorsal and anal fins are as long as the pectoral fin. There is a dark spot on the gill cover.

Janna Nichols

LIFE HISTORY

This is a schooling, open-ocean species that ranges from the Aleutian Islands in the Gulf of Alaska to southern Baja California and out into the tropical mid-Pacific. It usually occurs from the surface down to 183 m, but can be as deep as 400 m. There are migrations north into the Gulf of Alaska in the summer, particularly in El Niño years. The maximum size is 81 cm, but most are 20 to 50 cm. Females can mature in their second year at lengths of 20 to 24 cm. These fish can live 30 years or longer. Jack mackerel are batch spawners, releasing eggs every five days and up to 25 times a year. Spawning can occur throughout the year, but most takes place from February to August, with a peak from March to June. This is a large, plankton-feeding species that also preys on squid and small fishes. It is consumed by a large number of predators including sharks, seals, marine mammals, billfishes and cormorants.

PACIFIC POMFRET

Brama japonica VISITOR ▼

IN THE STRAIT OF GEORGIA

There were no reports of Pacific pomfret in the Strait of Georgia until two recent records in the groundfish database at the Pacific Biological Station in Nanaimo. Both were from commercial bottom trawl catches in the southern strait, one in October 2015 and one in January 2020. There are no other species in the Strait of Georgia that have the oval shape and appearance of the Pacific pomfret, except for the rarely occurring Pacific pompano. The Pacific pompano is a smaller fish that has a bright, iridescent, greenish-blue body with fine scales, and is easily distinguished from the larger Pacific pomfret, which has a grey-black colour when dead and prominent scales that are firmly attached. Thus, it is likely that the two records of Pacific pomfrets in the groundfish database were correctly identified.

Image Professionals GmbH / Alamy

DESCRIPTION

Pacific pomfret are oval-shaped with a laterally compressed body. The dorsal and anal fins are long. There is a large, deeply forked tail and a long, narrow pectoral fin. The head is distinctively blunt. The sides of a live fish can be silvery white, but the colour quickly changes to a dark grey or black after death. The body has prominent overlapping scales, which are difficult to remove, at the base of the fins and on the posterior of the head.

LIFE HISTORY

The Pacific pomfret is one of the most abundant and widely distributed fishes in the pelagic waters of the North Pacific. It is an open-ocean species that occurs near Japan and off the Russian Commander Islands to the Aleutian Islands and Gulf of Alaska, and south to California and off the coast of Peru. This is a schooling species that will migrate vertically to the surface at night and down to about 400 m during the day. There is a seasonal migration into the northern feeding areas in the late spring and summer, and south to spawning areas in the winter. Spawning occurs in batches, mostly from January to March, but some occurs throughout the year. Females mature beginning in their second year at lengths of about 30 to 40 cm. Individuals grow rapidly, reaching about 50 cm after five years. The maximum size reported is 61.4 cm and 2.7 kg; the maximum age is nine years, but most are three to five. Pacific pomfrets were the major bycatch species in the former high-seas drift net fishery for squid. They were not retained commercially, to reserve space for squid and tuna, but a former executive chef at the Hotel Vancouver considered them to be a gourmet fish. There are distinct size groups that have preferred prey, but in general, prey includes squid, fishes, amphipods and other crustaceans. The Pacific pomfret has a variety of predators, including sharks, swordfish, daggertooths and marine mammals.

BIGMOUTH MANEFISH

Caristius macropus VISITOR ▼

IN THE STRAIT OF GEORGIA

The first and only record of a bigmouth manefish in the Strait of Georgia was in February 2020. The 22.5 cm specimen was caught by the vessel *Arctic Ocean* while it was fishing for Pacific hake at a depth of 170 m, 10 m off the bottom, west of Texada Island. John Morgan and Michael Lester from Vancouver Island University picked up the frozen specimen from Seadrift Seafood Market and stored it in a freezer at the university. This is also the first specimen reported from British Columbia.

DESCRIPTION

The bigmouth manefish is immediately recognizable by its body profile and the appearance of its fins. The ovoid body shape results from an almost vertical

Michael Lester

anterior margin of the head, with a large eye that is mostly below the midline of the body immediately above the mouth. The body colours are mostly solid black, grey or brown; the fins are black. The peritoneum is soft grey and the flesh is soft. The dorsal fin is high and long, originating near the anterior of the head above about the middle of the eye. Pelvic fins are long, originating anterior to the base of the pectoral fin. The dorsal and anal fins have a base that forms a fleshy shallow groove. The dorsal fin rays are 32 to 36 in number and the pectoral fin rays 16 to 19. A species-specific character is the several short bristles along the distal third of the gill rakers that may also have a small bristle at the tip. The lateral line is not distinct.

LIFE HISTORY

This species is generally rare throughout its distribution. The bigmouth manefish (which once had the common name "veilfin") is found off Japan, in the western and eastern Bering Sea, along the Aleutian Islands, in the Gulf of Alaska and down to central Baja California. It is a mesopelagic to benthopelagic species occurring at 200 to 1,450 m, but the specimen from the Strait of Georgia was captured at a depth of less than 200 m. The first record in the Gulf of Alaska was in 2006. There is no mistaking a fish this unusual, which probably means that specimens will in future be recognized as such and kept. There is almost no information on the species' biology and ecology.

Croakers

The white croakers and queenfish that have been reported in the Strait of Georgia belong to the family Sciaenidae. The family comprises between 270 and 300 species that have a worldwide distribution in temperate and tropical coastal oceans but also occur in estuaries and fresh water. These are bottom-dwelling species that are found inshore, often over muddy bottoms. They are well known because many are important commercially, and because they produce sound.

WHITE CROAKER

Genyonemus lineatus

VISITOR ▼

Tom Laidig

IN THE STRAIT OF GEORGIA

This is another species that is rare in the Strait of Georgia. It has only been reported once in the strait, in March 1983 west of Pender Island in Swanson Channel. We have no catches in any of the several thousand on- and off-bottom research trawls since the mid-1990s. There are historical accounts of white croakers being caught by the Musqueam People.

DESCRIPTION

The white croaker is silvery to brassy yellow on the back, and a lighter silver ventrally. The fins have a silvery to light-yellow colour. There is a darker spot above the pectoral fin and a small barbel on the chin. Like other croakers, this species is moderately elongate and has a pointed pectoral fin, a tail fin without any fork and an anal fin that is shorter than the second dorsal fin.

LIFE HISTORY

This is a schooling species, commonly found off Central California but ranging from Barkley Sound to southern Baja California. It has a maximum size of about 41 cm and a maximum age of around 12 years. About one-half are mature at age one and a length of 13 cm. The white croaker is known to move north to off Oregon during El Niño and was reported there during the 1983 El Niño, which seems to coincide with the 1983 report in Swanson Channel. Thus, it is likely there will be more observations in the Strait of Georgia as oceans warm and warm-water events become more common and more extreme. In California, this is a popular sport fish that can be caught off piers. It may also be fished commercially.

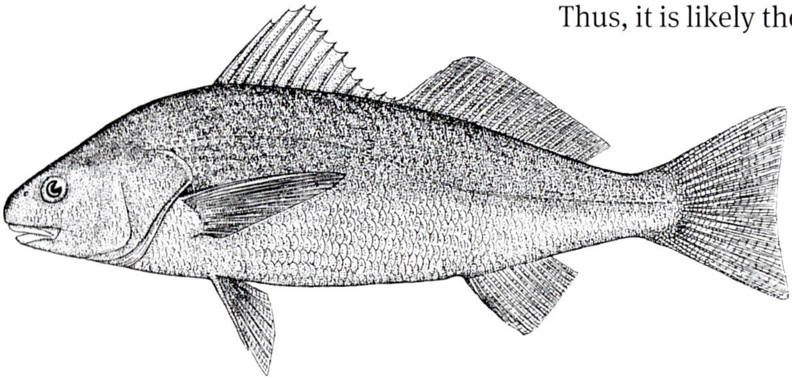

QUEENFISH

Seriphus politus

VISITOR ▼

IN THE STRAIT OF GEORGIA

Only one queenfish has ever been reported in the Strait of Georgia. It was a 63 cm fish caught in Burrard Inlet in 1981. This most likely was an immature individual that probably came from off California. The queenfish is a schooling species, so it is likely there were more individuals that entered the Strait of Georgia from Juan de Fuca Strait that year. There have been no catches in the thousands of research trawl sets in the last 25 years, which indicates this is a rare visitor to the Strait of Georgia.

DESCRIPTION

Queenfish are an elongate, large-mouthed, herring-like fish. They are silver to white on their sides, with some blue to tan shades dorsally and fins that have shades of yellow to green.

LIFE HISTORY

Queenfish belong to a group of fishes known as "croakers" or "drums" because of their ability to make sounds. They are common in estuaries in tropical waters. In California, schools are found over sandy bottoms in bays and around pilings and piers, where they are caught in recreational fisheries. They range from the Strait of Georgia to the southern Gulf of California. The maximum length is 30.5 cm and they live up to 12 years. This is a species that could occur more frequently in the Strait of Georgia as the oceans off the west coast warm.

Theo Moddler

Perches

There are four species of perch in the Strait of Georgia. Perch are in the family Embiotocidae (surfperches), which is reported to have about 24 species. All perches are shallow-water, inshore, coastal species in the northeastern Pacific and a presence near Japan and Korea. All are marine, except for one species in fresh water in California. Perches are small and have a deep, laterally flattened body and small mouth. These are live-bearing (viviparous) fish with internal fertilization; they produce fully developed juveniles that can be 2 to 5 cm long. The thickened anterior of the male anal fin inseminates the female, who stores the sperm for several months before the eggs are fertilized.

KELP PERCH

Brachyistius frenatus COMMON ●

IN THE STRAIT OF GEORGIA

Kelp perch are distributed throughout the Strait of Georgia from Campbell River to the Gulf Islands, but although they occur widely and often in small schools, they are not commonly seen. The kelp perch is not common in our trawl net capture records, but scuba divers find them abundant in shallow kelp bed communities of the Strait of Georgia.

DESCRIPTION

This is an attractive copper-brown, brassy-coloured fish that lacks bars and may have blue spots and pearly patches along the sides. Its colour is similar to that of kelp, which makes the fish hard to spot among kelp fronds, where they pick off small grazers like amphipods, isopods and copepods. It has a distinctive upturned snout. The tail end of the body trunk (caudal peduncle) is notably broader than in other surfperch species.

Kelp perch always have a kelp colour, and the trunk by the tail (called a caudal peduncle) is thick from top to bottom. Danny Kent

LIFE HISTORY

This is a pelagic species that occurs among kelp, as the name implies, and in subtidal areas from the surface to about 75 m, but mostly to about 15 m. It occurs from Sitka, Alaska, to Baja California. This is a smaller surfperch species, about the same size as the more abundant shiner perch, reaching 22 cm and 130 g. Kelp perch mature at young ages and live for only two to three years. Individuals mature in one year and mate in the fall; the young are born fully developed the next year in the spring to early summer. Up to 50 perfectly formed babies may be released, at lengths of about 3.3 cm. These newborns are smaller than those of other surfperch species, and in a commercial aquarium they are more difficult to rear, perhaps because they need smaller live foods. Most other perch species migrate into deep water during the winter and can then be taken in trawl nets, whereas the kelp perch is usually at shallow depths as it does not tolerate water cooler than 8°C. It is not known where kelp perch hide during the winter, when storms cause decay of seaweed canopies, so it remains to be discovered if this species prefers to hide in shallow rock piles during the winter. One reason they are not seen as often in display aquariums is that they usually have to be collected from the wild, whereas the other surfperch species reproduce like clockwork and have very high survival rates in an aquarium setting.

The kelp perch has a strong association with giant kelp and bull kelp, where it feeds on small crustaceans and breeds. It is one of the Strait of Georgia species that will on occasion act as a cleaner fish, picking crustacean parasites and leeches off larger fish and eating them. Since kelp perch pick small crustaceans off seaweed when feeding, it probably comes naturally to them to pick off crustacean parasites and leeches from the bodies of larger fish. The larger fish will hold still and be patient to achieve some relief from the discomfort of small creatures adhering to and eating into their skin. With the declining abundance of bull kelp in the Strait of Georgia, it is possible that kelp perch may also be declining in abundance in the strait.

SHINER PERCH

Cymatogaster aggregata

ABUNDANT

IN THE STRAIT OF GEORGIA

Shiner perch are very common throughout the Strait of Georgia and are one of the most abundant fishes in the strait. In the summer, they can be seen in schools around docks, pilings, floating wharves and eelgrass beds. They can be abundant in bottom research trawl catches and in the bycatch of shrimp trawls, as well as in pelagic trawl catches in water shallower than 30 m. This is a popular fish with children fishing from docks. A small hook baited with something living around the dock usually will catch the attention of shiner perch, though they are very good at pecking at the bait while avoiding the hook.

Shiner perch are the most abundant live-bearing perch in the Strait of Georgia. Danny Kent

DESCRIPTION

The shiner perch body is short, with a compressed head and large eye. The mouth is small and upturned, and the teeth are pointed. The shiner perch has one dorsal fin on the top, and an anal fin and two pelvic fins ventrally. A large caudal peduncle separates the terminus of the fins from the tail, which is forked and pointed. The scales are quite large and the lateral line is highly visible. This species varies significantly in colour from silver to black, but it is always shiny. The dorsal side may be darker than the rest of the body or green in colour. Breeding males are often black. Silver longitudinal bars extend the length of the body, interrupted by three yellow vertical bars. The species can be distinguished by the yellow bars as well as by its large scales, of which there are 50 or less along the lateral line. Old-timers sometimes call this species a yellow shiner.

LIFE HISTORY

The range of this species is from southern Alaska to Baja California. The maximum length is approximately 20 cm. In general, they are a shallow-water fish in the summer, but in the winter they occupy deep water down to 303 m. They have been caught in British Columbia on La Pérouse Bank at 128 m. Shiner perch are often found near the shore around wharves and pilings and can be found in low-salinity environments in wetlands, and even in fresh water. They are also common along sand beaches.

The eggs are fertilized internally, where they develop and grow, leading to live birth. In British Columbia, the breeding season extends from April to July. Prior to insemination, courtship behaviour involves the male pursuing the female and swimming in front of her snout in a criss-cross fashion until she appears receptive. He then moves parallel to her, aligning his anal fin with hers, and they swim slowly for several seconds. During this time, the male sperm is transferred to the female using the male's modified anal fin. The female may carry sperm from up to eight males. Females store sperm in an ovary compartment for as long as six months, or until November or December, at which time the sperm penetrates the ovary and initiates fertilization. Depending on the size of the female, up to 36 eggs may be incubated at a time. Incubation lasts from December until the following June or July, when young are born, tail first, into the warm coastal waters. Young are born at between 5.6 and 7.8 cm long.

Like many other species, the shiner perch exhibits sexual dimorphism in terms of length. Females are normally larger than males, beginning at birth. Males are mature either at birth or shortly after. Females mature in their first or second year at between 4 and 6 cm in length. The maximum age is six years, but females live longer than males. Food preference differs with age, sex and season. Young shiner perch consume mostly copepods before they graduate to a more substantial diet of mussels and algae. Adult shiner perch have been seen consuming barnacles. Shiner perch are carnivores, known to feed on worms, mussels, brittle stars, crustaceans and herring eggs. They have specialized, hard tooth plates that allow them to crush hard-shelled invertebrates such as crabs and mussels. They are often observed feeding in the evening around piers, where they pick away at items such as barnacles or mussels attached to pilings or floating structures. In California, there is a small commercial fishery for shiner perch (approximately 270 kg per year), as well as a vibrant recreational fishery. Shiner perch are good to eat, though a little bony.

PILE PERCH

Phanerodon vacca ABUNDANT

IN THE STRAIT OF GEORGIA

Pile perch are abundant throughout the Strait of Georgia and are commonly seen by scuba divers but are not particularly common in estuaries. They occur mostly around underwater structures close to shore. They are occasionally captured in research trawls on and off bottom, particularly in the southern strait.

The recent book *Fishes of the Salish Sea* includes the white seaperch (*Phanerodon furcatus*) in the Strait of Georgia from one record off Nanaimo. We have no records from any of the research catch databases or the groundfish database over the past 50 years and so are not considering it to be in the strait. In general, the white seaperch is like a pile perch without any bars, white in colour with yellow on the fins.

DESCRIPTION

The body of the pile perch is short, with a compressed head and large eye. The mouth is small and upturned, and the teeth are pointed. The fish have one dorsal fin, an anal fin and two pelvic fins. A thick, large caudal peduncle separates the posterior of the fins from the tail, which is forked and pointed. Pectoral fins behind the operculum are large and point toward the dorsal side of the fish. The scales are quite large and the lateral line is highly visible. The dorsal

Andy Murch

surface is dark grey or brown with silver on the sides and ventral surface, with a dark vertical bar mid-body. The pile perch is distinguished by its short head, small mouth with fleshy lips and strongly forked tail. The midpoint of the dorsal fin is higher than the anterior portion. Newborns are pink.

LIFE HISTORY

The range of this species is from southern British Columbia to central Baja California. The maximum length is approximately 44 cm. The maximum age is unknown, but research scientist Milton Love recorded a 40.4 cm fish as being 13 years old. They are often found near the shore around wharves and pilings.

All species in this family are viviparous: eggs are fertilized internally, where they develop and grow, leading to live birth. Like many other non-commercial species, little is known about their life history. As adults, they prefer habitats with dense algal growth and easy access to prey. In British Columbia, the young are born in August. Pile perch reach sexual maturity between the ages of 2 and 4, and have been recorded to be in breeding condition until at least age 10. The number of offspring is quite small, with a fecundity at age 4 of approximately 12; it can exceed 60 in fish 8 to 10 years old. In California, breeding has been recorded in July. In Oregon, male pile perch were running ripe (meaning their sperm was mature) in September and October, leading to the belief that this was the most significant time for breeding activities. As with many other species, the brood size is correlated with the size and age of a pile perch. The mating behaviour of this fish is somewhat different than that observed in other perch species. A male and female fish swim in the same direction, then suddenly turn to their sides or upside down to link their urogenital pores.

Pile perch are benthic feeders and are classified as “commuters” because they school in midwater when not foraging. They are ravenous mussel eaters and can crush the shells and encapsulate them in mucus in order to excrete them in the feces. They also consume other hard-bodied animals such as decapods, isopods, amphipods, gastropods and echinoderms. Juvenile pile perch are small and vulnerable to considerable predation by piscivores (fish-eating predators). Pile perch are born when kelp cover is greatest, which aids in juvenile survival. There are few predators of the adults. Pile perch are a favourite with youngsters learning how to fish. Many years ago, Bill Ricker, who was perhaps Canada’s most famous fisheries scientist, would fish for pile perch with his granddaughters off the dock at the Pacific Biological Station.

STRIPED SEAPERCH

Embiotoca lateralis

COMMON ●

IN THE STRAIT OF GEORGIA

Striped seaperch are somewhere between abundant and common in the Strait of Georgia. This is a nearshore, shallow-water species commonly seen around piers. Throughout the strait it is frequently observed by scuba divers and is occasionally found in bottom and pelagic research trawls. This fish can be caught by children using the natural bait found around wharves and piers.

DESCRIPTION

The body of the striped seaperch is short, with a compressed head and large eye. The mouth is small and upturned, and the teeth are pointed. A large caudal peduncle separates the terminus of the fins from the tail, which is forked and pointed. The pectoral fins behind the operculum are moderate in size and point toward the dorsal side of the fish. The scales are quite large, and the lateral line is highly visible. This species has a memorable coloration. Overall, it is copper to dark brown with horizontal blue stripes below the lateral line. The head may also have blue spots or stripes. The coloration of brilliant blue and orange lines with changing golds, yellows and greens makes this fish almost tropical in appearance. This is clearly one of the most beautiful fishes in the Strait of Georgia.

Striped seaperch have conspicuous horizontal blue stripes on a reddish-brown body. Danny Kent

LIFE HISTORY

The range of the striped seaperch is from southeast Alaska to Northern California. The maximum length is 38 cm and the maximum age at least nine years old. Striped seaperch mature at two to three years old and at lengths of 19 to 22 cm. In general, the striped seaperch is a midwater fish that occupies the same depths both during the day and at night. It is an inshore species, preferring rocky subtidal areas of the shoreline with considerable seaweed growth. The seaweed provides refuge from predators for adults and juveniles; it is also a good source of invertebrates for food.

Striped seaperch give live birth to their young. In British Columbia, this is believed to occur in June and July, with up to 44 fish being born at a time. The number of young produced increases with the age of the female. In general, a three-year-old striped seaperch spawn will produce approximately 18 offspring, increasing to 32 by age seven. The largest number of young is known to be 92, but anything over 44 is uncommon. Breeding behaviour of the striped seaperch involves one member of the pair orienting in a vertical swimming position while the other stays horizontal, keeping their anal fins close together. The horizontal fish, presumably the male, vibrates for several seconds and fertilization occurs. As with other live-bearing perches, the embryonic individuals have specialized organs called spatulate extensions on the anal and dorsal fins, which, when in contact with the ovarian wall, allow for the absorption of nutrients and oxygenation of the blood. These extensions are absorbed shortly before birth. Females are known to release young from March to September. Newborns resemble their parents, apart from the orange tinge on their fins and body.

Striped seaperch, in general, are benthic carnivores and known to feed on worms, mussels, brittle stars, crustaceans and herring eggs. They have specialized hard tooth plates that allow them to crush hard-shelled invertebrates such as crabs and mussels. They are inactive at night and feed in the daytime, preferring to graze on food found on the bottom. As adult fish, striped seaperch have very few predators, mainly seals and sea lions; as juveniles, they fall victim to rays, sharks, seals and sea lions. In the 1960s and 1970s, California had a commercial perch fishery that included this species; although it was a small fishery, it did range in annual value from $35,000 to $57,000.

Ronquils and Eelpouts

There is one species of ronquil in the Strait of Georgia. Ronquils are small cold-water species in the family Bathymasteridae, a poorly studied group of seven species found only in the North Pacific and Arctic Oceans. Ronquils occur on or near the ocean bottom from the intertidal and nearshore rocky areas to about 150 m. They are elongate and have large eyes, terminal mouths and long dorsal and anal fins that are continuous with the tail fin.

Eelpouts belong to the family Zoarcidae, which has a worldwide marine distribution ranging from very shallow to very deep, with a recent depth record of 4,250 m in an ocean trench. There are about 240 to 300 species, most of them occurring on or near mud bottom. They are elongate, are less than 46 cm long, and have an ovoid and somewhat flattened head. The tapering body has long dorsal and anal fins joined to the tail fin. The pelvic fins are small or absent. Some have internal fertilization, resulting in females giving birth to live young. There are eight species of eelpouts in the Strait of Georgia, some of which are very common, particularly in commercial shrimp trawl catches.

NORTHERN RONQUIL

Ronquilus jordani

COMMON ●

IN THE STRAIT OF GEORGIA

Northern ronquil are found in shallow water throughout the Strait of Georgia. They are commonly caught by shrimp trawlers in 18 to 37 m of water, occasionally caught in bottom research trawls and frequently seen by scuba divers throughout the strait.

DESCRIPTION

This species has an elongate, eel-shaped body with a large, compressed head. The eyes and mouth are both large and the teeth are small. The species has one tall, long dorsal fin extending from behind the head to the end of the body, terminating at the caudal peduncle. It has a similar anal fin extending from behind the anus and terminating before the end of the body. The tail is medium sized and rounded. The pectoral fins are large and fan-shaped, similar to the tail; the pelvic fins are also big and extend below the operculum. The northern ronquil is distinguished by the unbranched first 20 to 30 rays of the dorsal fin and by the pectoral fin extending well past the origin of the anal fin.

Northern ronquil live in moderately deep water and use their pelvic fins to prop their heads up above the bottom. Light-yellow markings are the most common colour pattern, but variations exist. Danny Kent

The northern ronquil exhibits sexual dimorphism, with the two sexes having distinctive colour patterns; ichthyologists would call this species dichromatic. The males are orange on top and olive green below, with dark bars extending down the sides. Males also have yellow bars below the eye and across the cheek. The fins of males also exhibit different colour patterns from those of females. The males have yellow pigment on the upper parts of the pectoral fins and a blotch at the anterior end of the dorsal fin. The dorsal and caudal fins have black blotches encircled with yellow, and the anal fin is iridescent blue outlined in black with golden spots at the base. The females are olive green on the dorsal surface and lighter on the ventral surface. The dorsal fin has a blotch on the anterior end, and the dorsal and caudal fins have bright-orange coloration on the edges. The anal fin is pale blue outlined with brown, and the pelvic fins are similar but outlined with white.

LIFE HISTORY

The northern ronquil is found from Central California to the Aleutian Islands and Bering Sea. It occurs throughout British Columbia and has been captured in waters as shallow as 18 m and as deep as 165 m. The first record of a northern ronquil in British Columbia was in 1891 off Victoria at 88 m. Members of this family are all bottom dwellers found in shallow coastal areas from 3 to 275 m throughout their range, but most are found at less than 150 m. They are inhabitants of both rocky reefs and associated sandy areas, where they are territorial. Their maximum length is approximately 20 cm.

It is believed this species reproduces from late winter to early spring, as in March males are observed in full breeding colours, which include increasing amounts of blue pigments. There are reports that breeding may even extend into early summer. Eggs of the northern ronquil resemble those of salmon, with dark-pink to red coloration. Eggs are laid in a mass under a rock and fertilized by the male. In general terms, the demersal eggs are tended, but it is not certain which of the two parents provides this care. Once the eggs hatch, the larvae become planktonic. Juveniles approximately 1.1 cm in length have been observed in the surface waters of Saanich Inlet and the Fraser River estuary. Northern ronquil have been identified in the diet of pigeon guillemot chicks in Alaska, as well as in arrowtooth flounder from the Gulf of Alaska. The northern ronquil feeds mostly on benthic crustaceans.

A male northern ronquil shows increasing blue pigment when mating season occurs.
Danny Kent

BLACKMOUTH EELPOUT

Lycodapus fierasfer RARE ★

IN THE STRAIT OF GEORGIA

Blackmouth eelpout are rare in the Strait of Georgia. We have catches from research shrimp trawls and bottom trawls that fished just off the bottom in deep water. In *Pacific Fishes of Canada*, J.L. Hart reported the first British Columbia record to be from Howe Sound, where a specimen was taken from the mouth of another fish.

DESCRIPTION

The blackmouth eelpout is a small fish with no remarkable colour pattern to help with recognition. Pelvic fins are absent, there are no scales, the pectoral fin is small and placed low, and the body is slender with a translucent skin that is pearly in colour. There are long, pointed gill rakers and two small pores on the top of the head between the eyes. The name "blackmouth" refers to the lips, mouth and gill cavity being black in live specimens.

Michael Oliver

LIFE HISTORY

The range of the blackmouth eelpout is from the western Bering Sea and on Shirshov Ridge in the eastern area of the sea, to off Kamchatka, and down the Pacific coast of the Americas to Peru at depths of 102 to 2,189 m. Blackmouth eelpout are mainly on bottom along the continental slope but are also captured in nets that fish off the bottom. A maximum length of 14.1 cm has been recorded. Very little is known about their life history.

PALLID EELPOUT

Lycodapus mandibularis ABUNDANT

IN THE STRAIT OF GEORGIA

Pallid eelpout are common throughout the Strait of Georgia at depths below 75 m and particularly below 150 m. Because of their small size and elongate shape, they would readily escape through the mesh of our research trawls, suggesting the abundance is greater than our catches indicate. Based on the frequency of catches in deeper waters throughout the strait, the species probably is one of the more abundant fishes in those areas.

DESCRIPTION

The body of the pallid eelpout is elongate, almost eel-like, with a long tapering tail. There is a terminal mouth with jaws about equal in length. Some specimens have large, protruding teeth. The fish does not have pelvic fins or scales. The skin is a pale, translucent cream and silver in colour and the body is a strong silver. There is a dark lining in the mouth, inside the gill cover and inside the body cavity. The gill opening extends upward past the base of the small pectoral fin. The gill rakers are stout, and the upper rim of the eye extends above the slope of the head.

Bill Pennell

LIFE HISTORY

This is a pelagic eelpout that is found in the midwater and down to 800 m in offshore areas. The maximum length is 19.8 cm. In a survey performed in 2011, the average length of 335 fish in the Strait of Georgia was 7.9 cm, and the range in size was 3.8 to 11.8 cm. The species is known to migrate toward shallower water at night, which might explain why pallid eelpout are occasionally captured in shrimp trawls. Catches outside of the deep pelagic areas are rare. In one study, most fish were found to mature at age three or four and did not live much longer than age five. However, fish from the Strait of Georgia have been neither aged nor examined for their maturation and spawning times.

Pallid eelpout have been observed spawning in a laboratory. The eggs were relatively large, adhesive and negatively buoyant. The fish in the laboratory dug U-shaped burrows about 1.5 cm in diameter and 10 to 20 cm in length.

Dick Beamish

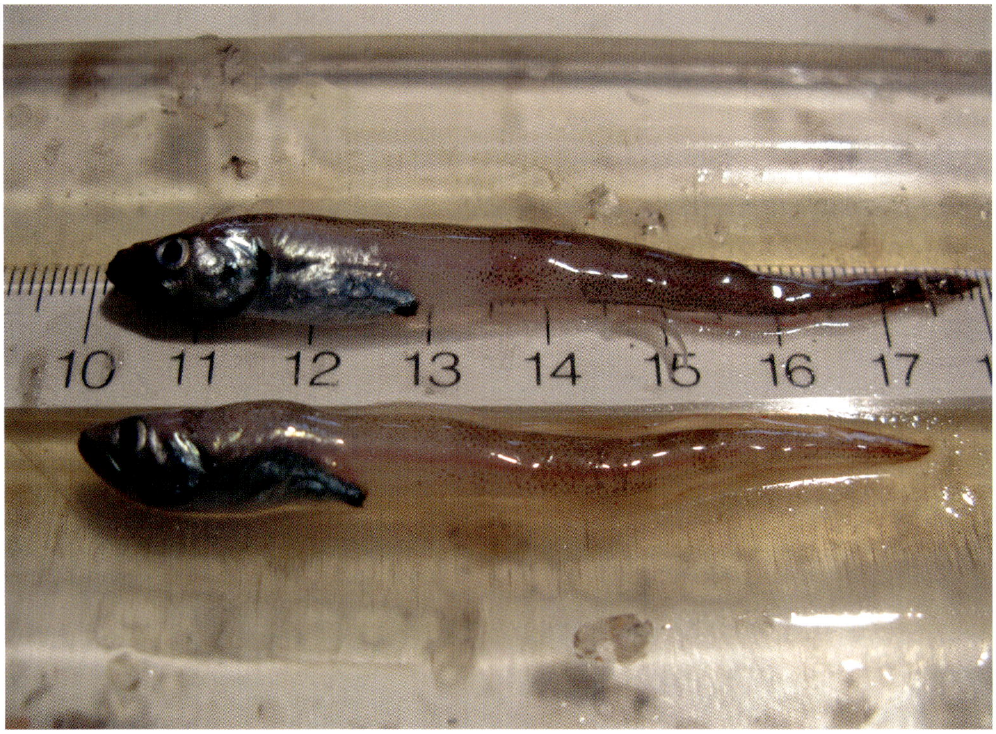

It is possible that as the pallid eelpout mature, they leave the pelagic areas and reside near or on the bottom, where they spawn. A benthic area of reproduction for a mesopelagic species would be most unusual. This could also mean the pallid eelpout we capture are at the immature, pelagic stages of development. It is also possible that once they mature, they spawn in batches and not just once a year. Pallid eelpout feed on copepods, euphausiids, amphipods, squid and other smaller organisms within the nekton (open-water) community, including deepwater pelagic fishes.

SMALLHEAD EELPOUT

Lycodapus parviceps RARE ★

IN THE STRAIT OF GEORGIA

The smallhead eelpout was collected in Howe Sound in the late 1970s. There are no other records for the Strait of Georgia, including any pelagic or on-bottom research trawls or in the groundfish database.

DESCRIPTION

In contrast to the pallid eelpout, the smallhead eelpout has a light lining in the mouth and gill cavity. Smallhead eelpout are elongate, with long dorsal and anal fins that are attached to the caudal fin. There is no pelvic fin. The gill slit is generally below the base of the pectoral fin, unlike the gill slit of the pallid eelpout, which extends above the base. The translucent cream to silvery-white skin is loose. The gill rakers are very reduced and stubby compared to those of the pallid eelpout, which are stout, curved and pointed.

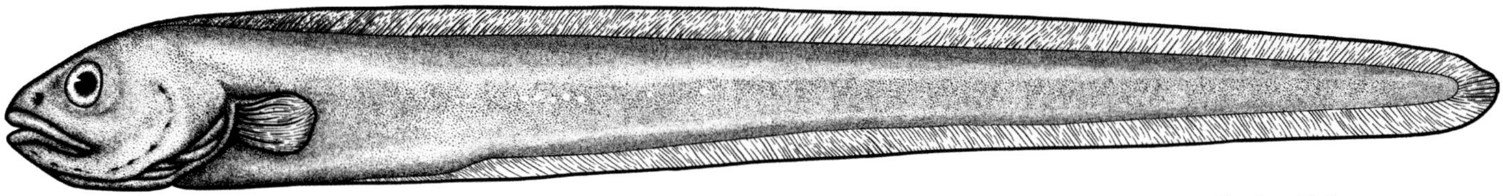

Meghan Walley

LIFE HISTORY

Smallhead eelpout range from southeastern Alaska to southern Oregon. They are found close to the bottom at depths ranging from 81 to 457 m. The maximum length is 16.6 cm. There is no information about the maximum age or the age at maturation. This is a pelagic species that is found in deep water and feeding on plankton. Because this fish is at the southern limit of its distribution in the Strait of Georgia and occurs in cooler, deeper water, it likely would be one of the fishes affected by the warming occurring in the strait.

SHORTFIN EELPOUT

Lycodes brevipes UNCOMMON ◆

IN THE STRAIT OF GEORGIA

Shortfin eelpout probably are uncommon throughout the Strait of Georgia. They have been caught in research shrimp trawls in English Bay and Burrard Inlet at depths between 27 and 119 m. They are a minor part of the catch in research shrimp trawls at shallower depths in areas of sand or mud bottom in the southern strait.

DESCRIPTION

These fish have the eelpout's typical highly elongated, eel-like body. They have a long, compressed head with a blunt snout, a large mouth with small teeth, and an upper jaw that hangs over the lower jaw. The shortfin eelpout has one long dorsal fin and one anal fin along the ventral margin that extends from the anus to the end of the body, where it meets up with the dorsal fin at the tail. This species of eelpout is light brown with bands of white running through the dorsal fin, down the back and over the side of the fish. The dorsal and anal fins both have dark margins. It is likely that this species is named the shortfin

TINRO

eelpout because of its very tiny pelvic fins. It is distinguished by the 13 narrow light bands on the body, the short pelvic fins and the unnotched pectoral fins.

LIFE HISTORY

Shortfin eelpout have been caught at depths ranging from 2 to 973 m and reach a maximum size of 32.8 cm. The range of the species is from Central California to the eastern and western Bering Sea and off the east coast of Kamchatka, Russia. Shortfin eelpout are benthic (bottom-dwelling) fish. In the Bering Sea, they have been captured on bottom where water temperatures were −1.4°C. The species is found as bycatch in flatfish and rockfish research trawl fisheries. Studies in the Gulf of Alaska have reported stomach contents of shortfin eelpout to include clams, crustaceans, shrimp, mysids and euphausiids. In the Bering Sea, similar types of studies reported brittle stars as the predominant prey item, with some consumption of polychaetes, amphipods, shrimp, crabs, clams and fishes. Shortfin eelpout have been identified in the stomach contents of walleye pollock, Pacific cod, wattled eelpout and arrowtooth flounder in surveys in Alaska. Bering Sea bottom trawl surveys estimated the relative abundance of the shortfin eelpout at 20 of their survey stations to be greater than 190,000 t, equating to 1.9 billion individuals. As is demonstrated by the type of information available on this species, the little we do know is often a result of work done on catches of other species. Much of the general life history of the shortfin eelpout remains a mystery.

BIGFIN EELPOUT

Lycodes cortezianus RARE ★

IN THE STRAIT OF GEORGIA

In *Pacific Fishes of Canada*, J.L. Hart reports that the bigfin eelpout is fairly common in the Strait of Georgia; this is currently not the case, as we have no records in any of the research trawl studies in the strait since the mid-1990s. Because it was previously considered to be common in the strait, we are including the species in our list to ensure there is an awareness that it is important to look for evidence of its existence.

DESCRIPTION

This is a big eelpout, growing to 49 cm. The head is large and broad and the pectoral fin is large and rounded. The body is light brown and the fins are black. The high position of the broadly rounded pectoral fin makes the common name "bigfin eelpout" an appropriately descriptive name.

Rick Starr

LIFE HISTORY

The range of the bigfin eelpout is from the Gulf of Alaska to San Diego, California. It is caught in trawl catches off the west coast of British Columbia from depths of 100 to 450 m, usually on mud, sand or gravel. The record depth of capture is 800 m. Reports from the Bering Sea are considered to have been misidentifications, so it will be important for voucher specimens from the Strait of Georgia to be deposited in fish taxonomy museums for detailed study. Features like the presence of palatine teeth but absence of vomerine teeth are details that call for expert examination. This fish is found in the guts of large flatfish, skates and sea lions.

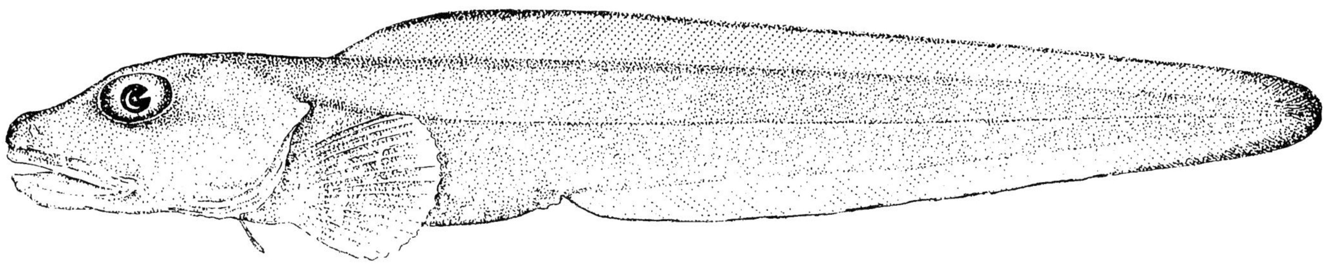

BERING EELPOUT

Lycodes beringi COMMON ●

IN THE STRAIT OF GEORGIA

In 2009 a new species of eelpout, named the Bering eelpout, was identified. It was originally considered to be the black eelpout (*Lycodes diapterus*). Importantly, the taxonomic study found there was very little overlap in the distributions of the Bering and black eelpouts. All of our records for the Strait of Georgia previously reported that the species in the strait was the black eelpout, but a recent examination of 11 preserved specimens from two locations in the strait indicated the specimens were the newly described Bering eelpout. The 11 samples had head scales only to the nape and not to the eyes as in the black eelpout. Assuming there is very little overlap between the two species, we now consider the species in the Strait of Georgia to be the Bering eelpout. However, because we are at the range extremes for both species, this merits closer examination and the use of molecular genetic samples. The Bering eelpout may be common in the deep water of the Strait of Georgia, where it occurs mostly on or near the bottom, but may be caught in pelagic trawls fishing from 75 to 200 m.

Dick Beamish

DESCRIPTION

The Bering eelpout is distinguished from the black eelpout by the scale pattern on the head and by meristic differences. It is recognized by the solid black colour and the notched pectoral fin.

LIFE HISTORY

The Bering eelpout is distributed from the Gulf of Alaska to the Strait of Georgia and Puget Sound, but specimens off the west coast of Vancouver Island are reported to be the black eelpout. The Bering eelpout is also found off Russia from Kamchatka to the Kuril Islands. The maximum length is about 37 cm. The recognition of the new species provides a good opportunity to study the biology and life history of the Bering eelpout in the Strait of Georgia in comparison to the closely related species off the west coast of Vancouver Island.

BLACKBELLY EELPOUT

Lycodes pacificus ABUNDANT

IN THE STRAIT OF GEORGIA

The blackbelly eelpout is a very common species throughout the Strait of Georgia and the most abundant of all the eelpouts. It is common at depths of 50 to 100 m in the strait. Scuba divers commonly see it at night in deeper water.

DESCRIPTION

The blackbelly eelpout has an elongated, eel-shaped body with a maximum length of 46 cm, but most fish in the Strait of Georgia are less than 20 cm long. It is recognized by the black lining (peritoneum) of the gut cavity, which is faintly visible through the skin of the belly. There is also a black area at the anterior edge of the dorsal fin and black along the margin of the dorsal fin. An index finger placed at the top of the mouth will not feel any teeth (vomerine and palatine teeth are absent). The skin colour is grey to reddish brown. Older males may be larger than females of the same age, and they have different body proportions.

LIFE HISTORY

This species is found from Baja California to the Aleutian Islands at depths from 9 to 399 m. It is not important commercially but is commonly caught as bycatch in commercial trawl fisheries. Of the relatively few fish that have been aged, the oldest was age five. It is likely there are older fish, but the maximum age is unlikely to be much greater. In the Strait of Georgia, all males and females are mature at about 17 cm long and about age three.

Blackbelly eelpout have a smooth snout with the upper lip overhanging the jaw, and a black trim to the dorsal fin. Lee Newman

Most spawning occurs from November to December, with females being ripe from August to December. Females produce surprisingly few eggs, averaging about 30. Spawning fish have not been observed, but fish that produce so few eggs would be expected to have some form of special protection for them, such as burying or guarding. As well, the eelpout's very large eggs typically hatch as fully developed juveniles rather than as a distinct larval form, thus eliminating an early life stage with high mortality.

The blackbelly eelpout is common on soft, muddy bottoms, where it feeds extensively on bivalve molluscs as well as amphipods, polychaete worms and small crabs. It is reported to be prey for larger fish predators. It does not appear to be eaten by seals, perhaps because it hides during the day and moves into shallow water to forage during the night. Despite the importance of the blackbelly eelpout in the Strait of Georgia and other ecosystems, little is known about its biology, life history and ecology. An important study was published by fisheries scientist Colin Levings in 1969 when he was still a student. We hope that universities will again recognize the importance of these life history studies.

WATTLED EELPOUT

Lycodes palearis UNCOMMON ◆

IN THE STRAIT OF GEORGIA

The wattled eelpout is another big eelpout that is uncommonly caught in the Strait of Georgia. There are records from the Gulf Islands and southern strait but no catches from the thousands of on- and off-bottom research trawls since the mid-1990s.

DESCRIPTION

The name "wattled eelpout" relates to the species' highly developed chin crests, which project forward. The pectoral fin is large and long, and the pelvic fin is shorter than the eye diameter but longer than the eye pupil. The colour is brown, light under the head and belly, with light vertical bars on the body and dorsal fin; two of the bars are in front of the dorsal fin. These bars become indistinct on larger fish, and a black blotch at the anterior margin of the dorsal fin can also disappear with growth.

Catherine Mecklenburg

LIFE HISTORY

Wattled eelpout occur from the Bering Sea and Sea of Okhotsk down to Oregon and are caught at depths from 25 to 925 m, but usually from waters shallower than 150 m. Wattled eelpout grow to 54 cm. Catches are from mud or sand bottoms. They are reported to spawn large, 7 mm ellipsoid eggs during the winter in Puget Sound. Food includes small clams and shrimps. This eelpout has very large eggs, and the fish hatch as juveniles rather than a larval form.

and Gunnels

Pricklebacks, wrymouths and gunnels in the Strait of Georgia represent three families. Pricklebacks are in the family Stichaeidae, which includes cockscombs, warbonnets and shannies. Fishes in this family are found in the northern hemisphere, mostly in the North Pacific, with some in the Arctic and North Atlantic Oceans. A total of up to 76 species is estimated. These are small fish, mostly less than about 25 cm long and none longer than 50 cm. All are marine fishes that are generally found inshore, including in tide pools and often under rocks. They have a small mouth and are elongate with a long, low spinous dorsal fin and long anal fin, both of which are attached to the tail fin in some species. Species in the family are known to be able to survive long periods out of water, and some guard nests. There are eight pricklebacks, two cockscombs, two warbonnets and one shanny in the Strait of Georgia.

Wrymouths are in the family Cryptacanthodidae, which has only four species. Two of these are in the Strait of Georgia. Three species are in the coastal areas of the North Pacific and one is in the North Atlantic. These are very elongate fishes that are rounded anteriorly and flattened posteriorly. The eyes are high on the head, which has an oblique jaw that appears almost vertical. The long dorsal and anal fins join the tail fin. Wrymouths are found in soft sediments, where they can be buried in tunnels they dig.

Gunnels are small elongate fishes in the family Pholidae that are mostly in the North Pacific, with some in the North Atlantic. These are nearshore species that commonly occur under rocks or in tide pools but have been reported at depths up to 200 m. There are about 15 species in the family, and 5 of these occur in the Strait of Georgia.

SLENDER COCKSCOMB

Anoplarchus insignis RARE ★

IN THE STRAIT OF GEORGIA

This is a rare species in the Strait of Georgia. We have records from the Cowichan River estuary and scuba observations from Active Pass, Howe Sound and Burrard Inlet. There also are records from Pender Harbour and Malaspina Strait. We have no records from the several thousand research trawl studies since the mid-1990s.

DESCRIPTION

The slender cockscomb gets its common name from the small size of its dermal crest (or "cockscomb") above the eye, in contrast with the higher crest on the high cockscomb. A fast and sure way to distinguish the slender cockscomb is by looking for prominent zebra stripes on the lower jaw. Slender cockscombs have distinct tail fins; they are not joined with the dorsal and anal fins, as in other intertidal prickleback species. The slender cockscomb and high cockscomb are recorded in the intertidal zone down to 30 m deep; the slender cockscomb tends to be found subtidally, whereas the high cockscomb is almost always intertidal. On rare occasions, a male slender cockscomb will be entirely orange in colour, but breeding efforts at the Vancouver Aquarium suggest this is a recessive trait, as all offspring of an orange male had normal pigment patterns.

Slender cockscombs occur in deep water and tend to be seen on silty bottoms, where their body pattern blends in with the surroundings. Lee Newman

LIFE HISTORY

The slender cockscomb occurs from Northern California to the west end of the Aleutian Islands and the southeastern Bering Sea, and it reaches about 12 cm in length. It is an intertidal and subtidal species, mostly in areas where there is current with rocks and algae for hiding, as would be found in Active Pass. The biology remains to be studied and reported.

HIGH COCKSCOMB

Anoplarchus purpurescens COMMON ●

IN THE STRAIT OF GEORGIA

In the Strait of Georgia, scuba divers report that high cockscomb are distributed around the strait, but their abundance is not known. They are common in intertidal areas. The high cockscomb is not found in fisheries or research trawls, as the species' intertidal preference keeps them away from the commonly fished areas.

The male high cockscomb has orange pigment on the pectoral, dorsal and anal fins during the winter mating season. Andy Murch

DESCRIPTION

This species was first recorded in British Columbia waters as the cockscomb prickleback in 1861. High cockscombs have a very slender, elongate body with a relatively small, pointed head. The eel-like body is accentuated by one long dorsal fin extending from behind the head to the short, fan-like tail. The anal fin begins after the anus and extends to the end of the body by the tail. Of particular note for this species is the tall and prominent fleshy crest that extends from the middle of the top lip to the crest of the head, hence the name "high cockscomb."

LIFE HISTORY

The range of this fish extends from Southern California to the Aleutian Islands. Interestingly, the number of vertebrae and fin rays increases going from south to north. This species has been recorded at up to 20 cm in length. It is primarily

Female high cockscomb have a salt-and-pepper pigment pattern, with the high fleshy crest as a ridge over the eyes and two warpaint stripes extending downward from the eye. When the female is paired with a male under a rock, the large white ovaries are visible through the gut wall during early winter, prior to spawning. Andy Murch

an intertidal species found no deeper than 3 m. Not surprisingly, given the shallow water it inhabits, this species can tolerate influxes of fresh water into its marine environment. It is commonly discovered by turning over small- and medium-sized rocks in the intertidal zone. It may be found in hollowed-out burrows underneath these rocks at low tide or among the rocks when they are covered with water. The high cockscomb has a home range and may be territorial, particularly during the spawning season. Its coloration is highly variable; it may be brown, olive, reddish, grey or black. Females and males usually have different colour patterns, especially during winter and spring courtship and spawning seasons. The females tend to have a more reticulated colour pattern, usually with a pale ventral surface. They have a mottled cockscomb, while the males' is pale yellow. During spawning season, males have bright-orange anal and pectoral fins and an orange-red dorsal fin and tail. Both males and females mature at two to three years of age and at lengths of 8 to 9 cm. The females are slightly larger than the males but in general grow more slowly. In British Columbia, both males and females are approximately 11.4 cm at three years old.

The breeding season extends from January to March. During spawning, males perform displays and compete with other males for access to females. In captivity, males have been observed spawning with more than one female. Females have been recorded producing up to approximately 3,000 eggs, with a diameter of about 1.4 mm. The number of eggs a female can deposit increases with her size. The egg mass is found between craggy rocks and under shells, usually in a cone shape. The female tends the eggs by wrapping herself around the adhesive egg mass and fanning it with her body until the eggs hatch. Newly deposited and fertilized eggs begin as white in colour and later change to grey; this makes the oil globule within the egg highly visible. It takes approximately three weeks for eggs to hatch, at which time the larvae are about 7.5 mm in length. They grow quickly, reaching a length of 3 cm by June, completely resembling the adult form. This species is known to consume nematodes, molluscs, polychaetes, crustaceans and echinoderms, as well as algae and eel grass. One predator of the high cockscomb is the common garter snake, which catches the fish in water. Other predators include greenlings and land animals such as mink and black bears.

DECORATED WARBONNET

Chirolophis decoratus COMMON ●

IN THE STRAIT OF GEORGIA

Decorated warbonnets are distributed throughout the Strait of Georgia. Scuba divers see them throughout the strait, usually when inspecting sponge openings, giant barnacles or crevices in cliffs. They are occasionally caught in bottom research trawls throughout the strait.

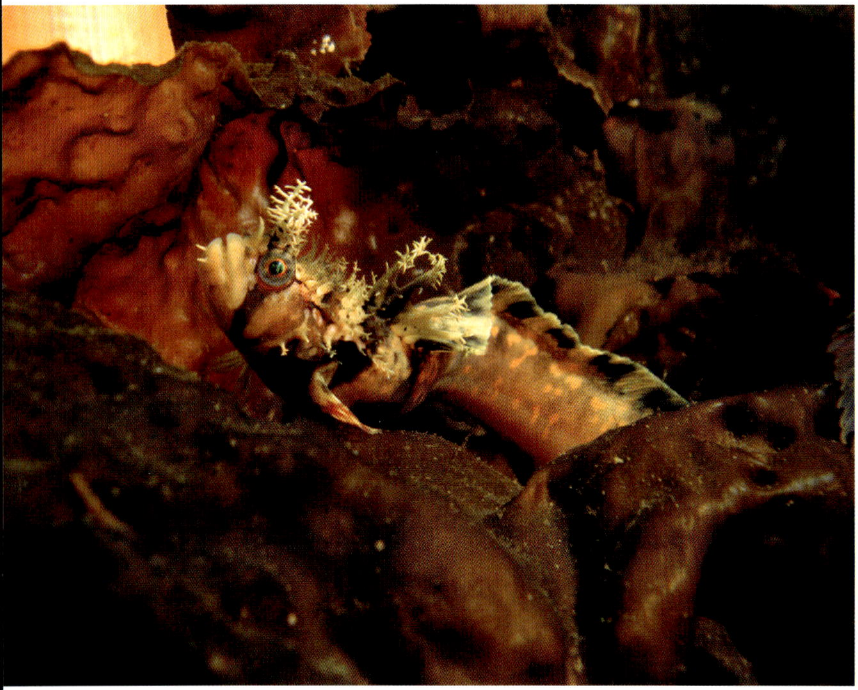

Janna Nichols

DESCRIPTION

The pectoral and tail fins have patterns of broken bars that appear to be like fingerprint patterns unique to each fish. This was noticed when large numbers of warbonnets were reared at the Vancouver Aquarium, but the FBI will never put the "fingerprints" to use because these patterns change with growth. An adult individual is not recognizable from its appearance in youth; in fact, none of the fish could be traced from one age group to another based on tail photographs since they all changed their patterns continuously as they grew. The species likes to peek its head out of hiding spaces, and its appearance is very cryptic because of the complex branching cirri arising between the eyes and extending back onto the first four or more dorsal fin spines. The body outline is also broken up with vertical dark bars that extend into the fins from the body.

LIFE HISTORY

Decorated warbonnets occur from Northern California to the Bering Sea and live on rocky bottoms at depths of up to 90 m. They grow to 42 cm in length, considerably larger than the mosshead warbonnet. They spawn their egg mass against the inside of their hiding place so that it usually takes on a semicircular appearance, with the guarding male occupying the centre of that curve. This is in contrast to related species, which all wrap their bodies around a more or less spherical egg mass.

MOSSHEAD WARBONNET

Chirolophis nugator COMMON ●

IN THE STRAIT OF GEORGIA

Mosshead warbonnets have been reported by scuba divers throughout the Strait of Georgia. Thus, they are widely distributed and probably common.

DESCRIPTION

The mosshead warbonnet's head is covered with dense, thread-like cirri of comparable size back to the first dorsal fin spine, giving the fish a crewcut look. The dorsal fin has 12 or 13 evenly spaced ocelli (eye-spots), which are black spots with white outlines, tending to disrupt the appearance of the actual eye and the location of the head. The usual interpretation of eye-spots is that they misdirect the lunge of a predator and give the eye-spot bearer a chance to barely escape instead of getting swallowed whole.

LIFE HISTORY

Mosshead warbonnets live in relatively shallow water, from the intertidal zone to about 80 m but usually less than 20 m deep, on rocky shores where there are hiding holes. They grow to about 15 cm long, much smaller than the decorated warbonnet. This species has been found in large tide pools at the lowest tide levels, living in rocky crevices within the pool. Mosshead warbonnets

Mosshead warbonnets have a crewcut of dense cirri on top of the head. Danny Kent

Danny Kent

were originally thought to occur only from Northern California to the Strait of Georgia, but over the last half century additional collections documented the species from Southern California to the end of the Aleutian Island chain in Alaska. The Strait of Georgia is thus at the centre of distribution for this species, not its northern end—but it is a tough animal to see, which adds to the sense of rarity and probably results from the species' successful dedication to blending in and hiding. They mature in their second year, and spawning occurs in the late winter to early spring. The mosshead warbonnet probably has a maximum age of at least five years.

DAUBED SHANNY

Leptoclinus maculatus UNCOMMON ◆

IN THE STRAIT OF GEORGIA

The daubed shanny has been caught in experimental shrimp trawls off Comox, in Howe Sound and at the mouth of the Fraser River. It has also been captured in bottom research trawls in the southern Strait of Georgia. It is not common, likely because it is at the southern limit of its distribution. The trawl catches in the strait were all in areas with sandy bottoms.

DESCRIPTION

The daubed shanny is an elongate fish in the prickleback family. The body is cream to yellow brown with irregular dark blotches that give the species its common name. The lower five to six tips of the pectoral rays are free from the membrane, distinguishing it from related species. The tail fin is completely separate from the dorsal and caudal fins.

Dick Beamish

LIFE HISTORY

Juvenile daubed shannies are pelagic for the first two to four years, when they are about 8 cm long, and then settle on the bottom, where they feed on marine worms and crustaceans. Recorded depths range from 2 to 475 m, but most are at less than 170 m. Females mature at age seven and lengths of 12.5 to 13 cm. Males mature at age six and at about the same size as females. The maximum length is about 20 cm, and the maximum weight is 129 g. A 13.6 cm female was found to be 11 years old, and the maximum age is probably about 12. Spawning occurs in the winter. The daubed shanny occurs in the North Pacific from Puget Sound to the Beaufort Sea and west to the Sea of Okhotsk and the Sea of Japan. It also is found circumpolar in the Arctic to Labrador and south to Cape Cod. It can be abundant in the Arctic, where it can be a major prey for a number of species, including diving birds.

LONGSNOUT PRICKLEBACK

Lumpenella longirostris UNCOMMON ◆

IN THE STRAIT OF GEORGIA

This species has been collected in the Campbell and Nanaimo River estuaries, as well as in Burrard Inlet and Howe Sound. It was one of the more common species in bottom trawl shrimp surveys in the southern Strait of Georgia but probably is generally uncommon in the strait, which is the southern limit of its distribution. Interestingly, it is not reported in the extensive observations of scuba divers.

DESCRIPTION

This is an elongate, bottom-dwelling fish with dark fins. It is bluish-brown dorsally and dark silvery-blue ventrally. The snout projects over the lower jaw and there are scales over all of the head. The dorsal fin spines have free tips and the anal fin usually has three to five spines.

LIFE HISTORY

The longsnout prickleback occurs over soft bottoms at depths from 25 to 1,140 m, but usually at 100 to 200 m. The maximum length is 42 cm, with individuals maturing at 27 cm. Very little is known about its biology, but Russian scientists report that the maximum age is eight years. It is an abundant forage fish in the Bering Sea, and one of the fishes consumed by bearded seals and a variety of other fishes. It occurs from the Strait of Georgia to the Aleutian Islands, across the North Pacific to the Sea of Okhotsk and the Sea of Japan. There is also a record in the Atlantic off the coast of Greenland. The diet includes small bottom organisms such as polychaetes, as well as shrimp, squid and larger species of zooplankton.

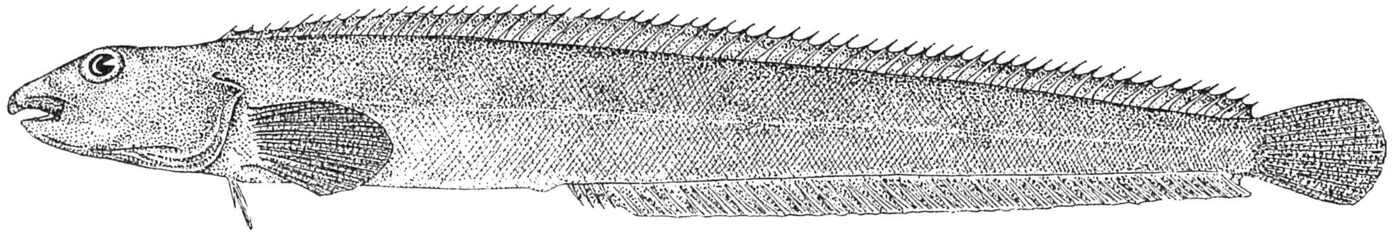

Y-PRICKLEBACK

Lumpenopsis hypochroma RARE ★

IN THE STRAIT OF GEORGIA

There is only a handful of records of the Y-prickleback in the Strait of Georgia, most of them old, and even fewer in British Columbia. The older records are for Departure Bay, Baker Passage in the north and Saanich Inlet in the south. There is a recent observation for Willis Point in Saanich Inlet between about 45 and 50 m by wildlife photographer Andy Murch, who produced the photograph in this book. We have no records from any of the numerous research trawl surveys since the mid-1990s or from any scuba observations. There also are no records of this fish in Puget Sound.

Andy Murch

DESCRIPTION

The Y-prickleback is distinctive and should be easily identified. It is the only prickleback with spots on the body that form a *Y*. There is a long dorsal fin with about 45 spines and a long anal fin with one anterior spine and about 31 rays.

LIFE HISTORY

This fish is reported to occur in southern British Columbia and possibly south to California, but very little is known about it. It is a bottom-dwelling species that may be up to 7.6 cm long and living in water 30 to 100 m deep. This is a fish that scuba divers could look for on ocean bottoms that are sandy or with scattered smaller rocks.

SNAKE PRICKLEBACK

Lumpenus sagitta COMMON ●

IN THE STRAIT OF GEORGIA

The snake prickleback is very common in the Strait of Georgia, occurring in most estuaries. Adults and juveniles are commonly observed by scuba divers throughout the strait, particularly at night, but the fish are absent in the winter, possibly because they move to deeper water. It has been caught in on- and off-bottom research trawls in areas of soft bottom throughout the strait. They are mostly found on soft bottom and often in eelgrass, but can also be found in the water column. Children enjoy fishing for this species and have had success with worms as bait. Juvenile snake pricklebacks have been recorded in abundance off the mouth of the Fraser River in April and May.

Andy Murch

DESCRIPTION

This species has a highly elongated, moderately compressed snake-like body with a small, short head, a small mouth, large eyes and scaleless cheeks. It has one long dorsal fin ending just before the slender, rounded tail. The anal fin is much shorter but similar in height to the dorsal fin. The dorsal fin rays extend beyond the margin of the fin tissue, resembling spikes. The pectoral fins are large and broad and come to a point. Typical body coloration includes light green on the dorsal surface and cream on the ventral surface. Of note are the green- to brown-coloured streaks on the sides of the fish and the brownish-green bars or dots on the dorsal fin and tail. There is a line of dark oval marks along the mid-body.

LIFE HISTORY

The snake prickleback is found from Northern California to the Bering Sea, across the Pacific to the Sea of Okhotsk and the Sea of Japan. It was first recorded in British Columbia in 1882. This species is quite common in the ocean off the coast of British Columbia and has been found from shallow waters to depths of 206 m. It may occur as deep as 425 m but is most common from the intertidal zone to 200 m. Specimens have occasionally been collected from tide pools as well. It reaches a maximum size of 51 cm long, but most are between 5 and 22 cm.

Studies in Alaska have shown that the snake prickleback is a common forage fish in the inshore marine areas. Interestingly, the adults were most commonly found in areas that were most recently glaciated; that is, the substrate was primarily glacial silt. This information is leading researchers to believe this was an important early colonizer of nearshore habitats after the ice age. This would be another excellent species for a student project.

Great blue herons, Caspian terns, double-crested and pelagic cormorants, red-throated loons and pigeon guillemots are known to prey on snake pricklebacks. The snake prickleback consumes amphipods, harpacticoid copepods, clams and other small organisms found on the bottom. Juveniles consume mostly copepods.

RIBBON PRICKLEBACK

Phytichthys chirus RARE ★

IN THE STRAIT OF GEORGIA

There are records of the ribbon prickleback in the Strait of Georgia from Departure Bay, off Texada Island and off the Cowichan River estuary. However, it has not been found in any of the numerous research trawl surveys since the mid-1990s or seen in hundreds of scuba observations. Thus, while it is reported to be common in the San Juan Islands area, it is rare in the Strait of Georgia.

Ribbon pricklebacks have a sort of warpaint striping behind the eye, and the pectoral fin is larger than the eye. Danny Kent

DESCRIPTION

The "warpaint" behind the eye consists of four parallel light and dark bars of rather indistinct outline compared to other prickleback species. The pectoral fins are slightly larger than the eye, and there are four lateral line canals. The body markings vary, but tend to occur in linear arrays, like a line of dark dots halfway down the body or light markings along the posterior dorsal fin, toward the tail, which tends to have a vertical light bar at its base. The body colour grades from an olive green to brown.

LIFE HISTORY

The ribbon prickleback occurs from the east coast of the Bering Sea and along the Aleutian Islands to Southern California from the intertidal zone to 12 m deep. It is most likely to be encountered in the more exposed rocky shoreline areas of the Strait of Georgia that have high relief, meaning lots of ledges, crevices and boulders. It can also be a shallow-water species, most often seen during close inspections of tide pools or under rocks at low tides. It is commonly found on mussel beds in the San Juan Islands area. This fish grows to about 21 cm in length. Spawning occurs from mid-winter to spring, with males and females guarding the nest. The ribbon prickleback, like some of its close relatives, eats small crustaceans, as well as the red and green seaweeds that such small animals often live on.

BLUEBARRED PRICKLEBACK

Plectobranchus evides RARE ★

IN THE STRAIT OF GEORGIA

Until recently, the bluebarred prickleback was rarely observed in the Strait of Georgia. Three specimens were collected in Ladysmith Harbour in 1951, and scientific surveys for shrimp captured several in the Fraser River plume area. More recently, scuba divers have reported them in Saanich Inlet, on the bottom at depths of about 30 m. Amazingly, a remotely operated undersea camera in Saanich Inlet photographed 109 over a period of one year. Footage from remotely operated cameras also contained a number of images from the Sunshine Coast. Also, a recently published fish plankton study reported a number of larval bluebarred pricklebacks in their samples. Thus, although the species has been considered rare, there may be more of them in areas not commonly sampled by traditional methods.

DESCRIPTION

This is a colourful fish. There are at least 25 narrow blue bars projecting dorsally and ventrally, some of which are continuous from top to bottom. It is dark olive-brown dorsally and paler ventrally. The dark spaces between the bars may be about twice the width of the blue bars. There are distinctive white margins on the pectoral, caudal and anal fins. The long dorsal fin has two to three dark blotches posteriorly. The caudal fin is distinctly separated from the dorsal and anal fins.

Andy Murch

LIFE HISTORY

The bluebarred prickleback is a deepwater fish that is found on sandy slopes and soft bottoms from central British Columbia to San Diego, California, at depths from 57 to 368 m. Very little has been reported about the biology of this prickleback species except that the maximum length is about 13 cm.

WHITEBARRED PRICKLEBACK

Poroclinus rothrocki UNCOMMON ◆

IN THE STRAIT OF GEORGIA

The whitebarred prickleback was first recorded in British Columbia in 1929, when a specimen was caught in a shrimp trawl in Burrard Inlet at a minimum depth of 91 m. This is not a common species in the Strait of Georgia, but it is occasionally found throughout the strait in bottom research and shrimp trawls, particularly in the Northern Gulf Islands and around the mouth of the Fraser River. The only observation from an estuary was at the mouth of the Cowichan River. It has not been reported by scuba divers.

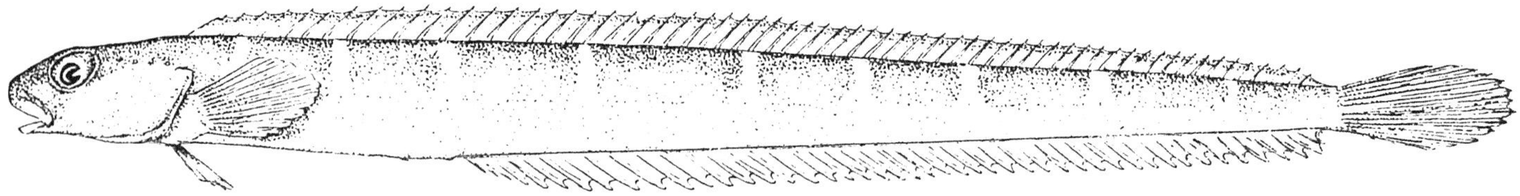

DESCRIPTION

The whitebarred prickleback has a highly elongated, moderately compressed, eel-shaped body. Large, vertical white bars edged with brown pigment on the dorsal surface spread and merge with the ventral colour. It has a small, short head with a small mouth and moderate-sized eyes. Its dorsal fin ends just before the slender, rounded tail. The anal fin is much shorter but similar in height to the dorsal fin. The dorsal and anal fins are not connected to the tail. The anterior of the anal fin has three spines. The dorsal fin rays extend beyond the margin of the fin tissue, resembling spikes. The pectoral fins are large and broad and come to a point. The front of the pelvic fin has a short, sharp spine. The typical body coloration is light brown on the dorsal surface and lighter underneath.

LIFE HISTORY

This species is distributed from the southeast Bering Sea and Alaska to Southern California. It is reported on rocky bottoms at depths from about 35 to 160 m. The maximum length is approximately 25 cm. It is mostly nocturnal and feeds on bottom organisms such as clams, snails and other invertebrates. The species has been identified in the stomachs of walleye pollock, Pacific cod and arrowtooth flounder from the Gulf of Alaska.

BLACK PRICKLEBACK

Xiphister atropurpureus UNCOMMON ◆

Andy Murch

IN THE STRAIT OF GEORGIA

Black pricklebacks are uncommon in the Strait of Georgia. There are reports from Departure Bay, off Saturna Island, Burrard Inlet and Howe Sound. They have been reported from exposed boulder beaches that receive high levels of wave action in Montagu Channel in eastern Howe Sound. We have no records from our various research trawls over the past 25 years. We have one scuba observation from off the Sunshine Coast and records from the Nanaimo and Cowichan River estuaries.

DESCRIPTION

The head on this purple-black fish can be easily recognized by the two distinct, warpaint-like black stripes located behind the eye and outlined with fine white margins. These two large stripes go down from the eye, while a third patterned stripe, much smaller but otherwise identical, points upward from the back of the eye. In addition, if the fish already has its head under the next rock and is wriggling away fast, you can positively identify the species by noting the conspicuous white vertical bar with a black outline (the opposite pattern from the cheek markings) at the base of the tail fin.

LIFE HISTORY

The black prickleback is found from the western Gulf of Alaska to northern Baja California and in depths of up to 12 m. It occurs in exposed rocky areas, under rocks or in algae. It hides under boulders that themselves lie on other boulders or pebbles, so that it can swim around in tight spaces under the rocks without risk of capture by predators or being swept away by wave action. From December to May, a male and female pair will lie together under a rock as the female grows larger and larger with developing eggs. When they finally spawn, the male fertilizes the eggs as the parents coil around them to cause them to stick together in a spherical mass. The male will stay wrapped around to hold the egg mass in place under the shoreline boulder while waves sweep strong currents under the rock and keep the eggs well aerated. After hatching, the larvae spend up to two months in the plankton until all their fins and organs are developed, then they settle into fine gravel on the beach and turn black overnight. They are then miniature versions of the adult form; after creating their dark camouflage pigment, they are measurably smaller in cross-section than they were just hours before while swimming in the water column. Once the settled fish start growing, they gradually choose increasingly large rocks to reside under.

The black prickleback grows to nearly 32 cm in length, maturing at an age of 2 to 3 years and lengths of 12 to 13 cm. Research scientist Milton Love has reported a 26 cm fish that was 12 to 13 years old, so we know they probably live longer. The species' diet includes invertebrates such as crustaceans, polychaetes and snails, as well as algae. There is evidence that the Strait of Georgia black pricklebacks are genetically distinct from other populations.

ROCK PRICKLEBACK

Xiphister mucosus RARE ★

IN THE STRAIT OF GEORGIA

The rock prickleback is rare in the Strait of Georgia. There is a scuba observation in Howe Sound, but no records in any of the several thousand research trawl surveys from the mid-1990s until the present. There also is a report from off Gabriola Island. The species is reported to be commonly found in the San Juan Islands, however.

DESCRIPTION

The rock prickleback is the larger cousin of the black prickleback, and it has yellowish-green blotches that resemble the colour under a little kid's runny nose—hence the species name *mucosus*. The facial markings are the reverse of black pricklebacks, with the eye bands light inside and outlined with dark pigment. The vertical tail band is the same as on their cousin—light with a dark

Andy Murch

outline—but rather than white and black, it is a light, mucus-like colour inside and dark olive outside. The marbled colour of this fish is most pronounced toward the tail end. There is no pelvic fin or anal spine. The dorsal fin originates above the very small pectoral fin.

LIFE HISTORY

This species is found from the western Gulf of Alaska to Southern California in the intertidal zone to 18 m deep. It is a big species, up to 58 cm long, maturing at 20 to 30 cm and an age of four to five years. The maximum age is at least 11. It can breathe air for nearly a full day, so is well adapted to its intertidal life. Spawning occurs from late October to May. The spawning habits are very similar to those of the black prickleback, except that the females lay much larger egg masses. In both species, a nesting male will spawn repeatedly with different females, guarding two or even three egg masses on occasion, but the egg masses will always be in different stages of development because of the courtship required to stimulate egg-laying in each successive female.

Rock prickleback become herbivorous as adults, eating mostly green and red seaweeds, but on occasion they will eat small crustaceans that live on the seaweeds. Those crustaceans are the main diet of younger, smaller juveniles. The boulders that these fish choose to live under lie on cobbles and pebbles, providing spaces that match their body size. The rock prickleback is capable of wrestling its way through cobbles, moving them about. The lateral flexure capability of these big eel-like "blennies" is too great for humans to contain the fish in their hands; these fish are hard to pick up unless they are quickly tossed into a bucket filled with seaweeds that the fish will relax in and hide under. Otherwise, they tend to get away, but it is always good beach etiquette to return boulders to where they were found and put the fish back where they can hide after you have had a good chance to examine and identify them. The rock prickleback was eaten traditionally by First Nations people and more recently has been marketed live in Asian communities.

DWARF WRYMOUTH

Cryptacanthodes aleutensis

UNCOMMON ◆

IN THE STRAIT OF GEORGIA

The dwarf wrymouth is found in research trawls and in the bycatch from shrimp trawls throughout the Strait of Georgia but is uncommon. In *Pacific Fishes of Canada*, J.L. Hart reports that it was found in the southern strait, but more recent research catch records show it is found occasionally throughout the strait in flatter, soft-bottom sediments. It is more common in the waters around the San Juan Islands.

DESCRIPTION

The species is bright red or pink as a consequence of the loose and transparent skin that allows the blood colour to be seen. It is often buried in the bottom at depths from 28 to 350 m. It reaches a maximum length of 31 cm. The eyes are positioned on top of the head and the mouth is oblique to almost vertical, clearly an adaptation to hiding in the mud and waiting for prey to come too close. The dwarf wrymouth does not have pelvic fins, and unlike the giant wrymouth, it does not have scales.

Lindsay Dealy

LIFE HISTORY

Dwarf wrymouths are found from the southeastern Bering Sea and the Aleutian Islands to Northern California. The dwarf and giant wrymouth species in British Columbia probably spawn in the winter to summer, as indicated by larval stages observed near the surface. These larval fish settle on the bottom when they reach about 3 cm in length. Very little is known about the biology of the dwarf wrymouth, though it is known to be eaten by skate. This is another species that needs some basic biological information and would be a good study for a student considering a career in marine biology.

GIANT WRYMOUTH

Cryptacanthodes giganteus

RARE ★

IN THE STRAIT OF GEORGIA

The giant wrymouth is rarely observed in the Strait of Georgia and seen much less often than the dwarf wrymouth. The giant wrymouth is reported in *Pacific Fishes of Canada* to be in the strait off Nanaimo, Comox and Victoria, and in English Bay and Pender Harbour. Several specimens have recently been collected in research trawls in the southern strait, and there is a photograph online of a specimen on the bottom at about 12 m deep off Hornby Island.

DESCRIPTION

The giant wrymouth has a highly elongate body, with one dorsal fin beginning just posterior of the head and merging with the small, fan-like caudal fin. The anal fin begins posterior to the anus, which is more than halfway down the abdomen from the head, and it too merges with the caudal fin. The pectoral fins are small and pelvic fins are absent. The giant wrymouth has an eel-shaped body, similar to that of gunnels but much longer. The head is large and compressed

Danny Kent

with a strongly upturned mouth and relatively small eyes. The mouth is outfitted with small, sharp teeth and thick lips. The overall coloration is pale brown with yellow and purple tinges. The ventral side of the fish is similar, but coloration is normally lighter. This species may exhibit two different colour phases. The first phase is a drab beige colour with a dark band along the lateral line; the second is much more colourful, with dark bands above and below the lateral line, a pale area in between, and a dark stripe along the base of the dorsal fin. It is distinguished from the dwarf wrymouth by its larger size, pale-brown colour and scales on the posterior body.

Danny Kent

LIFE HISTORY

The giant wrymouth occurs from the southeastern Bering Sea to Northern California. It is an eel-like fish that is found on soft bottoms, where it is buried in the mud or sand, often in tunnels. The giant wrymouth is found at depths from 6 to 128 m, but usually shallower than 20 m. The maximum length is 123 cm. It is not a commercially valuable fish in British Columbia, but in 1998 the Oregon Department of Fisheries and Wildlife reported that 652 kg of giant wrymouths had an ex-vessel value of $236 USD. Trawl surveys in Alaska report catching giant wrymouths in modest amounts.

The giant wrymouth feeds on invertebrates, and mostly at night. The dwarf and giant wrymouths probably spawn in the winter to summer, as indicated by larval stages observed near the surface. These larval fish settle on the bottom when they reach about 3 cm in length. There is a report of the giant wrymouth being in the diet of harbour seals feeding in the San Juan Islands. This is a species that needs more study to identify diets, behaviour and ages at maturity.

PENPOINT GUNNEL

Apodichthys flavidus COMMON ●

IN THE STRAIT OF GEORGIA

The penpoint gunnel is most common in tide pools and intertidal areas in the Strait of Georgia. It is only rarely caught in other areas, but is widely distributed within the strait and frequently seen by scuba divers.

DESCRIPTION

Like other gunnel species, the penpoint gunnel is elongate and eel-like in shape. It has a small head with small eyes and a small, upturned mouth with small teeth and thick lips. It has a long dorsal fin extending from behind the head to the end of the body, where it merges with the small rounded tail. The anal fin extends from behind the anal opening to the end of the body and also merges with the tail. Pelvic fins are absent, and the anal fin spine is large and grooved, resulting in the common name "penpoint." The maximum length for this species is approximately 46 cm. The dorsal and ventral surfaces of the fish are almost identical in colour. Normal body colours can include green, yellow, brown or red, but the colour is highly variable depending on the individual's

Penpoint gunnel are often a bright green, as in this photo, but different species can have similar colours. Aaron Baldwin

diet and habitat. A line of dots is present along the side of the fish and can vary in colour but is normally lighter than the body colour. A distinguishing feature of this species is the narrow, dark-coloured line typically extending from the corner of the eye back and downward, ending near the operculum.

There has been discussion about body colour, with authors presuming that the pigment comes from eating small animals that in turn have been feeding on seaweeds. In 2013 this notion was put to rest when a large number of penpoint gunnel were reared to settlement at the Vancouver Aquarium. Subsequently, they were housed in two tanks with seaweed habitats of green or red or brown. Within a couple of months the fish had come to match their habitat seaweeds, all of them having received identical diets.

LIFE HISTORY

Penpoint gunnel are sometimes referred to as "blennies." In general, they are small fish with long, slender, eel-like bodies, small mouths and teeth, and short fins. The penpoint gunnel is one of the largest gunnels. It is not a commercial species, but is an important ornamental fish for private and public aquariums. The distribution of this species is from Kodiak, Alaska, and throughout coastal British Columbia to Southern California. The penpoint gunnel has been shown to be present in subtidal eelgrass beds during the summer but not once the water temperatures drop in the winter, coinciding with the seasonal disappearance of eelgrass. Additionally, the species has been shown to be more active at dawn and dusk.

In British Columbia, penpoint gunnel spawn from January to March, when breeding pairs have been observed under rocks in the intertidal zone. The female lays a single cluster of eggs that will normally be tended by both the female and male until they hatch, after approximately two and a half months. Larvae are about 1.3 cm long when they hatch, with little yolk sac remaining. After a month and a half of development, they reach 2.5 cm in length and settle as juveniles. Underwater observations have led researchers to believe that growth is rapid in the first year of life; fish were reported at between 2 and 4 cm in April to May, and between 10 and 12 cm by the end of summer. At age six, they are approximately 30.5 cm long.

This is an abundant forage fish and an integral component of the diet of many marine bird and groundfish species. The penpoint gunnel consumes epibenthic and benthic organisms, including amphipods, crab larvae and polychaetes. In Juan de Fuca Strait and northern Puget Sound, known predators include a variety of birds.

ROCKWEED GUNNEL

Apodichthys fucorum

RARE ★

IN THE STRAIT OF GEORGIA

The rockweed gunnel is rare in the Strait of Georgia. We have only one scuba observation from Howe Sound. In *Pacific Fishes of Canada*, J.L. Hart reports observations from off Comox and Courtenay.

DESCRIPTION

The rockweed gunnel is a small cousin of the much larger penpoint gunnel, and both species have the marked tendency to match the seaweeds that they live in—whether brown, green or red—so they might initially be hard to tell apart, aside from an obvious difference in body size. On closer inspection, though, the rockweed gunnel can be distinguished easily by noting the very small size of its pectoral fin, which is no bigger than its eye. The penpoint gunnel has a more normal, relatively big pectoral fin, as well as more prominent markings.

Jeremiah Degenhardt

LIFE HISTORY

These fish tolerate outer estuaries, where salinity is low. They occupy shallow shoreline habitats and are found by intertidal explorers who look under rocks or search among seaweeds. They wrap their bodies around seaweeds when the tide is in, but often hide under rocks at low water, when the tide is out. Rockweed gunnel eat the small crustaceans that feed on seaweeds, which is where assumptions about the origin of their body colours arose. Once a fish develops a colour that matches one type of seaweed, it will actively seek to remain amid that type and colour, thus reducing the chances that a shorebird like a great blue heron might spot it and eat it.

Spawning takes place in spring. The eggs adhere to each other in a spherical mass that the male parent will wrap around under a rock. A parent is required to actively keep the egg mass from being swept out from this nesting area. The egg mass is tended for over a month until hatching occurs, at which time the transparent little eel-shaped larvae swim out from under the rock and up toward the surface, where they will remain for up to a couple of months, feeding on tiny planktonic crustaceans. During this time they gradually develop all their fin rays, but they remain transparent until the time of settlement, after which they develop an adaptive body colour.

The rockweed gunnel occurs from central Baja California to northern British Columbia and reaches 23 cm in length.

LONGFIN GUNNEL

Pholis clemensi UNCOMMON ◆

IN THE STRAIT OF GEORGIA

Longfin gunnel are commonly seen by scuba divers throughout the Strait of Georgia. It also occurs in catches of bottom research trawl surveys in shallower areas, which may indicate it is a subtidal species, living deeper than the other gunnel species in the strait. It is captured in Pacific herring surveys, which are pelagic trawls at night. This species has not been documented in any of the estuaries around the strait. It is caught occasionally in vertical fishing plankton nets at night. Catches at the surface at night probably indicate that migration to the surface is a feeding response.

DESCRIPTION

This is the only gunnel species whose body up to the beginning of the tail is shorter than its tail section, which is to say that the anal fin is long (over 50 percent of the fish's total length). This is a character usually used to distinguish the prickleback family (short gut area) from the gunnel family (long gut area). A simpler and more immediate method to recognize the longfin gunnel is to examine the body markings: they always involve dark dots, usually against a reddish background. There is a series of dark marks evenly spaced along the base of the dorsal fin. Each has a light patch at the rear side, with dark dots inside each light patch, as well as dark dots along the lateral line, each inside a pale blotch. This fish frequently appears to be pink with purple polka dots, an appearance that can lead observers to consider whether they have been doing too much partying. Longfin gunnel have large pectoral fins, very small pelvic fins and an indistinct lateral line.

Danny Kent

LIFE HISTORY

Longfin gunnel occur from southeast Alaska to Point Lobos in Central California. They are found at depths of 6 to 64 m, but mostly from 7 to 18 m, under seaweed in shallow areas with rocky and sandy bottoms. Longfin gunnel are also found in intertidal area pools, where they are capable of breathing air. Very little is known about their life history, but a maximum length of 13 cm has been reported. They spawn in the winter, with males guarding the nest, which can be under rocks or shells. Prey includes crustaceans and worms. The species name *clemensi* recognizes W.A. Clemens, who was an ichthyologist and the director of the Pacific Biological Station from 1925 to 1940.

CRESCENT GUNNEL

Pholis laeta COMMON ●

IN THE STRAIT OF GEORGIA

Crescent gunnel are most common in intertidal areas, including tide pools, where they are commonly found under rocks at low tide. They are widely distributed around the Strait of Georgia and can also be found in estuaries. They are seldom found at intermediate depths.

DESCRIPTION

Like other gunnel species, the crescent gunnel is elongate and eel-like in shape. Its head is small, as are its eyes, upturned mouth, and teeth, and it has thick lips. It has a long dorsal fin extending from behind the head to the end of the body, where it tapers to the small, rounded tail. The anal fin extends from behind the anal opening to the end of the body, also tapering to the tail. The

Andy Murch

entire body is a yellow-green colour, with the dorsal side darker than the rest of the body. It is named the crescent gunnel for the crescent-shaped colour pattern along the back, near the dorsal fin. These crescents are yellow to orange with dark outlines. The head is ornately coloured as well, with light patches outlined in black above and below the eye, and a green spot on the operculum.

LIFE HISTORY

The crescent gunnel is sometimes referred to as the bracketed blenny. It is commonly distributed from Kamchatka, Russia, to the Bering Sea, along the Aleutian Islands and south to California. Its maximum size is approximately 25 cm. It spawns in the winter, probably mid-January, in intertidal and subtidal rocky habitats. Between the months of November and April, the species can be found in pairs under rocks. It is believed this is the courtship period before and during spawning. Immature crescent gunnel have been observed under the same rock that sexually active gunnel are found under. Adhesive eggs are tended by both males and females, who curl around the egg mass with their bodies. Eggs hatch in about two months and the larvae are pelagic. Crescent gunnel attain sexual maturity at approximately 10 cm long and have a maximum life span of about six years. Males in reproductive condition are an orange-red colour along the cheeks, throat, pectoral fins, anal fin and ventral side of the fish. This is in direct contrast to females and immature crescent gunnel, which are pale green to cream coloured in the same body regions. During the summer, males are indistinguishable from other life stages of the crescent gunnel.

In Juan de Fuca Strait and northern Puget Sound, known predators include the double-crested cormorant, red-necked grebe, common merganser, Arctic loon, common loon, red-throated loon, pelagic cormorant, pigeon guillemot, red-breasted merganser and Caspian tern. Other predators include mink, river otter and subtidal fish species. The crescent gunnel consumes epibenthic and benthic organisms including amphipods, crab larvae and polychaetes, crustaceans and molluscs. Like other gunnel species, it has been shown to be more active at dawn and dusk, which corresponds to when the greatest quantity of food is found in the gut.

SADDLEBACK GUNNEL

Pholis ornata

COMMON ●

IN THE STRAIT OF GEORGIA

The saddleback gunnel is widely distributed in shallow areas throughout the Strait of Georgia and can tolerate low-salinity water. It is reported by scuba divers throughout the strait. This is another intertidal species, but it is more common in areas near the outlets of rivers, where it prefers muddy substrate. We have records of catches in research trawls on bottom at intermediate depths, but this is not common. Saddleback gunnel were first recorded in British Columbia waters in 1861.

In contrast to the crescent gunnel, the saddleback gunnel has a series of open V-shaped marks along the dorsal fin, which are dark with a light outline that does not close at the top.
Andrew Howe

DESCRIPTION

The saddleback gunnel is elongate and eel-like in shape. As with other gunnels, its upturned head is small, as are its eyes, and its upturned mouth has thick lips and small teeth. The saddleback gunnel has a long dorsal fin extending from behind the head to the end of the body where it tapers to the small rounded tail. The anal fin extends from behind the anal opening to the end of the body, where it too tapers to the tail. The saddleback gunnel is very colourful and is often misidentified as a crescent gunnel. The dorsal side of the body is olive green to brown, while the underside is yellow or orange. It is named the saddleback gunnel for the green or brown V- or U-shaped pigment pattern, outlined in white, that is visible on the dorsal fin and dorsal margin. A series of dark bars is found along the side of the fish. The head of the saddleback gunnel has a distinct colour pattern: a dark line extends vertically through the eye toward the chin, and two dark lines stretch vertically to the top of the head.

LIFE HISTORY

This is an intertidal, nearshore species found from Vancouver Island to California. It is present in coastal areas of British Columbia and has been found at depths down to 60 m. Scuba divers find these fish under solid structures or in dens during the daytime. However, in the evening the fish tend to venture out where they can be seen by a patient diver. The maximum length of this species is 30.5 cm. Saddleback gunnel spawn in the winter, but little is known about most of their biology and ecology. This is another species that needs the attention of a student looking for a good project. Like other gunnels, the saddleback gunnel consumes epibenthic and benthic organisms including amphipods, isopods and polychaetes. Coho salmon are known to prey on them.

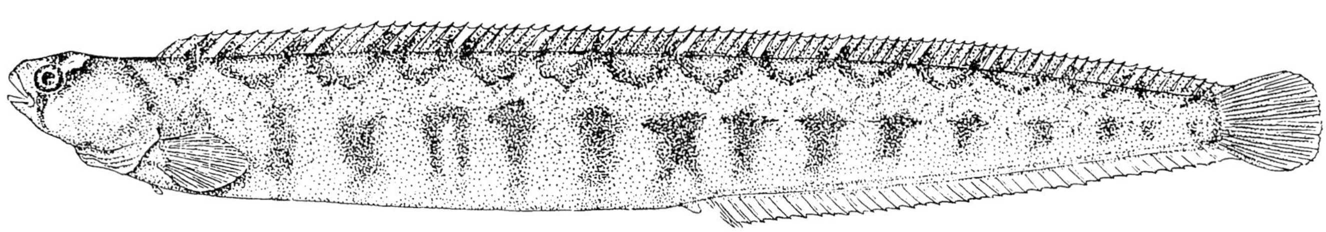

Wolffishes, Quillfishes, Prowfishes and Sandfishes

Wolffishes, quillfishes, prowfishes and sandfishes are each represented by a single species in the Strait of Georgia. Two of these four species—the quillfish and the prowfish—are in families comprising only one species, and one, the sandfish, is in a family of only two species. Quillfishes are in the family Ptilichthyidae. They are found in the North Pacific from the Bering Sea to Oregon. They are very long and thin, but are poorly studied. Prowfishes, in the family Zaproridae, occur in the North Pacific from Japan to California. They have a robust body and head and no pelvic fins. The distinctive head has a convex shape and a projecting snout, resulting in the name "prowfish." Sandfishes, in the family Trichodontidae, are found in the North Pacific. These are also very distinctive with their laterally compressed, deep, scaleless body and an upturned mouth. Wolffishes, in the family Anarhichadidae, have five or six species distributed in the North Pacific and North Atlantic Oceans. These are robust fishes with strong canine and molar teeth and no pelvic fins. They are found on bottom at depths down to 600 m, and grow up to 2.5 m long.

WOLF-EEL

Anarrhichthys ocellatus COMMON ●

IN THE STRAIT OF GEORGIA

Wolf-eels are commonly found throughout the Strait of Georgia. Studies that focus on their distribution, preferred habitats and general ecology are badly needed. Larval or juvenile wolf-eels can be a common species in the surface waters in the strait in the late summer.

DESCRIPTION

This species will not be confused with any other in the Strait of Georgia. Wolf-eels are readily distinguished by their elongate body and the numerous dark spots, called ocelli, surrounded by light pigment on the body. Adults are dark blue, grey and brown, and juveniles are an attractive orange to gold with tones of brown and black. The skin appears scaleless, but small scales are buried

Andy Murch

Jon Bryan

deep in the skin, and there is no lateral line. The skull is well developed with powerful canine teeth and strong molars in the back of the mouth that crush food before it is swallowed.

LIFE HISTORY

Wolf-eels are called eels because of their elongated body, not because they are true eels. They occur from the southeastern Bering Sea, along the Aleutian Islands and Gulf of Alaska to Southern California to depths of 225 m. The maximum length is 240 cm. Wolf-eels are frequently seen in commercial aquariums and by scuba divers. Despite their threatening appearance, they are friendly fish—possibly even too friendly if they are offered food by scuba divers. They occur in rocky areas, where they occupy crevices or caves, with most of their elongated body remaining in their den. They are occasionally caught in trawls or traps, but there is no commercial fishery for them. Consequently, there is very little information on their biology. It seems a shame to kill these magnificent fish to obtain information on their life history; however, fish that are killed in the bycatch of commercial species could be accumulated so that biologists could produce the basic information needed to understand their population dynamics.

There are no available age and growth studies and thus the maximum age is unknown, but it could be over 20 years. Research at the Vancouver Aquarium has made a significant contribution to our understanding of their reproductive biology. As part of these studies, specimens collected in the Strait of Georgia in 1995 were still spawning in 2011 and looked young to the biologists. The research indicated that at about age four and lengths of 100 cm the males and females are sexually mature. Males and females apparently mate and maintain a monogamous relationship for life. Detailed long-term monitoring by scientists at the Washington Department of Fish and Wildlife reveals that whenever a mate is lost, another will quickly be found to replace it. Therefore, it is more like serial monogamy without any long period of single widowed status. There are both den and mate fidelities, but the duration of the fidelity appears to vary. There is a report of wolf-eels at a popular dive location remaining in the same den for up to 25 years. Sailfin sculpin have been observed in and around the dens of wolf-eels, possibly indicating a sharing of resources. It also is not uncommon to observe the Pacific giant octopus in an occupied wolf-eel den. Lingcod and rockfish have also been observed near and even inside occupied dens.

Spawning occurs as early as October and as late as April, but most spawning may be in winter. The self-adhesive eggs are guarded by the female wrapping her body around the egg mass and the male then wrapping his body around the female. Eggs hatch in 13 to 16 weeks. During the incubation period, the female uses her body to rotate the egg mass. She also attends to the egg mass by removing dead eggs. Just prior to hatching, the female begins to fan the egg mass with her body. The young wolf-eels maintain a planktonic existence up to an age of about three months or a length of 15 to 30 cm, after which they settle to the bottom. Between four and five months old, or lengths of 25 to 40 cm, the juveniles start to find shelter. After about one year, the juveniles begin to take on the adult coloration and their molars develop. Prior to one year, the juveniles depend on their canine teeth to capture prey.

Mature fish will aggressively defend their territory against other wolf-eels, though they seem to enjoy the company of humans. The large, fleshy area on the head and fleshy jaws may be damaged in fights between males. Bite scars on the bodies are common in larger male wolf-eels in the wild. Wolf-eels are sluggish swimmers and have predominantly sedentary behaviour. They feed on other relatively slow-moving prey, such as sea urchins, clams, mussels, snails, crabs and sand dollars. Their abundance may be limited by available habitat as the monogamous pairs are both large and territorial. Taking wolf-eels in Puget Sound is prohibited; it should also be prohibited in the Strait of Georgia.

QUILLFISH

Ptilichthys goodei UNCOMMON ◆

IN THE STRAIT OF GEORGIA

We have captured several hundred quillfish in our pelagic research trawls from 1995 to the present in the open waters of the Strait of Georgia. There are scuba diver records from off Gabriola Island, in Howe Sound and in Burrard Inlet, including Indian Arm. There also is a very old record from off Porlier Pass. There is no doubt that this fish is uncommon in the Strait of Georgia, but once seen, it is unlikely the observer will ever forget the sighting.

DESCRIPTION

The quillfish has isolated, curved spines along two-fifths of its body length behind the head, followed by soft dorsal rays to the point at the rear end of the body where the dorsal fin joins the anal fin at a thread of a tail fin. That thread is frequently broken short or absent. The small head ends in a jaw that is longer on the lower jawbone, ending in a tip that protrudes and often has a downward, fleshy appendage. Adults have a dark racing stripe lengthwise along the side of the body.

Quillfish have a very elongate body with horizontal stripes, and their swimming is characterized by undulating, snake-like movements. Lee Newman

LIFE HISTORY

The quillfish is unique to its monotypic (one-species) family. It is unmistakable for its extremely long, thin body and snake-like swimming undulations. Scuba divers in the Strait of Georgia tend to see quillfish in shallow water near soft sediments. It is interesting that some major fish compendiums report a belief that these fish bury themselves in mud during the day, but no references were found to verify it. This species has not been kept successfully to adulthood in public aquariums, preventing confirmation of the burying reports. It is usually the pelagic juveniles that are seen near the sea surface or near the bottom in shallow water. Juvenile quillfish swim in surface waters at night. These fish are attracted to lights and will approach the side of a dock where people are using night lights to witness plankton that often includes larval or juvenile fishes. This is one species that is really easy to identify once you know about them. They occur from the Seas of Japan and Okhotsk across to the Bering Sea and south to central Oregon and grow to 39 cm in length.

PROWFISH

Zaprora silenus RARE ★

IN THE STRAIT OF GEORGIA

There is one old record of a prowfish caught off Nanaimo in 1895, and a second from the groundfish research database at the Pacific Biological Station that was caught in a bottom trawl off the mouth of the Fraser River in 1979. We have no observations from any research trawl studies or scuba divers in the Strait of Georgia, except one unconfirmed scuba observation from Saanich Inlet. Obviously, the species is very rare. We include it in our list of species as an indication of the importance of reporting any observations of it.

DESCRIPTION

The prowfish's body colour can be grey, green or brown, often with blotches of bright yellow, especially near the head. The head pores are obvious, in part because of their white interior colour and a ring of white, then black, around them, as well as their large size. This species has no lateral line and no pelvic fins.

LIFE HISTORY

The prowfish is a big, heavy fish with solidly built fins and huge pores on its blunt head surrounding the jaws and eyes. This is a crevice- or cave-dwelling fish that swims out from its hiding places into the water column to eat sea jellies (jellyfish). It may use its head pores as a sense of distant touch in the dark

Catherine Mecklenburg

to detect movement of potential prey items, like the pulsing of jellies, to locate them and bite chunks off. It has sharp little teeth around the edge of the jaw for slicing pieces of jellies. Prowfish tend to emerge with a fearless attitude from their daytime hiding places to inspect scuba divers. Unfortunately, most captive specimens have come in as bycatch from trawl nets, and while these tough fish seem to be healthy upon capture, they usually have internal injuries and require veterinary care that is typically not successful in the long term. Juvenile prowfish take shelter among the tentacles of large jellies. Prowfish occur across the North Pacific from Hokkaido, Japan, to the Sea of Okhotsk, the Bering Sea and Southern California. The prowfish is in a monotypic family, meaning it is a unique species at this time in the Earth's history, but it had relatives that lived around five million years ago.

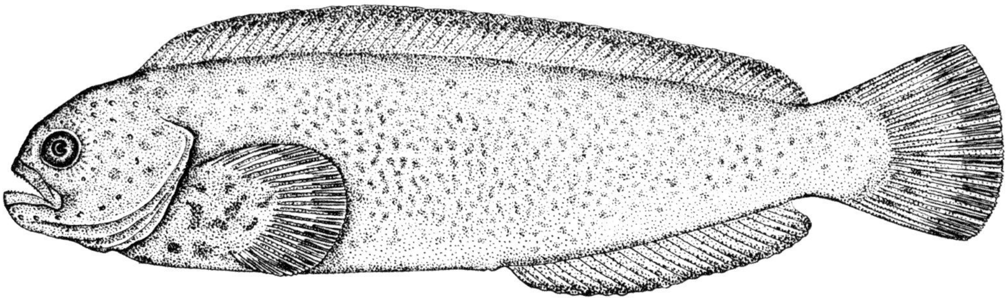

PACIFIC SANDFISH

Trichodon trichodon UNCOMMON ◆

IN THE STRAIT OF GEORGIA

Pacific sandfish are uncommon in the Strait of Georgia. They are occasionally caught in both pelagic and bottom trawl surveys throughout the strait, but mostly in its southern areas. From 1998 to 2010, 44 were caught in pelagic trawl surveys. They have been caught with recreational fishing gear in the Gulf Islands. It is possible that their migration into the Strait of Georgia is increasing.

DESCRIPTION

The Pacific sandfish will not be confused with any other species in the Strait of Georgia. It is distinguished by its upturned jaw, which is surrounded by fringed lips, and by the large eyes at the top of its head. The silvery body is dark brown dorsally with dark spots or short bars and is tapered vertically like a wedge, thick along the back and narrow at the base of the belly. It does not have scales. The two dorsal fins are high and the tail fin is forked, indicating its capability for rapid swimming in the water column, where it preys on pelagic fish and krill. The large pectoral fins that curve out at the edge are used to backpaddle while the fish is wriggling down into sand, where it buries itself so that only the eyes, jaw tip and dorsal fins are visible. The fleshy fringe around the jaw probably enables them to inhale water through a nearly closed jaw without getting sand inside the gills.

LIFE HISTORY

These fish occur around the North Pacific from northern Japan and off the coast of Russia to the Bering Sea, Aleutian Islands and south to Central California. They are abundant in the nearshore areas off southeastern Alaska. The name "sandfish" relates to the fish's behaviour of burrowing in the sand or mud, leaving only its eyes, the top of its head and perhaps the top of its body and dorsal fin exposed.

Danny Kent

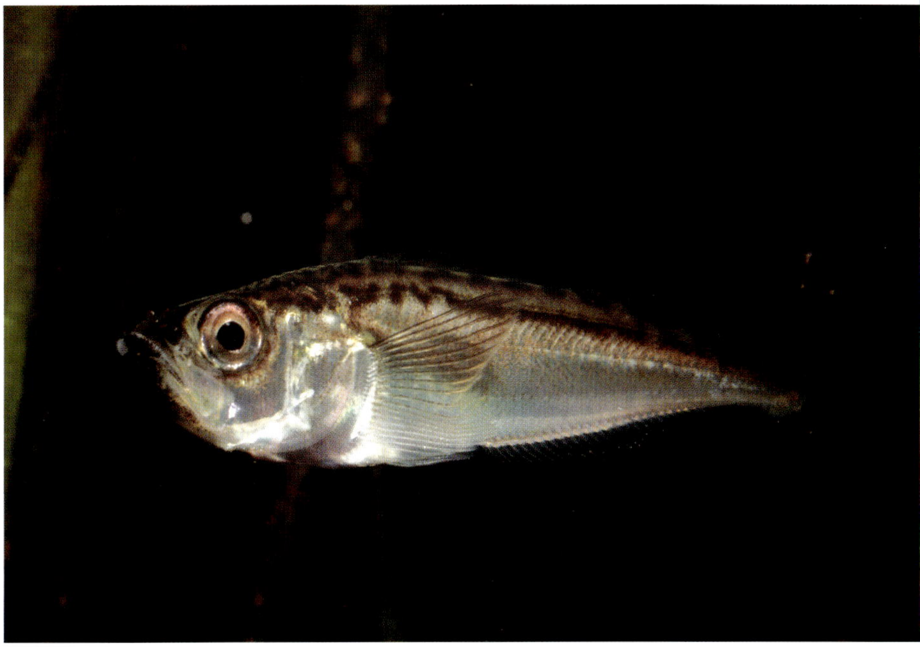

Individuals bury themselves in the sand only after they have metamorphosed into miniature versions of the adult form, at about 5 to 6 cm in length. The fleshy fringes along the edge of the mouth may help ensure sand and mud are excluded. Pacific sandfish are reported to be ambush predators when they are buried in the sand, but larger fish also occur in schools in the midwater, where they use their needle-like teeth to catch a variety of other fishes, such as small cods, sand lance and Pacific herring. Smaller individuals, less than about 10 cm long or in their first two years of life, feed mainly on invertebrates such as shrimp, euphausiids and crab larvae. The maximum age of nine years is probably a reliable estimate, though a relatively small number of fish have been aged. One-year-old fish average about 7 cm in length, and two-year-old fish about 10 cm. Most adults seem to be age four to six and about 14 to 23.5 cm long. The maximum size found in a recent study was 33.3 cm for a nine-year-old fish. The Pacific sandfish occurs intertidally to 375 m, but is most abundant in shallower areas no deeper than 100 m. The average age at maturity has not been determined, but females grow larger than males.

Bill Pennell

This fish spawns in rock crevices along wave-swept beaches and produces batches of up to 15,000 eggs in masses that are attached to rocks or seaweed. The eggs incubate for a full year before hatching into larvae with advanced developmental features like fully formed caudal fin rays that enable rapid swimming. Their larvae are so rare in samples from the open North Pacific as to lend to the argument that these larvae are typically non-planktonic, inhabiting the extreme nearshore areas with the help of their well-developed swimming capabilities. After hatching, the juveniles develop their adult appearance in two to three months, which is also when they begin their burying behaviour. This species swims in schools right at hatching and is capable of cruising continuously at about ten body lengths per second—something that fish physiologists have determined that salmon species can achieve only in brief bursts. Pacific sandfish tend to school among pink salmon fry along cliff faces in southeast Alaska. This is an important prey for a number of other fishes and birds in areas where it is abundant.

and Clingfishes

Sand lances, ragfishes and clingfishes are each represented by a single species in the Strait of Georgia. The ragfish is the only species in the family Icosteidae and occurs only in the North Pacific. This is a distinctive species because of the limp, flabby body that is mostly cartilage. It is another poorly studied species and has a larval form very different from the adult. Sand lances, in the family Ammodytidae, include about 23 species and are pretty much worldwide. Sand lances are known to occur in very large schools and as prey for a number of species. Clingfishes are in the family Gobiesocidae, in which new species are being described regularly. Currently there are 150 to 170 species that are found worldwide in marine and some in fresh water. These are distinctive tadpole-shaped fishes that are scaleless, have a ventral adhesive disc and are found from the intertidal zone to 500 m.

PACIFIC SAND LANCE

Ammodytes hexapterus COMMON ●

IN THE STRAIT OF GEORGIA

The Pacific sand lance is a common species in the Strait of Georgia. It is one of the species caught most often in research trawls that are fished at the surface in the strait. Our study catches have occurred throughout the strait and are relatively small, in part because the shape of the fish allows them to escape through the mesh of the net. However, on some occasions, we may catch several thousand in one set, indicating that large schools occur throughout the strait.

Jackson W.F. Chu

DESCRIPTION

This is an eel-shaped fish that is metallic blue or green dorsally and silvery ventrally. There are long, low dorsal and anal fins, no pelvic fins, a prominent caudal peduncle, a deeply forked tail, a high lateral line canal and a projecting lower jaw. There is a lateral skin fold ventrally along the length of the body.

LIFE HISTORY

Pacific sand lances grow to about 28 cm long and occur from the intertidal zone to about 100 m. The genus name *Ammodytes* means "sand diver" and relates to the fish's habit of diving into the sandy bottom, where it may remain during the day with only its head protruding. Pacific sand lance can occur in large schools. In some areas offshore, these schools contain literally hundreds of thousands of fish. The otoliths are used for age determination and are relatively easy to interpret, probably because there is no feeding and growth during the winter. The maximum age appears to be seven years and the maximum length 28 cm. However, fish up to one year old are most common, and fish older than three years or larger than 20 cm are rare. Pacific sand lances mature in their second year.

In Puget Sound, there are accounts of spawning behaviour at depths of 25 to 100 m in areas with strong currents. Spawning occurs in dense formations from early November to about mid-February. Females burrow into the sand or fine gravel when spawning intertidally, but fertilized eggs may be exposed to air. Females release eggs in one batch within dense spawning aggregations. Despite the importance of sand lances and their spawning behaviour, which allows them to be detected in the intertidal areas, we do not know where the spawning areas are in the Strait of Georgia. Locating these areas would be a significant contribution.

Studies show that Pacific sand lance eggs are about 1 mm in diameter and adhere to the bottom. They hatch in about two to four weeks, depending on the water temperature. Larval fish remain in the sand until the yolk sac is absorbed. The larvae are about 5 mm long when they hatch and become free-swimming at 1 cm. Larvae of 2.5 cm in length have been observed in the early summer off the mouth of the Fraser River. After metamorphosis, which occurs at lengths of 3 to 4 cm, the young fish resemble adults. There are large fluctuations in recruitment, an indication that their desired prey may experience similar fluctuations in abundance.

Areas with flat, sandy bottoms and strong tidal currents are the preferred habitat. Pacific sand lances appear in the Strait of Georgia from the spring to late summer, and accounts of them in the late fall and winter are rare, which may confirm that winter is a period of dormancy. If this is true, it should be possible to find the burrowed individuals. Copepods are the major prey, but they will also consume euphausiids, crustaceans, and herring and eulachon larvae. A study in 1967 found that larval Pacific sand lances were 6 mm to 3.7 cm in length and adults were 8 to 10.5 cm. In early June, larvae were 1.3 to 2.6 cm in length and metamorphosed individuals were 2.7 to 4.4 cm. Adults ranged from 6.4 to 11.5 cm. In early July, there were no larvae caught, but the adults were 6.8 to 11.5 cm.

Adults feed at dusk and dawn, schooling during the day and burrowing in the sandy bottom at night. The absence of a swim bladder facilitates the burrowing. The availability of the preferred bottom type is believed to limit their abundance. Whether Pacific sand lances have a dormant period, just prior to spawning, is currently speculation regarding the fish in the Strait of Georgia, as the species' biology is poorly understood. This lack of understanding of their natural history is almost inexcusable, as sand lances are a key item in the food chain. In fact, around the world, they are a common prey of over 100 species, including 40 species of birds. Because studies of the population ecology of this important species are so limited in the Strait of Georgia, we do not know if the abundances are increasing or decreasing.

RAGFISH

Icosteus aenigmaticus VISITOR ▼

IN THE STRAIT OF GEORGIA

There are 23 reports of ragfish from the Strait of Georgia in the groundfish research database at the Pacific Biological Station. There also were catches from the Pacific hake fishery in the central strait in December 2016, February 2018 and January 2020. It would be very difficult to misidentify ragfish because of the shape and colour of juveniles and the limp body and appearance of larger fish, so we consider the reports in the database to be valid. We hope that future catches will be retained for museum collections. The database also includes one midwater research trawl catch in the northern strait in February 1999. There were catches from bottom trawls in May and December 2015 in the southern strait. In general, ragfish would be a rare visitor to the Strait of Georgia.

DESCRIPTION

Ragfish differ from other fishes taxonomically, in their development and in their appearance. This is a flattened and flabby fish whose juveniles have an appearance completely different from the adults'. Additionally, the adult can be very large. Juveniles up to about 40 cm are ovoid and deep-bodied. They are

Danny Kent

brown and yellow with big spots and blotches, a small head and a rounded caudal fin. As the juveniles mature, they become elongate and dark brown with some purple hues, and they lose all scales and the juvenile pelvic fins. Adults retain the limp, flabby body that is mostly cartilage, but the shape is now elongate with a large forked tail.

LIFE HISTORY

Juvenile ragfish will be in shallower water or closer to the surface, while adults can be as deep as 1,420 m. They range from Japan to the Sea of Okhotsk, Bering Sea and Gulf of Alaska and to Southern California. Adults can grow to lengths of 213 cm and tend to be close to the bottom. Females release pelagic eggs throughout the year, but probably more during the autumn to winter period. The diet includes squid, fishes and octopuses.

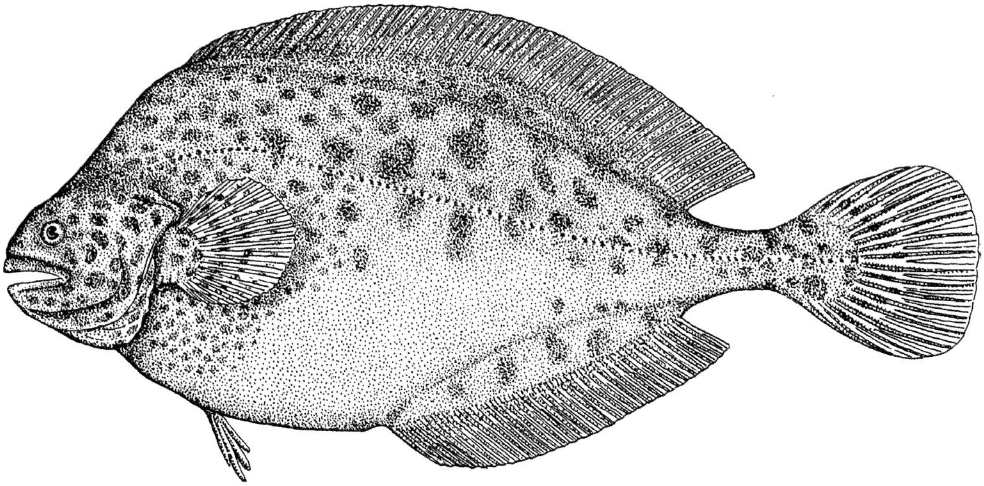

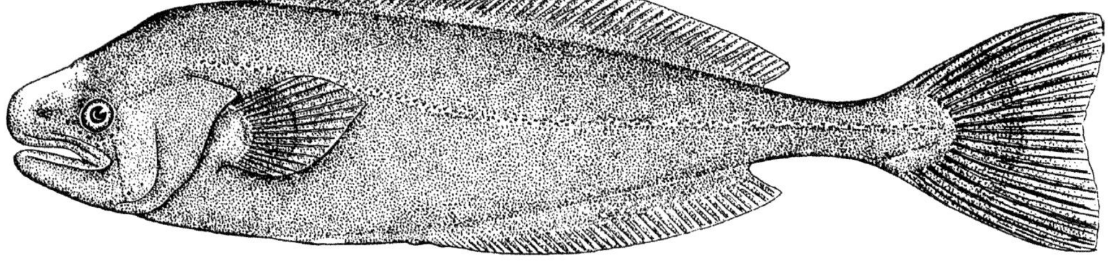

NORTHERN CLINGFISH

Gobiesox maeandricus UNCOMMON ◆

IN THE STRAIT OF GEORGIA

Northern clingfish are frequently reported by scuba divers from rocky shoreline areas around the Strait of Georgia. While this species occurs in the intertidal and shallow nearshore areas, it may not be common throughout the strait. It is able to breathe out of water, indicating it should be found in the intertidal areas in low tides. The species is often found associated with the penpoint gunnel, black prickleback and high cockscomb.

Scott Stevenson

DESCRIPTION

The northern clingfish is a uniquely shaped fish. The body is elongate with a huge head, which is flattened down and out to the sides, accommodating a very large mouth and oral cavity. There is one dorsal fin and a similarly sized anal fin below. The body ends with a large, fan-shaped tail. The pectoral fins, which are located behind the operculum, fan out and meet up with the pelvic fins on the bottom of the fish. The northern clingfish, as one would expect with such a name, has a prominent adhesive disc cupped by the pelvic fins on the underside of the fish. When the disc is engaged, negative pressure creates suction to allow it to adhere to a surface. Interestingly, the disc of the northern clingfish is able to support a weight of 454 g/cm^2. The efficiency of the disc in this species, as in other fish with adhesive discs, is related not only to the structure of the disc itself but to the body form of the entire fish. The adhesive disc plays a significant role in highly volatile environments, allowing the fish to cling to rocks under high current and wave action. It also allows the fish to remain in place while eating crustaceans and small molluscs such as limpets and chitons. This species is beautifully coloured. The overall colour can be from light brown to red with motley patches of a lighter shade. Dark lines radiate from the eyes across the side of the face to the operculum.

LIFE HISTORY

This species inhabits intertidal and subtidal areas from southeast Alaska and through coastal British Columbia to Southern California at depths from the intertidal zone to 8 m. It commonly clings to the underside of rocks, rock walls, kelp or anything flat. The maximum length is 16.5 cm. This species is a springtime spawner, depositing adhesive eggs on the undersides of rocks in April and May. In February, northern clingfish that are 8 cm in length and larger move from the subtidal to intertidal area in preparation for spawning. It is believed that females begin maturing at approximately 6.4 cm in length. Once the eggs hatch and larvae emerge, they become planktonic and move into the water column to feed. The larvae school within a couple metres of the seabed near shallow seaweeds that they hide under during darkness, so larvae are rarely taken in offshore plankton samples. Northern clingfish consume a variety of prey, including crustaceans and molluscs, as well as fish. The larger northern clingfish are able to chisel limpets off rocks with their teeth, using their adhesive disc as a lever. Both small and large northern clingfish have been reported to be cannibalistic. The fish in the Strait of Georgia are genetically distinct from those on the west coast of Vancouver Island.

Gobies

There are about 2,000 species of gobies in the family Gobiidae, making it the largest family of marine fishes. Gobies are found worldwide in oceans, and some in fresh water, at shallow depths mostly with sandy to muddy bottoms and a variety of habitats, including burrows and caves. Most are less than 10 cm long, but some can be as large as 50 cm long. The world's smallest fishes are gobies that are only 8 mm to 1 cm long, making them among the world's smallest vertebrates. Gobies have eyes near the top of their head and a unique structure of their pelvic fins, which form a cone shape used for attachment. They move in short swimming bursts followed by rests on their fins. Most species guard their nests. There are only three species of gobies in the Strait of Georgia.

ARROW GOBY

Clevelandia ios UNCOMMON ◆

IN THE STRAIT OF GEORGIA

There are reports of the arrow goby throughout the Strait of Georgia, but generally it is uncommon. Arrow goby are abundant on Spanish Banks, just downhill from the University of British Columbia. They can be found around ghost shrimp burrows. One of the reported locations is Taylor Bay on Gabriola Island (Taylor Bay is named after Reverend George Taylor, who was the first director of the Pacific Biological Station). The Strait of Georgia is the northern limit of this species.

DESCRIPTION

The arrow goby has an elongate body with a large head, big fins and very small scales. The mouth is extremely large and directed upward and has fine, sharp teeth. The fish has two dorsal fins. As with other gobies, this species' pelvic fins are fused together to form a cone to prop up the body and aid in attachment. Overall coloration is grey, green or tawny with black and orange speckling, and green and white spots on the sides. All fins except the pelvic fins have dark spots aligned into bars.

Danny Kent

LIFE HISTORY

In general, the arrow goby is a bottom dweller. It is commonly found in estuaries and lagoons, where it spends the majority of its adult life in or around muddy sediment and in burrows. The species is highly tolerant of changes in salinity and temperature. Arrow goby are known to live commensally in burrows with the ghost shrimp, the mud shrimp and the fat innkeeper worm to avoid both predation and drying out during low tide. Typically the fish is found from the intertidal area to 10 m and may even be found in fresh water. The maximum length of this species is 5.1 cm. It lives to at least three years old and is found from the Gulf of California to the Strait of Georgia. It has been reported that the arrow goby in California spawns in the winter to spring, mostly between March and June. Females spawn more than once, and there are reports that spawning may occur throughout the year. Their 0.8 mm eggs are demersal and laid in clumps of over 1,000. Whether these eggs are laid in nests and then sink into the sediment or are laid in the shared burrows has not been determined. The eggs are club-shaped with an attachment at one end. Eggs hatch quickly, after approximately 12 days at 15.5°C.

Larvae are pelagic and grow quickly, consuming eggs, copepods and nauplii in the water column. Adults feed on diatoms, green algae and marine eggs. They have been known to work co-operatively with the ghost shrimp, placing food too large to eat near a shrimp and then waiting for smaller pieces to be broken off. Arrow goby are food for the whitespotted greenling, the Pacific staghorn sculpin, terns and members of the rockfish family. A short report in a local paper by anyone finding a location of large numbers of arrow goby would be a useful contribution.

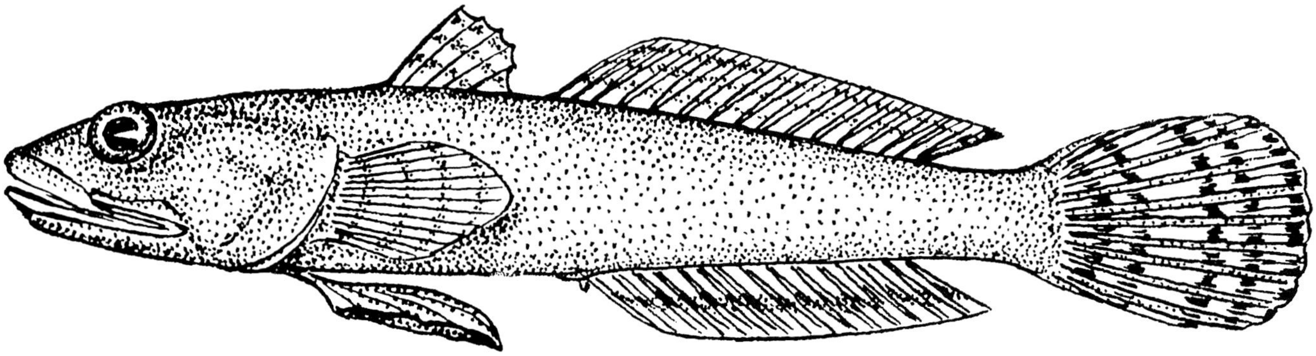

BLACKEYE GOBY

Rhinogobiops nicholsii COMMON ●

IN THE STRAIT OF GEORGIA

The blackeye goby is one of the fish most commonly seen by scuba divers exploring shallow, rocky seabeds in the Strait of Georgia. There are reports throughout the strait, but it appears more common in southern areas such as the Nanaimo and Cowichan River estuaries, Departure Bay, Howe Sound, Burrard Inlet and Horseshoe Bay.

DESCRIPTION

The male blackeye goby has black pelvic fins to match its black eyes and the black edge on top of the first dorsal fin. The pelvic fins are fused to form a pedestal that props the head up off the bottom. One-year-old fish have the black eyes but do not yet have other black markings.

LIFE HISTORY

Growing to 15 cm and occurring from southeast Alaska to central Baja California, this goby functions as a protogynous hermaphrodite, spawning as a female and then becoming a male after achieving full size. This species can crowd areas where there are abundant small rocks in shallow water, with one goby per rock. Newly settled juveniles about 2 to 3 cm long often sit upon very large boulders,

The blackeye goby props its body up with its fused pelvic fins. Lee Newman

The male blackeye goby digs a nest under a flat rock, where the more mottled female lays eggs that hang from the roof of the nest. Lee Newman

right out in the open, in the late summer. The males and females all guard territories around hiding places under small rocks, usually burrowed out by the fish from sand under the edge of the rock. The male will clean the underside of his rock den and then attract successive females into the nest to spawn; the eggs hang in a monolayer from the underside of the rock. Spawning is in early summer. Larvae and early juveniles of this goby swim in the freshened surface layers of the Fraser River plume between the mainland and the Southern Gulf Islands, and these fish have been found to have crustacean plankton and clam and mussel larvae in their guts. The settled juveniles and adults eat small crustaceans, snails, worms and other bottom-dwelling food, but they also rise up off the bottom to eat plankton when it is abundant. Another feeding behaviour is to suck up a mouthful of sand, then spit it out and ingest the small animals that are uncovered.

BAY GOBY

Lepidogobius lepidus

COMMON ●

IN THE STRAIT OF GEORGIA

The bay goby is distributed throughout the Strait of Georgia from Comox to Burrard Inlet. It most likely is somewhere between common and uncommon. It occasionally occurs in bottom research trawls in the Baynes Sound area.

DESCRIPTION

The bay goby has an elongate body with a large head, big fins and small scales. The mouth is large and directed upward and has fine, sharp teeth in several irregular rows. It has two dorsal fins; the posterior one is similar to the anal fin in size, shape and relative position. As with other gobies, this species' pelvic fins are fused together to form a cone to prop the body up and aid in attachment. The maximum length is approximately 10 cm, but most are less than 7 cm. Overall coloration is pale green with darker areas dorsally. Dark bands can be seen on the dorsal fin and tail. The fish is distinguished by the wide space between the dorsal fins, and the pectoral fin does not reach the anal fin.

The pattern of grey blotches and the black trim to the first dorsal fin help identification of the bay goby, which tends to be seen on open, soft bottoms. Lee Newman

LIFE HISTORY

This species is found from southeast Alaska to central Baja California on shallow muddy bottoms to a maximum depth of 305 m, but mostly less than 70 m. Bay goby apparently do not stray too far from a protective burrow to which they can retreat when disturbed. This behaviour makes them difficult to observe or to catch in nets. The early juveniles, however, school in shallow water near the seabed during late spring and early summer. A study in California that involved counting otolith annuli (ear bone annual growth zones) reported that the bay goby can attain an age of seven years. Half of the entire growth of the animal appears to occur in the first two years. From that age onward, growth appears to be constant until five years of age, when it slows down remarkably; mortality rates also increase with age. In general, it was determined that females reach maturity at approximately 4 cm in length. Bay goby are prey to the Pacific staghorn sculpin, the California flounder and sea slugs. Some marine birds are also believed to prey on bay goby, including marbled godwits, willets and terns. Bay goby consume copepods, polychaetes, amphipods, molluscs and other crustaceans.

and Butterfishes

The Pacific barracuda, Pacific pompano, Pacific chub mackerel and Pacific bonito are all rare visitors to the Strait of Georgia. There is one record of a Pacific barracuda in the strait. The species is in the family Sphyraenidae (barracudas), which has recently been reported to have 28 species. These fishes are tropical to subtropical and found worldwide near the surface. There is no mistaking their elongate, round body anteriorly or their large mouth full of prominent caniniform teeth of varying sizes. Younger fish are in small schools; older fish up to 183 cm in length may become solitary and migrate north in the summer. There are no records of barracudas in the Pacific attacking humans.

The family Scombridae includes mackerels and bonitos and has a total of 51 species. These are schooling marine fishes that are mostly near the surface. Two species—Pacific bonito and Pacific chub mackerel—are occasional visitors to the Strait of Georgia. The Pacific pompano is in the family Stromateidae (butterfishes), which comprises about 15 species. These fishes are found near the shore in most tropical and temperate seas worldwide. They are deep-bodied with a silver, iridescent appearance and a round head with a small mouth. They are very flat sideways and have long pectoral fins, no pelvic fins and a deeply forked tail.

PACIFIC BARRACUDA

Sphyraena argentea VISITOR ▼

IN THE STRAIT OF GEORGIA

A Pacific barracuda was caught in August 1998 in a location that is now within the northern boundary of the Salish Sea. The 77.5-cm-long specimen was captured in a gillnet during test fishing for sockeye and pink salmon about 25 km north of Campbell River in Johnstone Strait, opposite Quadra Island's Kanish Bay. There is a video of the capture, and the specimen hangs on the wall of the Department of Fisheries and Oceans office in Nanaimo, uphill from the Pacific Biological Station.

DESCRIPTION

There is no mistaking the identity of this long, slender and cylindrical fish. Pacific barracuda are greyish black dorsally, shading to silvery white ventrally with some blue and even green on the head area. The tail has a yellow tinge with a dark-blue margin. Our specimen has a series of vertical dark-blue bars on the body. The mouth is large, with the lower jaw extending beyond the upper jaw, and full of different sizes of sharp, fang-like teeth. There is a uniform margin of smaller teeth on each side of the lower jaw. The tail is forked, and the widely spaced first and second dorsal fins are positioned directly above the abdominally located pelvic and anal fins.

Bill Pennell

LIFE HISTORY

The Pacific barracuda is a highly migratory species that occurs from the Gulf of Alaska at Kodiak Island to Baja California and the Mexican Revillagigedo Islands. It is rare north of California but occurs occasionally farther north, particularly during El Niño events. The specimen captured in Johnstone Strait in 1998 most certainly resulted from the 1997–98 El Niño, which was the most

The Pacific barracuda specimen in the hands of Dick Beamish is on the wall of the Department of Fisheries and Oceans South Coast offices in Nanaimo, up the hill from the Pacific Biological Station. Bill Pennell

powerful such event on record up to that time. Another specimen was reported in 1998 off Oregon. There are no historical reports from the Strait of Georgia or Johnstone Strait. The maximum size is reported to be 122 cm and 8.2 kg, but there may be larger specimens. There are estimates of ages up to 18 years, but there may be older individuals. Females are larger than males at all ages and mature at two to three years old and a length of 55 cm. Males mature at one to two years old and 35 to 45 cm long. The specimen from Johnstone Strait, at a length of 77.5 cm, could be older than 10 years. Older and larger fish are known to be solitary, which could explain its 1998 visit to the Strait of Georgia.

Spawning occurs from April to October, mostly from May to August. A female will spawn more than once in a year, releasing batches of eggs numbering from 42,000 to 513,000. Prey includes small fishes and squid. This is an important species for the recreational fishing community.

PACIFIC CHUB MACKEREL

Scomber japonicus VISITOR ▼

IN THE STRAIT OF GEORGIA

Pacific chub mackerel are rare in the Strait of Georgia, with only two reports in the last 50 years. Both reports are recorded in the groundfish database at the Pacific Biological Station in Nanaimo and come from commercial bottom trawl fisheries. There was one specimen caught in the southern strait in November 2000 and a second off the mouth of the Fraser River in January 2002. Pacific chub mackerel have not been found in the several thousand pelagic trawl sets made from 1995 to the present.

DESCRIPTION

There would be no mistaking this fish if you were to see one in the Strait of Georgia. It is elongate, streamlined and blue-green with 25 to 30 dark-green bars and spots across the back and irregular spots on the sides. There are five to six small finlets between the second dorsal fin and the tail.

LIFE HISTORY

This fish is most commonly known as the "Pacific mackerel" and is widely distributed throughout the Pacific and Indian Oceans. It can be found in large schools off the west coast of Vancouver Island, particularly in years of warm water in the summer. This is a coastal species that occurs in depths of up to 300 m, but is most often found near the surface. Juveniles begin to school at about 3 cm long, and the fish will school with other species, often staying near the bottom during the day. The maximum size is 64 cm and 2.9 kg, but most are about 30 cm. They mature at age two to four and spawn several times a year, mostly from April to August. The maximum age is about 18 years. Pacific chub mackerel are fished off piers in California, farmed in Japan and used as medicine in China.

TINRO

PACIFIC BONITO

Sarda chiliensis VISITOR ▼

IN THE STRAIT OF GEORGIA

There is only one record of the Pacific bonito in the Strait of Georgia. It was found near Yellow Point, and that was a little over 50 years ago. It has not been caught in any of the several thousand scientific trawl and seine studies since the mid-1990s. We list this species because it is found off the west coast of Vancouver Island, so it is possible there will be future visits to the Strait of Georgia as the waters warm here and offshore.

DESCRIPTION

This is a sleek, silvery blue to violet fish with a series of oblique black stripes on its posterior sides. There are eight finlets behind the second dorsal fin.

LIFE HISTORY

This is a small tuna that can occur in large schools off the west coast of Vancouver Island. The maximum length is 122 cm and the maximum weight is 11.3 kg. It ranges from the Gulf of Alaska to southern Baja California. A southern population occurs off Chile and Peru. Pacific bonito are fished commercially, but it is a popular sport fish too—particularly off California, where it can occasionally be caught off piers. The sport-caught record weight is 9.7 kg. Males can mature and spawn in one year. Females mature after two years and spawn several times a year, releasing a total of about half a million floating eggs from January to March that hatch in about three days. Individuals grow rapidly, feeding mostly on small fish such as the northern anchovy, and reach an average length of about 51 cm in the first year. The maximum age is about six years.

Ken Jones

PACIFIC POMPANO

Peprilus simillimus RARE ★

IN THE STRAIT OF GEORGIA

Pacific pompano have occasionally been recorded in the Strait of Georgia. In 2015 they were seen for the first time in 25 years in several locations within the strait. In June, one was captured in a beach seine in Cowichan Bay, and several were captured earlier in a trawl catch near the mouth of the Fraser River. They were abundant in the summer of 2015 in Juan de Fuca Strait. Children were catching them off the dock at Port Renfrew marina. Schools as wide as 200 m would show up in the early evening and were everywhere in the marina. It is known that there are occasional periods of abundance in the Juan de Fuca Strait area, and 2015 appears to have been one of those years.

Marc Trudel

DESCRIPTION

The Pacific pompano is easily recognized by the oval shape and laterally compressed body with an iridescent green or blue colour dorsally and silvery ventrally. There is a prominent pectoral fin, no pelvic fin and a small mouth.

LIFE HISTORY

This species is found from Queen Charlotte Sound to Mexico and is abundant in Southern California, ranging from mostly near the surface to 311 m. The maximum size has been reported to range from 28 to 36 cm in length. The specimen in Cowichan Bay was 20.5 cm. This is a poorly studied species despite its abundance off California and its reputation years ago as being a delicacy with the attractive name “butterfish.” It is known that spawning occurs in all months, but it seems that most of the pelagic eggs are released between the spring and fall. We were not able to find information for the maximum age or age at maturity.

Righteye Flounders

Sanddabs, flounders and soles are in the order Carangiformes, which includes flatfish species within families that have eyes on the right side or left side of the fish. All flatfish look like typical fish when they hatch, with an eye on each side of the head. As the larvae develop, at about 2 cm long, one eye migrates to the left or right side and other changes to the mouth and organs occur. When the eyed side of a flatfish is on top and the mouth is slanted upward, a left-sided fish will be pointing left and a right-sided fish pointing right. All flounders have long, continuous dorsal and anal fins, with the dorsal extending onto the head. All species have a flat, ovate body with a white or whitish colour under the side that has no eye, and a pigmented and slightly less flattened eyed side. The pigment pattern of the eyed side changes depending on the surroundings to provide camouflage. There are no swim bladders and no spines in the pectoral or pelvic fins. These flatfish species range in size from the small speckled sanddab to the extremely large Pacific halibut, and the total number of species is the third-largest in the strait, following sculpins and rockfish.

Sanddabs, with eyes on their left side, are in the family Cyclopsettidae. Estimates of the number of Cyclopsettidae species in the world range as high as 114. They occur worldwide in the Atlantic, Pacific and Indian Oceans, with rare occurrences in fresh water. They are mostly found on shallow, sandy or muddy bottoms along a continental shelf. Most are less than 40 cm long, but the maximum size is about 1.5 m. Like all flatfish, they swim on their sides. There are two species of sanddabs in the Strait of Georgia.

Righteye flounders, with eyes on their right side, are in the family Pleuronectidae. There are about 105 Pleuronectidae species worldwide, with rare occurrences in fresh water. Species in this family occur from near the surface to about 2,000 m deep. Many species are of commercial importance, including the Pacific and Atlantic halibut. There are 16 species in the Strait of Georgia, including the English sole, petrale sole and Pacific halibut. Petrale sole are not abundant, but they may be the best-tasting species of all fishes caught around British Columbia. The English sole was the principal species when bottom fishing started in the Strait of Georgia around 1912. Pacific halibut are now rare in the strait, but in the late 1800s they were so common that one government official wrote that their catch would be "unlimited."

PACIFIC SANDDAB

Citharichthys sordidus COMMON ●

IN THE STRAIT OF GEORGIA

Pacific sanddabs are likely common in most areas of the Strait of Georgia, but they may be more abundant in the southern areas around the Fraser River, Howe Sound and near the San Juan Islands. They are observed by scuba divers throughout the Strait of Georgia, but are not commonly seen during the day. Like some species of flatfish, they move into shallower water to feed at night. Pacific sanddabs may be more abundant in Puget Sound than in the strait.

DESCRIPTION

The Pacific sanddab has eyes on the left side of its head. The thick body is oval and has large scales. The lower eye is larger than the snout; the interorbital space is concave, and there is a high bony ridge above the lower eye. When the pectoral fin is bent forward, it reaches the middle of the lower eye. The eyed

Andy Murch

side is a dull tan or brown with mottled yellow or reddish-orange spots. The lateral line is almost straight. The blind side is a lighter tan to whitish, but not a distinct white.

LIFE HISTORY

This species is found on soft bottoms from the intertidal zone to depths of 549 m but is common from 50 to 150 m. It may move off the bottom when feeding. It is common from California to British Columbia, but it ranges from the eastern Gulf of Alaska to southern Baja California. It also occurs on offshore seamounts. The maximum size is 41 cm long and about 1 kg. Most are around 25 cm and 150 g. The maximum age is 13 years. Females grow faster than males after age 4, and grow to a larger size. Females first mature at age 3 at lengths of 19 to 23 cm and spawn from July to September off California; they may spawn earlier in the Strait of Georgia. Adults spawn near the bottom, in bays and open water, areas where the water temperatures may be cooler. This is another species that would benefit from a routine study of its biology, including an examination of its gonads to determine when the fish spawns and whether spawning occurs over a protracted period. Juvenile fish feed extensively on copepod nauplii. Older fish feed on euphausiids, mysids, polychaetes and even juvenile rockfish. This is an important commercial species off California, where it is considered a delicacy.

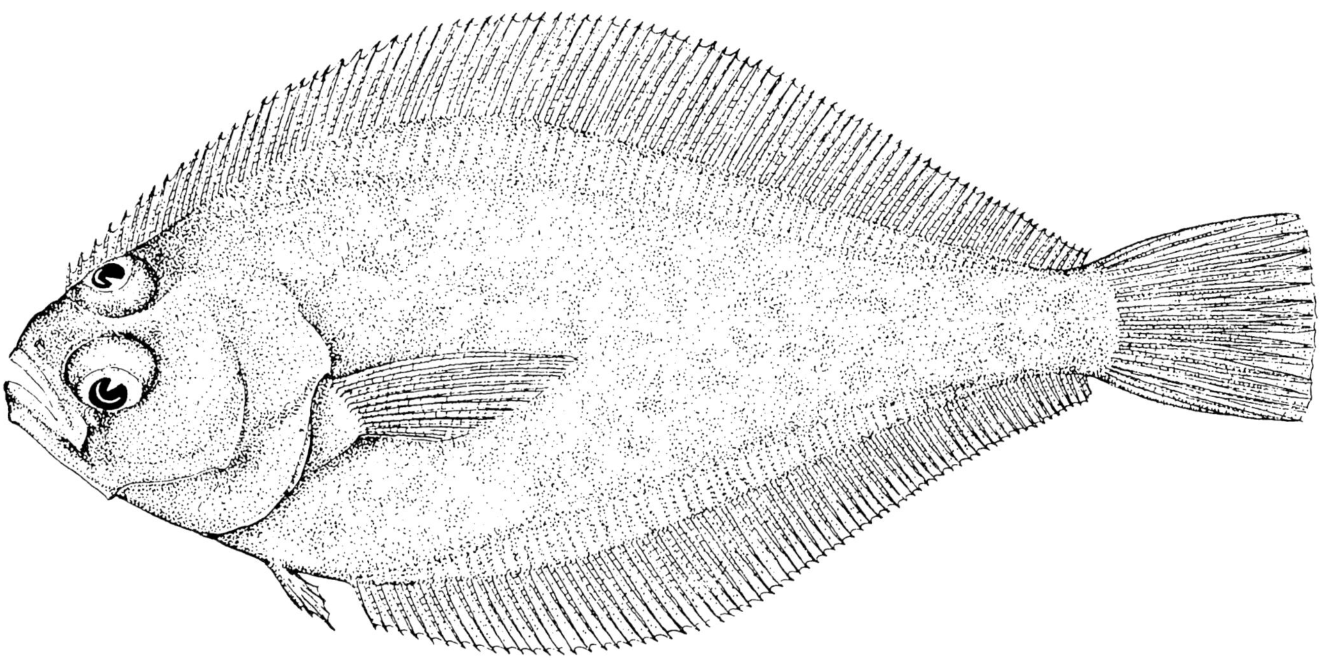

SPECKLED SANDDAB

Citharichthys stigmaeus COMMON ●

IN THE STRAIT OF GEORGIA

This small flatfish is found along sandy intertidal beaches and in trawls in shallow water in the Strait of Georgia. It is commonly seen by scuba divers all around the strait. Scuba divers have observed that speckled sanddabs will follow them, looking for food that becomes exposed. This fish may be more abundant in the southern areas of the strait, particularly in Burrard Inlet, Howe Sound and Nanoose Bay.

DESCRIPTION

Speckled sanddabs have eyes on their right side. They are olive-brown or tan with black speckling on their eyed side, and off-white to cream or light tan on their blind side. The colouring provides excellent camouflage, perhaps the best of all flatfish. This is a small fish, reaching a maximum size of about 17 cm long and 450 g. However, most are less than 10 cm long and rarely over 13 cm. They are distinguished from the Pacific sanddab by their small size, including a smaller head and eyes, and by the black speckling on the body. The pectoral

Jackson W.F. Chu

fin does not reach the eye when folded forward, the lower eye width is equal to or slightly less than the eye length, and the space between the eyes is narrow, with no ridge above the lower eye.

Gregory Jensen

LIFE HISTORY

Speckled sanddabs are found on sandy bottoms from the intertidal zone to 366 m, but most often to just 20 m and seldom deeper than 91 m. They occur from the northern Gulf of Alaska, in Prince William Sound, to Baja California. This is apparently a short-lived species, reaching maximum ages of three to four years. Females mature at age two and lengths of 7 to 8 cm. At age one, males and females are 6.7 cm and 6.9 cm, respectively. By age three, both sexes range from 12.1 to 15 cm.

These fish spawn from April to September. Fish larger than about 13 cm were found to have 27,500 eggs. It is possible that small amounts of sperm and eggs are released sporadically throughout the spawning season. Smaller fish feed on copepods, amphipods, cumaceans and mysids. Adults consume decapods, molluscs and small fish. Speckled sanddabs are eaten by marine birds, which is probably why their camouflage is so highly developed. Small flatfish that swim away when disturbed by people wading at shallow beaches likely are speckled sanddabs.

ARROWTOOTH FLOUNDER

Atheresthes stomias UNCOMMON ◆

IN THE STRAIT OF GEORGIA

Arrowtooth flounder are found throughout the Strait of Georgia, but they are probably uncommon. They are less abundant than the slender sole, English sole, rex sole, flathead sole and Dover sole. It is probable that they are considerably less abundant today than they were from the mid-1950s to 1960s, when they were an important bycatch in the commercial trawl fishery for use as mink food. Arrowtooth flounder may be found in shallow water of about 40 to 80 m, where they are caught in shrimp trawls, and rarely less than 12 m deep, where they are seen at night by scuba divers. They are more common in the deeper water of the strait and probably are more common in Juan de Fuca Strait than in the Strait of Georgia.

DESCRIPTION

The arrowtooth flounder is a large flatfish that can reach a length of 86 cm and a weight of 7.7 kg. An average length is about 50 cm. This is an elongated, diamond-shaped fish with a large mouth and two rows of sharp, arrow-shaped teeth that characterize this fish as a top predator in the ecosystem. It is distinguished from other flatfish by its larger size (though halibut are larger); the upper eye, which is visible from the blind side; more than 14 gill rakers on the first arch and 2 in the upper limb of the second arch; and the maxilla extending behind the eyes. Scales are large, coarse and easily removed from the fish. The caudal peduncle is long. The fish is brown to olive-brown; the blind side is greyish white and finely dotted with black.

Alexei Orlov

LIFE HISTORY

Arrowtooth flounder occur at depths from 12 to 900 m, but are most common offshore at depths of 100 to 300 m. The smaller fish occur on soft bottoms, but adults do not appear to have a preferred bottom type. The adults are fish feeders, which perhaps explains this lack of a preference. Arrowtooth flounder are more abundant at the northern part of their range. They are reported to be the most abundant of any groundfish species in the Gulf of Alaska, where their numbers have increased dramatically since the

1970s. They occur from the eastern Sea of Okhotsk in Russia, across the Bering Sea to the Gulf of Alaska and south to the coast of California.

The maximum age is about 30 years. Females mature at 4 to 8 years. Fifty percent of females are mature at age 7 and a length between 44 and 47 cm, and most females arc mature by age 10. Males mature at 3 to 7 years old. Fifty percent of males are mature at a length of 29 to 42 cm. Most individuals have a life span between 2 and 14 years. Females will grow to larger sizes than males, with maximum lengths exceeding 90 cm. Mature arrowtooth flounder migrate into deeper water in the late fall and winter, where they spawn over a protracted period in water deeper than 366 m. Off Washington, eggs have been seen as early as September, and by March females are finished spawning. However, in the Gulf of Alaska spawning begins in December and can last for eight months. Females produce 250,000 to 2.4 million eggs, depending on their size. The planktonic eggs hatch into larvae. The larvae go through metamorphosis, which can last anywhere from four weeks to several months, depending on growth and temperature. After metamorphosis the young settle on the bottom. Adult arrowtooth flounder migrate back into shallower water of several hundred metres during the summer, but in general the larger fish are found in deeper water.

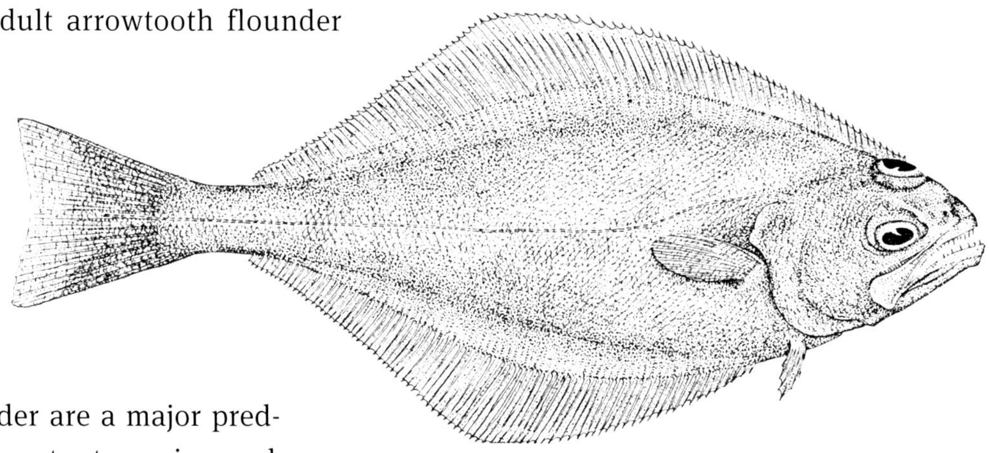

In the Gulf of Alaska arrowtooth flounder are a major predator of commercially important species such as walleye pollock, Pacific herring, capelins and sand lances. Adults prefer fish but consume shrimp, euphausiids, squid and a variety of benthic invertebrates. They also consume juveniles of various flatfish species. They are eaten by halibut, skate, Pacific spiny dogfish, sharks, orcas and toothed whales, and in some areas by Steller sea lions. They have been identified as prey of seals in the Strait of Georgia. They are not a sought-after commercial species, in part because the flesh contains an enzyme that causes the muscle to soften when heated, resulting in an unappealing paste-like product. Their dominant biomass and their preference for fish make them a key species in any effort to manage fisheries on an ecosystemic basis.

PETRALE SOLE

Eopsetta jordani UNCOMMON ◆

IN THE STRAIT OF GEORGIA

Petrale sole occur throughout the Strait of Georgia but are more abundant in the south, off the mouth of the Fraser River and just east of the Gulf Islands. They have also been found in the Fraser and Cowichan River estuaries. Petrale sole were rare in groundfish surveys of the Strait of Georgia from the 1970s up to the early 2010s, after which there was evidence of an increasing abundance.

DESCRIPTION

This fish has a broad head, double rows of teeth, a wide, flat space between the eyes and a bluntly rounded snout above the large jaw, which extends under the eye. None of the forward dorsal fin rays extend beyond the fin. The eyed side is either dark or light brown and the blind side is white. The lateral line gently slopes up toward the head and branches above and below the eyes on the eyed side. The spine at the forward end of the anal fin is sharp and visible, angled forward. The tail is pointed in the centre.

LIFE HISTORY

These are larger flatfish that prefer sandy or mud bottoms. Offshore, they occur from the Aleutian Islands to southern Baja California in nearshore areas to depths of 550 m. Most catches occur at depths of 50 to 300 m. The maximum size of females is 70 cm long and 3.6 kg. Males grow to about 53 cm. Fifty

Bill Pennell

percent of females are mature at age five and a length of 33 cm, while 50 percent of males are mature at age four and 32 cm. There is evidence of extensive migrations from the summer feeding grounds to winter spawning areas in deeper water. Spawning is from November to March. A large female can produce up to 1.5 million pelagic eggs. The larval fish remain in the midwater for about five to six months before settling on the bottom. The maximum age is about 40 years, but most fish live to no more than 25, with females slightly larger at any age. Few fish are older than age 17 in the fishery. Their prey is mostly fishes, but includes squid, octopuses and a variety of invertebrates.

Old-timers in British Columbia still call the petrale sole "brill." This is one of the best-tasting—if not *the* best-tasting—fish from the British Columbia commercial fisheries. Along with Pacific halibut, the petrale sole has always been considered to be among the top-rated flounders for seafood restaurants in the Pacific Northwest. Their flesh is worth more than that of any other flatfish except Pacific halibut.

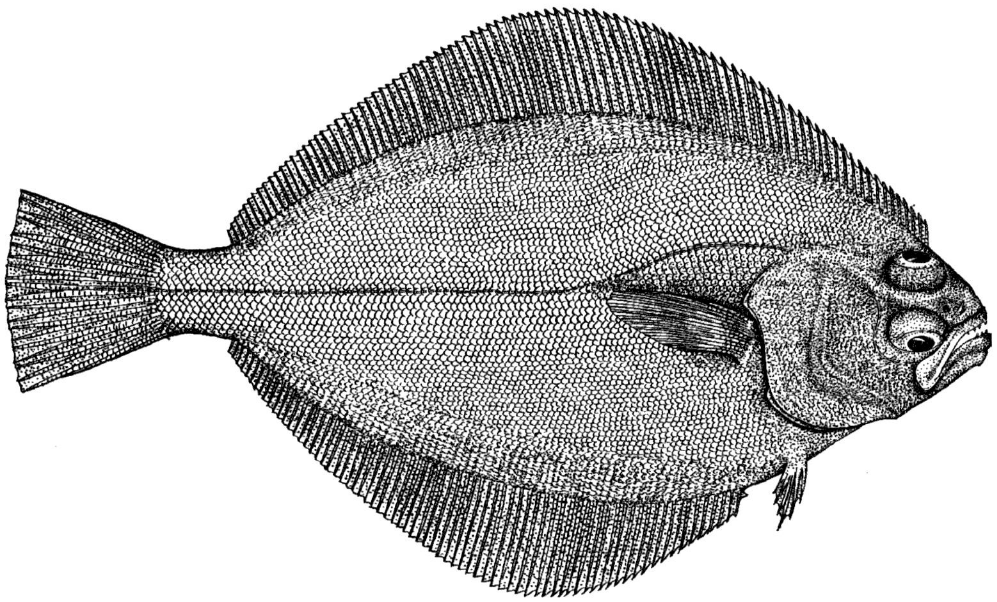

REX SOLE

Glyptocephalus zachirus COMMON ●

IN THE STRAIT OF GEORGIA

The rex sole is found throughout the Strait of Georgia on soft bottoms, mostly at depths from about 40 to 110 m. It is routinely caught in bottom research trawl surveys and probably is the third most abundant sole in the strait, behind the slender and English soles.

DESCRIPTION

This is a relatively slender fish. It has a small mouth and, on the eyed side, a long, mostly black pectoral fin that is longer than the head and the caudal fin. The lateral line is mostly straight. The caudal fin is pointed in the middle, and the dorsal and pelvic fins may be edged in black. The eyed side is brown to grey, and the blind side is a dusky or creamy white.

LIFE HISTORY

The rex sole is widely distributed from the southern tip of Kamchatka, Russia, around the rim of the North Pacific to Baja California. It has also been reported well offshore on seamounts. Off the west coast of North America it has been reported at depths of 850 m, but is common at depths from 50 to 450 m. The maximum size is 61 cm long and about 2 kg. Most fish are between 25 and 35 cm and about 200 to 300 g. In a study off Alaska, the maximum age reported was 27. Females matured at ages from three to nine, with 50 percent mature at age five and a length of 35 cm. They may mature at a smaller size toward the southern limit of their distribution. Males appear to mature at younger ages, with 50 percent mature at age three. Males also grow slower and to smaller sizes than females.

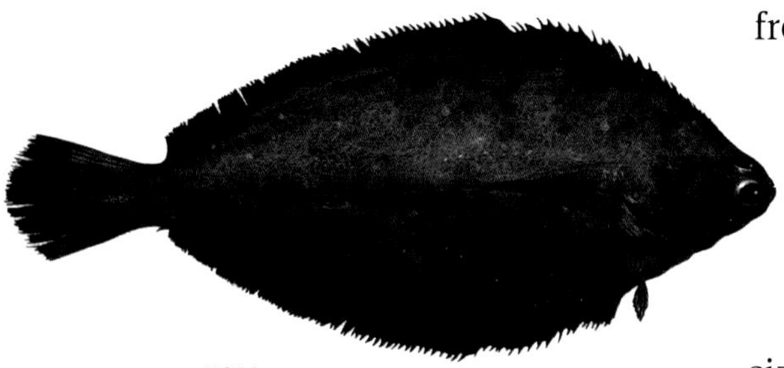
NOAA

Off Oregon, spawning occurs from January to June with a peak from March to April. Off Alaska the spawning period lasts eight months, from October to May. Rex sole are batch spawners, which means the females release ripe eggs in batches rather than all at once. The larvae have a prolonged pelagic period of about 12 months, which would allow them to become widely distributed throughout the Strait of Georgia. Rex sole feed on benthic invertebrates, including crustaceans such as gammarid amphipods, shrimp, crabs and polychaetes. The rex sole has long been an important commercial species, particularly off the west coast of North America. The largest commercial fishery has traditionally been in the Gulf of Alaska. It is an important species because of its abundance and pleasant taste. Its thinness makes it difficult to remove fillets, but the fish can readily be cooked whole by first removing the fins and the contents of the body cavity. This is the preferred method of cooking similar species of flatfish in some fine European restaurants.

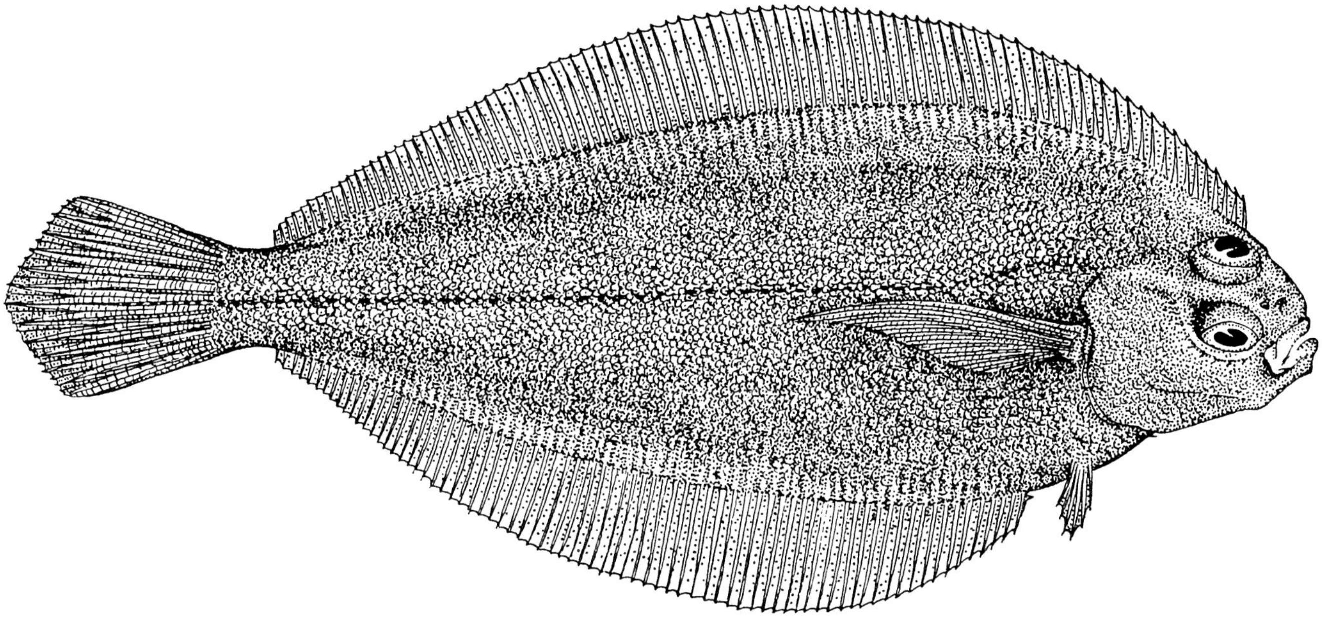

FLATHEAD SOLE

Hippoglossoides elassodon

COMMON ●

IN THE STRAIT OF GEORGIA

The flathead sole is commonly found throughout the Strait of Georgia, particularly in intermediate depths in the Baynes Sound–Cape Lazo area, the Gulf Islands, Howe Sound and the mouth of the Fraser River. It is not often seen by scuba divers, with reports only from Howe Sound and Burrard Inlet.

DESCRIPTION

This species is distinguished by its oval-shaped and elongate body. The mouth is large and the jaw is steeply angled. The dorsal and anal fins are blotched and the caudal fin membranes are mostly clear. The interorbital space is narrow with a sharp ridge, and there is one row of teeth on each side of both jaws. The lateral line curves slightly around the pectoral fin. The gill rakers on the lower limb of the first arch usually number more than 14. There is a row of pores under the eyes that may be difficult to identify. The eyed side is an olive-brown that is sometimes blotched, and the blind side is white with translucent areas.

Gregory Jensen

LIFE HISTORY

Flathead sole occur from the northern Sea of Japan to the Sea of Okhotsk, Russia, and through Alaska, British Columbia and south to Central California. Females grow to larger sizes than males. They are usually found in shallow water but sometimes down to 1,000 m. In the Strait of Georgia, they occur mainly at depths from about 10 to 100 m on soft, silty mud bottoms. The maximum size reported in recent studies in Alaska is 58 cm, with most samples in the 20 to 40 cm range, and an age range of about 6 to 18 years. The maximum age has been reported as 27 years for females and 30 for males. A recent study of the age at maturation off Alaska found that 50 percent of females matured at ages of 7 to 9 and lengths of 32 to 33.3 cm. Ages of fish in the fishery were reported as 4 to 21. Flathead sole in Puget Sound have been reported to reach maturity at 2 to 3 years old. In the Strait of Georgia, it would be most useful to sample spawning flathead sole, perhaps in the Boat Harbour area in the Gulf Islands, to determine their ages.

Alexei Orlov

Spawning has been reported in the Boat Harbour area in late March, while the major spawning period is March and April. However, flathead sole have been shown to be batch spawners, which means that they spawn quantities of eggs over a prolonged period. Eggs are buoyant, and larvae remain in the water column until they metamorphose and settle on the bottom in late summer. Nursery areas are in shallow bays or inlets at depths of 30 to 70 m. Young flathead sole feed heavily on small crustaceans such as mysid shrimps. Their diets change as the fish increase in size: they eat clams, polychaetes, euphausiids, shrimps and squid when they are smaller, and consume more fish as they grow. In the 1950s, the flathead sole was used for animal food, but recently it has become an accepted commercial species in Canada and the United States. The largest commercial catches are off Alaska, where it is taken mostly as bycatch in fisheries directed at other species.

PACIFIC HALIBUT

Hippoglossus stenolepis VISITOR ▼

IN THE STRAIT OF GEORGIA

It is possible to catch the occasional halibut in the Strait of Georgia, but they do not reproduce there. Experienced recreational fishermen have some secret spots around Campbell River, near Victoria and even near Nanaimo. Pacific halibut are not common in the strait, certainly less common than they were 100 years ago or at the time that the name Halibut Bank was chosen for a mid-strait area.

Jeff Sha

DESCRIPTION

There is no mistaking a Pacific halibut in Canadian waters, as it will be large and thick relative to other flatfish. These are elongate fish with a diamond-shaped profile. Pacific halibut also have an arched lateral line above the pectoral fin, and the tail is crescent-shaped, with the edges extending beyond the inside margin.

LIFE HISTORY

It is not known how large or how old the Pacific halibut in the Strait of Georgia are, as the fish are rarely sampled for length and for otoliths. However, it is likely the fish are about the same age as those found off the west coast of Vancouver Island. Most of these fish are 8 to 15 years old, with two-thirds being 9 to 12. Very few fish are older than 20. The average weight in the commercial fishery is 12 to 15 kg, and in the recreational fishery, about 8 kg. In Puget Sound the record halibut caught in the sports fishery was 134 kg. Pacific halibut in the Strait of Georgia probably migrate into the strait from offshore areas, most likely through Juan de Fuca Strait, but some will enter from Johnstone Strait. It would be useful if a program were started to determine the age and maturity of Pacific halibut caught in the Strait of Georgia.

Pacific halibut are distributed from Hokkaido, Japan, to the Sea of Okhotsk; east to the Chukchi Sea, Bering Sea and Aleutian Islands; and south along the coast to Northern California. The centre of abundance is in the central Gulf of Alaska. Spawning occurs in the winter, especially between December and February at depths of about 365 to 550 m. Males mature at ages of 5 to 11, and females at 8 to 16. Fifty percent of the females are mature at age 12 or 13 and a

Alexei Orlov

length of about 119 cm, while 50 percent of males are mature at age 8 and at a smaller size. A halibut claimed to be the largest ever caught is on display in the airport in Anchorage, Alaska; caught in June 1996, it had a length of 2.87 m and a weight of 208 kg. It lived to 31 years old. Unconfirmed reports have Pacific halibut exceeding 320 kg. Females grow to a larger size, and in recent years have been almost twice the size of males. Females spawning for the first time can produce 100,000 eggs, with larger females producing up to 4 million. The non-adhesive eggs hatch after about 15 days, depending on the temperature. Spawning probably occurs near the bottom, with eggs rising in the water column. The young larvae are carried north by the surface currents for six or seven months. After metamorphosis, juvenile fish settle on the bottom and some begin to migrate south. Virtually all halibut in the Canadian zone migrate through Alaskan waters at two to six years old. Pacific halibut generally are along the edge of the continental shelf, but some are at depths to a maximum of about 1,100 m.

Halibut are fast-growing and long-lived. The maximum age is 55 years, but few live longer than 30, and most are caught in the fishery before they are 15. The population size of Pacific halibut in the late 1990s was considered to be at historically high abundances, but numbers have been declining in recent years. The abundance has been shown to respond to natural trends in the ocean environment. Recruitment into the fishery at about age eight is influenced mainly by the survival of the larval fish, which is dependent on climate and ocean conditions during the spawning year. In recent years, Pacific halibut have not been growing to the previous lengths and may even be about one-half as large as they were in the 1970s.

It is tempting to summarize the diet by saying they eat whatever they want to, as they feed both near the bottom and pelagically. Smaller Pacific halibut feed on invertebrates, and larger individuals eat squid and octopuses, as well as fish. Some sharks and marine mammals feed on halibut.

There is a substantial longline fishery that is well managed by the International Pacific Halibut Commission. This commission was established in 1923 and was the first international agreement signed independently by Canada. The commission has a permanent staff and is managed by commissioners from Canada and the United States.

BUTTER SOLE

Isopsetta isolepis UNCOMMON ◆

IN THE STRAIT OF GEORGIA

The butter sole occurs in the Strait of Georgia, mostly in the southern strait, as well as in the San Juan Islands at depths of less than about 30 m. We have records from the southern strait, off Courtenay and in the Fraser and Cowichan River estuaries. There are no observations by scuba divers, probably because butter sole bury themselves in muddy areas not favoured by divers.

DESCRIPTION

The butter sole's upper eye is not visible from the blind side. The dorsal branch of the lateral line extends beyond the operculum. The eyes are small with a narrow, flat space between them. The eyed side is generally brown or grey with lighter mottling and yellow spots. The scales are rough on the eyed side and continue onto the dorsal and anal fins, which may be edged with yellow. The blind side is whiter.

Vicky Okimura

LIFE HISTORY

Butter sole are found from the southeastern Bering Sea and Aleutian Islands to Southern California. They occur on soft bottoms at depths of 2 to 425 m, but most are in shallow water of less than 150 m, usually less than about 75 m. Most are in coastal areas within 18 km of shore. Butter sole have a maximum length of 55 cm, with most commercially caught fish being about 20 cm. Spawning starts in February and extends into May in shallow water, about the same time that English sole spawn; hybrids of these two species have been found in the Strait of Georgia and Puget Sound. Off British Columbia, spawning occurs at depths of about 27 to 63 m. Females that are about 38 cm long produce about one million eggs. The eggs are planktonic, as are the larvae, which feed primarily on copepods of varying life stages. Metamorphosis occurs at lengths of 1.8 to 2.3 cm, with the juveniles settling to the bottom in shallow water from May to August. Yearling juvenile butter sole move offshore. They are about 6.3, 13.3 and 19 cm after age one, two and three, respectively. The maximum age of the butter sole is 12 years, based on very limited data. Almost all males greater than 10 cm in length are mature or maturing. Females 25 cm or larger (age three and older) are mature. Older butter sole feed on polychaetes, amphipods, molluscs, cumaceans and decapods. From the mid-1940s to the 1960s, butter sole were fished in the Skidegate Inlet area of Haida Gwaii, mostly for animal food. The fishery occurred from February to April, when the fish were spawning in shallow water.

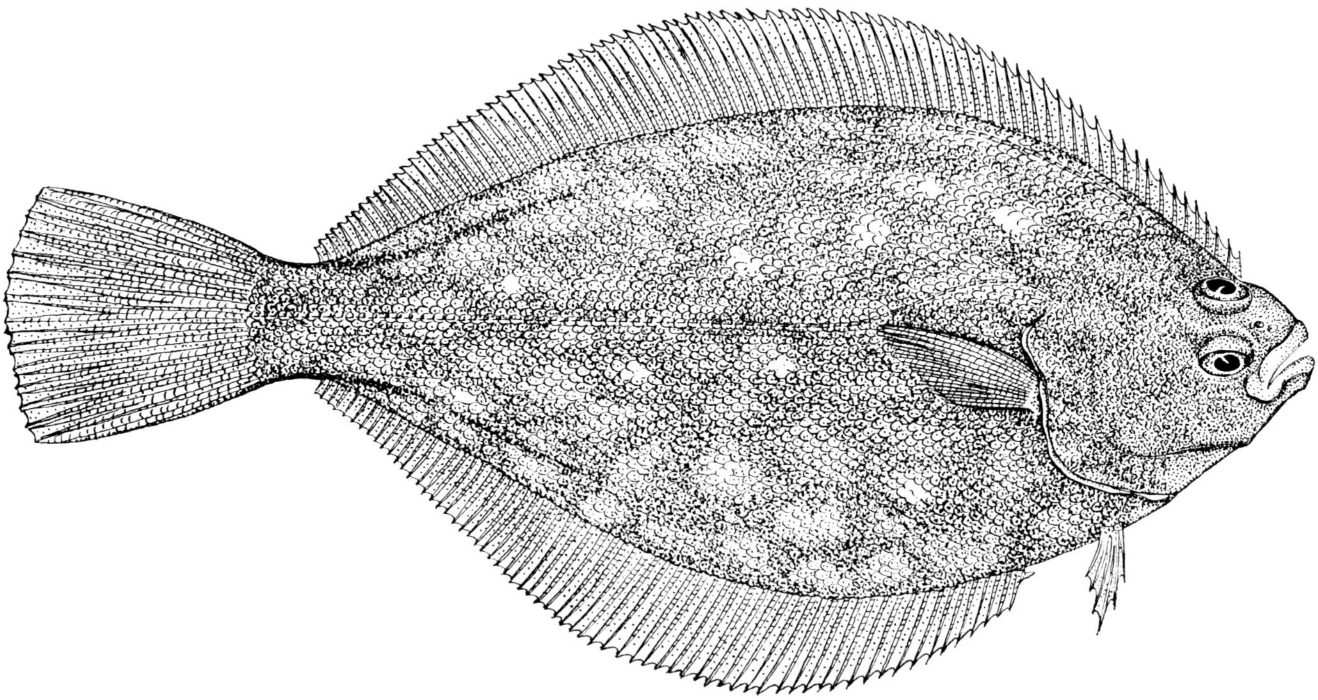

NORTHERN ROCK SOLE

Lepidopsetta polyxystra UNCOMMON ◆

IN THE STRAIT OF GEORGIA

The northern rock sole is a newly identified species that was distinguished from the rock sole (*Lepidopsetta bilineata*) in 2000. As early as 1985, biologists recognized there were larval rock sole that looked different from others. Biologists managing the rock sole fishery in the Gulf of Alaska in the 1990s could distinguish two different forms, and after a lot of detective work, two ichthyologists were able to show that what was thought to be one species was in fact two. The original species retained the common name "rock sole," while the new species was given the common name "northern rock sole" and the new scientific species name *polyxystra*, which roughly translates to "many gill rakers." Both species are reported in the Strait of Georgia, but a new study is needed to determine their relative abundances and distributions, as previous studies that caught large numbers of rock sole in the strait did not look specifically for northern rock sole. One recent study in the southern transboundary of the strait estimated that the two species were about equally abundant.

Gregory Jensen

Alexei Orlov

DESCRIPTION

Unfortunately, you really have to know what you are looking for to identify the species of rock sole when sorting through a catch of soles. The northern rock sole is distinguished by having one to three pores above the eyes (four to seven in rare cases); these pores are not easy to count. However, there are 10 or more slender gill rakers on the first gill arch, and three or more on the upper limb of the arch. There is some overlap between the two species on the upper limb, but on the lower limb, the northern rock sole has seven or more gill rakers, while the rock sole almost always has six or fewer. The northern rock sole also has a wide space between the bulbous eyes and thick lips and up to five rows of scales at the narrowest point of the interorbital space. The blind side is a creamy white, not a bright white. It is distinguished from other flatfish by the strongly arched lateral line with a short dorsal branch.

LIFE HISTORY

Northern rock sole occur from Puget Sound to the Bering Sea and along the Aleutian Islands to the Kuril Islands off the Sea of Okhotsk in Russia. They have a maximum age of 28 years, with 50 percent of the females mature at age 7 and all mature by 10. The maximum size is 69 cm long and 3.5 kg. Spawning occurs in winter to spring, with females producing eggs that adhere to each other and to the bottom. Both species of rock soles are important commercially. They are also occasionally caught in the recreational fishery in the Strait of Georgia, though the species remain to be confirmed.

ROCK SOLE

Lepidopsetta bilineata

ABUNDANT ○

IN THE STRAIT OF GEORGIA

The rock sole is one of the most abundant flatfish in the Strait of Georgia. It is found in numerous small concentrations throughout the strait. In 2000, it was determined that rock soles off the west coast of North America were actually two distinct species, a southern species (which retains the common name "rock sole") and a new northern species, the northern rock sole (*Lepidopsetta polyxystra*). One difficulty introduced by this finding is that the new northern species has also been found to the south in Puget Sound, indicating that there is overlap in the distributions of the two species. This means that the species of the rock soles in the Strait of Georgia need to be confirmed. Confirmation requires collecting rock soles from throughout the strait, but especially in the areas of greatest concentration near Chrome Island and between Neck Point and Icarus Point near Nanaimo. From the end of April to mid-June 2001, a bottom trawl survey in the transboundary area in the Strait of Georgia between British Columbia and Washington State recorded catches of both the rock sole and northern rock sole. In the Washington portion of the strait, both species were moderately abundant, with catches only slightly smaller than those of the Dover sole and starry flounder. In the San Juan Islands area, the rock sole was about as abundant as the Dover sole and about four times more abundant than the northern rock sole. The larvae of these two rock sole species look very different from one another and perhaps contribute to the ecological separation of the two species. Both types of larval rock sole occur in the Southern Gulf Islands of the Strait of Georgia. Thus, it appears that both species are in the strait, with the rock sole being more abundant than the northern rock sole.

DESCRIPTION

The rock sole is distinguished from other flatfish by the lateral line's high arch and short dorsal branch. The eyes are bulbous and the lips are thick. Scales on the eyed side are rough, and there are dark blotches on the dorsal and anal fins. The differences between the two rock sole species are not easily recognized, but this southern species typically has more than 10 gill rakers on the first arch and 3 or fewer on the upper part of the first arch, where the gill rakers are broad and robust. The interorbital space is narrow, with up to three rows of scales at the narrowest point. There are almost always more than four supraorbital pores, but they range in number from three to nine. The blind side is a bright white with distinct muscle bands.

Andy Murch

LIFE HISTORY

The distributions of the rock sole and northern rock sole overlap: the rock sole occurs from the southeastern Bering Sea to Baja California, and the northern rock sole from Puget Sound around the North Pacific to the Sea of Okhotsk, the northern coast of Hokkaido, Japan, and off Korea in the southern Sea of Japan. Adult rock soles are found from the intertidal zone down to 732 m, but they are not common below 300 m. They are found on sandy or gravel bottoms but have also been reported along the sides of steep rock slopes in Puget Sound.

A recent, updated maximum age of rock sole in the Bering Sea is 26 years; however, the species that was examined needs to be confirmed. The average size of mature rock soles in the Strait of Georgia is around 35 cm long, while the maximum size reported was a 60 cm female. In general, females mature at ages four to five and lengths of 33 to 36 cm. Males mature at ages three to four and a length of 30 cm. Fifty percent of females and males are mature at an average length of 34.6 cm and 30.3 cm, respectively. A 35 cm female produces about 400,000 eggs, and a 46 cm female about 1.3 million eggs.

The spawning season in British Columbia lasts from late winter to early summer. In Puget Sound, spawning occurs from December to April, with a maximum in March. The species migrates into deeper waters to spawn, then back to shallower waters in the summer. Fertilized eggs are demersal and stick to each other and to the substrate. Species in this genus are the only flatfishes to produce demersal and sticky eggs. In Puget Sound, rock sole eggs have been found from late December to early March in sand and gravel intertidal areas. The time it takes for the eggs to hatch depends on the temperature, but would be from 9 to 18 days in the Strait of Georgia. The larvae are 3.4 to 5 mm long when they hatch, and they continue to absorb yolk for about 10 to 14 days. Metamorphosis, during which the left eye migrates to the right side, occurs at lengths of about 1.7 to 2 cm. Most juveniles move into shallow sandy areas, but not into bays and estuaries. Juveniles will migrate into deeper waters in the winter but remain in shallower areas than adults. Older and larger rock sole are found in deeper water.

Rock sole larvae feed on plankton. Juveniles feed on cumaceans, carideans and gammarid amphipods. Adults consume a variety of benthic organisms, including crabs, shrimp, sand lances, polychaetes and molluscs. Some diet studies found a high percentage of clam siphons. Juveniles and adults reduce predation by matching their colour with the bottom, burrowing into the substrate and limiting activity in the presence of a predator.

The rock sole is a common species in the commercial fishery and is also the flatfish most likely to be caught in the recreational fishery. In the Bering Sea, the rock sole is the third most abundant commercial groundfish species.

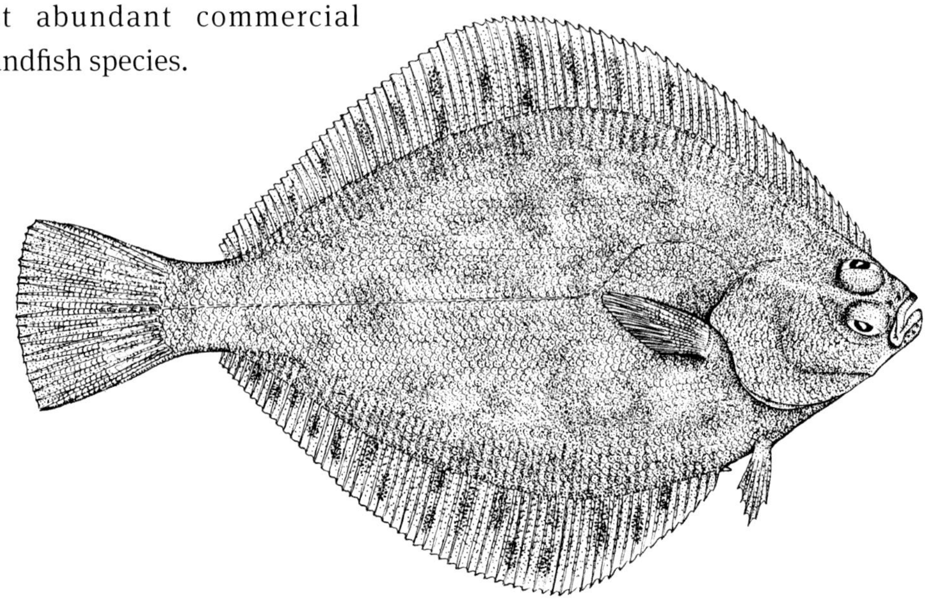

YELLOWFIN SOLE

Limanda aspera VISITOR ▼

IN THE STRAIT OF GEORGIA

There are records of the yellowfin sole in the Strait of Georgia in the groundfish research database at the Pacific Biological Station. Two were caught off Powell River in June 1985 using a handline. A total of 14 were caught in a bottom trawl at a depth of about 25 m in the Southern Gulf Islands area in July 1991. There is no question that this species is rare and an obvious visitor from the north.

DESCRIPTION

Yellowfin sole are readily distinguished by the yellow colour on both sides of the dorsal and anal fins, and by the high lateral line arch above the pectoral fin with no accessory branch. The deep body is covered with rough scales. The colour is brown to olive-brown with faint light mottling, and there may be faint dark bars on the fins. There is a narrow black line at the base of the dorsal and anal fins. The blind side is white.

Catherine Mecklenburg

LIFE HISTORY

The yellowfin sole is a dominant flatfish in the Bering Sea and support major commercial fisheries. It is found off Korea and Japan, in the Sea of Okhotsk, in the Bering and Chukchi Seas, in the Beaufort Sea off Alaska, in the Gulf of Alaska, along the Aleutian Islands and south to the Salish Sea. This is a shallow-water species, ranging from depths of 10 to 600 m, but mostly on soft bottoms around 150 m. The maximum size is 49 cm long and 1.8 kg. About 50 percent mature by age 10. The maximum age is 39 years. All females mature at age 18 and a size of 34 cm and 483 g. Males at age 18 average 32 cm and 366 g. Batch spawning occurs in the winter, with females producing one to three million eggs. The prey of yellowfin sole includes polychaetes and amphipods.

SLENDER SOLE

Lyopsetta exilis ABUNDANT

IN THE STRAIT OF GEORGIA

The slender sole is a small flatfish and perhaps the fourth most abundant fish in the Strait of Georgia in deep water. It has been a common bycatch in experimental and commercial shrimp trawls, particularly in depths of 10 to 30 m. It may be more abundant in the southern strait around the mouth of the Fraser River, including Vancouver Harbour and into areas of the strait that are United States waters. In shallower water the slender sole is much less abundant than the rock sole, Dover sole, starry flounder and butter sole, but on the bottom at 100 m and deeper it is a dominant species. Using an estimate of the relative abundances of Pacific herring and slender sole in our survey catches, we estimate there could be over 500 million slender sole in the Strait of Georgia.

DESCRIPTION

This species is distinguished by its elongate body, large, thin scales that are easily removed, a prominent ridge in the interorbital space, and a large angular mouth that extends to the middle of the lower eye, with small teeth and a double row on the upper jaw. The eyed side is olive to reddish brown. The

Gregory Jensen

deciduous scales may have dark edges. The fins are slightly dusky, and the dorsal and anal fins are lighter at the edges. The blind side is an off-white to pale orange-yellow.

Gregory Jensen

LIFE HISTORY

The slender sole is found in the Gulf of Alaska, east of Kodiak Island, but not in the Bering Sea. The distribution extends south to Baja California. It is found on soft bottoms at depths of 10 to 800 m, but mostly occurs between 50 and 350 m. Slender sole are known to bury into the sandy or silty bottom, leaving only their eyes visible. They are caught in trawls fishing off the bottom in the Strait of Georgia, indicating they may move off the bottom in search of food. The maximum length is 35 cm, but most adults are about 25 cm.

Spawning occurs in winter and into spring. Females in spawning condition have been found in the Strait of Georgia in February, and many fish spawn in April. The maximum age is 20 years. Males mature at ages two to three and females at three to five. Adults feed on polychaetes, small fishes and crustaceans like amphipods and shrimp.

This is another species of flatfish in the Strait of Georgia that needs to be studied, particularly because it is so abundant on the bottom at deep depths. A survey over the winter and spring in the vicinity of the Fraser River and extending south into Puget Sound would likely provide the specimens necessary to determine the average ages of the population, as well as spawning times and locations.

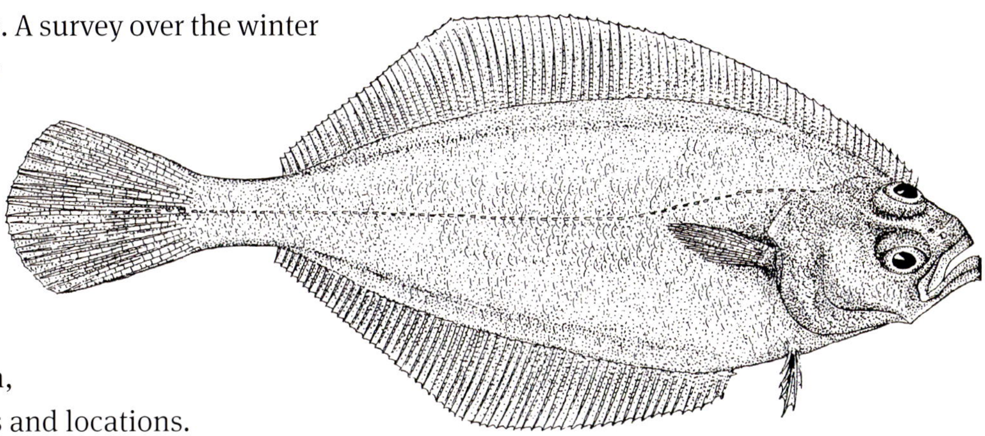

DOVER SOLE

Microstomus pacificus COMMON ●

IN THE STRAIT OF GEORGIA

Dover sole are commonly found throughout the Strait of Georgia, mostly in relatively deep water. They can be found as shallow as 10 to 40 m but are most common at depths below 80 m. They are occasionally seen by scuba divers in areas such as Howe Sound and Burrard Inlet. There are no abundance estimates of Dover sole in the strait, but catches in research trawl studies indicate they may have about the same abundance as starry flounder or rock soles and about 10 percent of that of the English sole. They are considerably less abundant than slender soles.

DESCRIPTION

The Dover sole has an elongate, soft and flaccid body with a small mouth, which is the origin of the genus name *Microstomus*, or "small opening." The eyed side, with large eyes, is brown and may have mottled dark or dull red spots. The blind side is off-white to dull smoky grey, or even a very light brown. The fin margins are black; there is a short, unconnected branch of the anterior or dorsal part of the lateral line, which is almost straight. The teeth are mostly on the blind side of the mouth. The gill opening is small. The fish produces large amounts of mucus, giving it the fishermen's common names "slime sole" and "slippery sole." The name "Dover sole" was created to help market the fish, which seemed too slippery and slimy. Few if any consumers were aware of the sliminess as almost all of it had been removed by shore filleters, who did not particularly like to handle the fish. The price originally kept demand down in Canada, but there was a higher demand in the United States.

LIFE HISTORY

This fish is found on the continental shelf and slope from the Bering Sea to Baja California at about 10 to 1,200 m on mud bottoms. Most commercial catches are from about 60 to 840 m. The largest specimen was 76 cm long and weighed 4.6 kg, but the average length is about 40 to 45 cm and the average weight is 450 g. The maximum age is 58 years for males and 53 for females. These ages identify the Dover sole as a long-lived species, which indicates it has evolved its life-history strategy to maximize its survival in a particular type of environment. In British Columbia and elsewhere, the commercial fishery occurs around the time that the fish concentrate for spawning.

Males and females mature at older ages and larger sizes in the northern areas of the species' distribution. In the Gulf of Alaska, females first mature at a length of 32.4 cm and age five, with 50 percent mature at a length of 43.9 cm and age seven. All females are mature by age 11. Males first mature at a length of 30 cm; 50 percent mature at 33 cm, and 100 percent at 40.6 cm and age seven. After age five, females grow faster than males and reach sizes that are about 4 to 5 cm larger than males of the same age. The largest specimens of this species occur in northern Hecate Strait and eastern Dixon Entrance, where females attain lengths of up to 72 cm and males up to 50 cm. In this area, length at 50 percent maturity is 57.4 cm for females and 47.2 cm for males.

Dover sole are reported to be batch spawners, which means individuals spawn a number of times. One report found an individual that spawned nine times in one season. However, this spawning behaviour has not been observed in British Columbia. In Puget Sound, spawning occurs from January to March. In the Gulf of Alaska, it is from February to May. Males and females migrate into deep water prior to spawning, which occurs at or near the bottom. A 57.5 cm female was estimated to produce 266,000 eggs. The eggs are pelagic or found near the surface and hatch in about 24 days at 8°C. Although spawning occurs in deep water, the eggs rise to the surface; most larvae are found in the top 50 m, but they can occur as deep as 600 m. This pelagic phase extends for up to two years. During this period, the left eye migrates to the right side. Metamorphosis occurs at lengths longer than 4 cm. When metamorphosis is complete, the juveniles settle on the bottom. The Dover sole's long life and batch-spawning behaviour, along with the extended pelagic phase of the larvae, all contribute to the dispersing of eggs and larvae over a large distance.

Gregory Jensen

As Dover sole grow, they move into deeper and deeper water. Off the west coast, they have been found as far as 840 km from shore. However, tagging studies confirm they tend to remain in the same general area. Adults migrate into water deeper than about 200 m in the fall for spawning and come back into shallow water for active feeding in the spring and summer.

Dover sole do not eat vertebrates. Their small mouth and teeth, mostly on the blind side, are adapted to feed on bivalves, benthic crustaceans, polychaete worms, scaphopods, shrimp and brittle stars. The Dover sole is a sought-after commercial species that should be managed in a way that recognizes the need to maintain a minimum spawning biomass, which some biologists consider to be about 30 percent of the abundance, before they are fished commercially. This is a long-lived species that is likely going to be difficult to manage in a changing ocean environment.

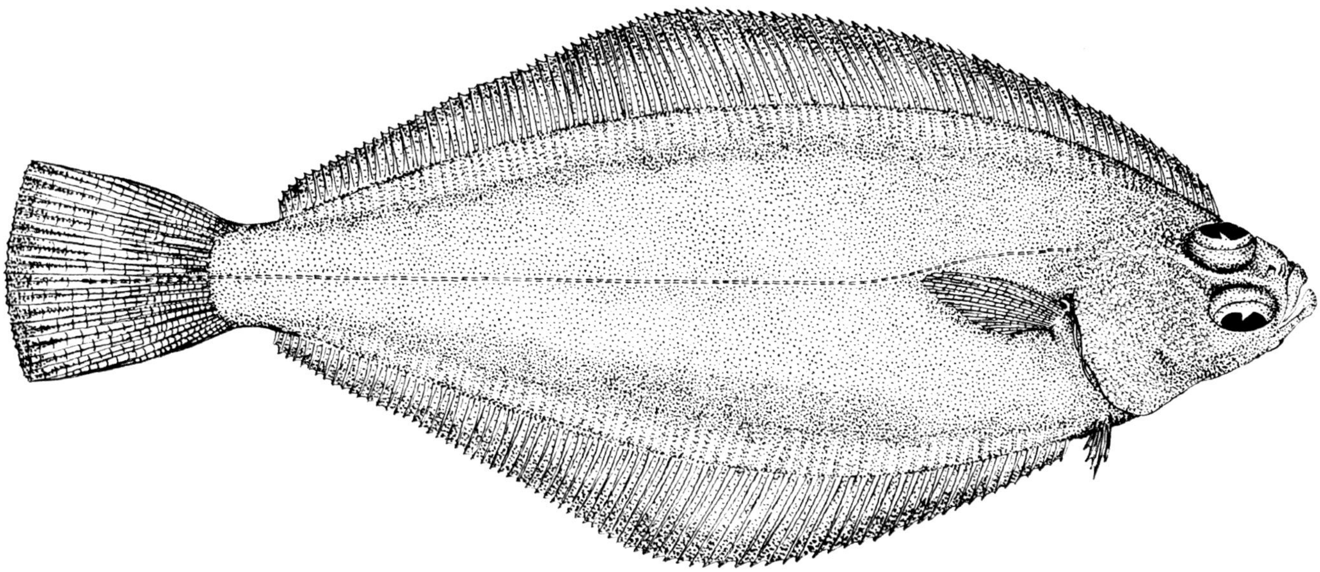

ENGLISH SOLE

Parophrys vetulus

ABUNDANT

IN THE STRAIT OF GEORGIA

The English sole is perhaps the second most abundant flatfish in the Strait of Georgia after the slender sole. Canadian fishermen call the species "lemon sole," and it was a major contributor to the strait's developing groundfish fishery in the 1940s and 1950s because of its commercial importance. Tagging studies have shown there were separate stocks in the Parksville–Comox–Baynes Sound area, the Gulf Islands and the Burrard Inlet–Fraser River estuary area. A fourth stock has appeared to enter the strait in the summer from the west coast of Vancouver Island and migrate back in the late fall. Most adults remain in one area, but some exceptional migrations have been documented: for example, a tagged fish from the Strait of Georgia was recovered in California. English sole are commonly found in bottom research trawl surveys throughout the strait. The populations in the Baynes Sound–Cape Lazo area and in the Gulf Islands were depleted by fishing during World War II. However, the abundance in the strait seems large as of 2024.

A diamond-shaped body and long, square tail, together with a bony ridge between the eyes and a small mouth, all help identify the English sole. The speckling is often absent from its brown body shade, however. Andy Murch

DESCRIPTION

The elongate body and pointed snout distinguish the English sole from other species. The dorsal branch of the lateral line is long, extending below the first 21 to 37 dorsal fin rays. A portion of the upper eye is visible from the blind side, and there is a narrow ridge between the eyes. The caudal fin is almost square, with slightly longer rays in the middle. The eyed side is a uniform reddish- or yellow-brown that may have light blotches or minute spots. The blind side is white to pale yellow and may have yellow colouring at the bases of the dorsal and anal fins.

LIFE HISTORY

English sole are found from Baja California to the Aleutian Islands and into the Bering Sea. In British Columbia, the highest abundance occurs in Hecate Strait. They are found on soft bottoms to depths of 550 m, but most are in nearshore shallows to 250 m. The maximum age is 23 years. In Hecate Strait, 50 percent of females are mature at 34 cm long, while the corresponding length for males is 29 cm. A recent study in the United States found that, on average, 50 percent of females are mature at age five and a length of 23 cm. Some females mature at age three and 26 cm long, and all females over 35 cm are mature. Males mature earlier, at two years old and a length of 21 cm. All males are mature at 29 cm long in Puget Sound. In general, females begin to spawn at age three or four and males at two or three. The growth rates of males and females are similar for the first two years, after which females grow larger than males; their maximum length is 57 cm, compared to 49 cm for males.

The spawning season for the English sole is protracted and variable. In the Strait of Georgia it is mainly from January to March, with peak spawning in February and early March. Some spawning fish have been observed as late as August on the Cape Lazo grounds. In Hecate Strait, spawning occurs from October to April, but mainly in late winter. Temperature seems to be an important determinant, as few fish spawn at temperatures below 7.8°C. Spawning occurs mainly in Baynes Sound, south of Cape Lazo, and in the Boat Harbour area among the Gulf Islands. Spawning in the Strait of Georgia occurs over mud bottoms at 50 to 70 m, but spawning fish have not been directly observed. A five- to six-year-old female that is 36 to 38 cm long produces about one million eggs, while a larger female may produce about two million. The larvae are pelagic and undertake daily migrations. They metamorphose over an 8- to 10-week period, after which they resemble the adult and settle on the bottom at lengths of about 1.8 to 2.6 cm. Young English sole appear to be dependent on shallow (2 to 20 m) depths for nursery areas. In the Strait of Georgia there are extensive rearing areas along the east coast of Vancouver Island and near the mouth of the Fraser River. Juveniles move away from the beach area after their first year, at lengths of 11 to 15 cm, to progressively deeper water as they increase in size.

Older English sole feed on polychaetes, harpacticoid copepods, amphipods, shrimps, cumaceans, tube worms and molluscs. They also appear to nip off the siphons of clams. English sole and starry flounder have reportedly produced a hybrid sole, sometimes called a forkline sole, which we do not include in our list. Water temperature may limit the viability of eggs, and temperatures above 18°C may affect the growth and survival of adults.

STARRY FLOUNDER

Platichthys stellatus ABUNDANT

IN THE STRAIT OF GEORGIA

The starry flounder is an abundant resident of the Strait of Georgia, with some migrating and remaining in rivers as juveniles. It has been found as far as 120 km up the Fraser River, between Agassiz and Hope. A few fish between 20 and 30 cm long have been found in the lower river, but generally fish in the river are shorter than 10 cm. They are common along sandy beaches and probably are the most abundant flatfish in estuaries around the Strait of Georgia. Starry flounder are more common in shallow water throughout the strait than in habitats where species like the English or rex sole are found. This is a species that can be caught by anglers fishing from piers and even from shore. There are no sport catch size records for the strait, but a recent fish caught off the coast of Oregon weighed 4.3 kg. Starry flounders were eaten occasionally by the Squamish People, who caught them in the shallow waters of Burrard Inlet and Indian Arm.

Andy Murch

DESCRIPTION

The starry flounder is an attractive fish that is easily recognized by the alternating light-orange and black bands on the dorsal, anal and caudal fins, as well as by the rough skin on the eyed side. The uppermost side is olive to dark brown to black. The blind side is a creamy white. Fish have adaptive coloration that provides important camouflage. The caudal fin is almost square, and the body shape is oval. There is no accessory or dorsal branch of the lateral line. The skin is rough because of the plate-like stellate scales; the skin sometimes has been dried and used as sandpaper. The flesh is tasty, but there is a dark layer of fat beneath the skin that detracts from its appeal. Like all flounders, the larvae have eyes on both sides. As they grow, one eye migrates to the other side, which can be either the right or left for this species. The proportion of fish with the eye on the left side varies from about 50 to 60 percent in the south off North America to a slightly higher percentage in the far north and almost 100 percent off Japan. The fish is commonly found in commercial aquariums.

LIFE HISTORY

These fish are found from Southern California to Alaska and in the Arctic into the Beaufort Sea in Canada, as well as off the east coast of Russia, into the Sea of Okhotsk, and in the Sea of Japan off Korea. They are found on sandy or muddy bottoms to depths of up to 375 m, but are mostly encountered shallower than 100 m. Their unconfirmed maximum size has been reported as 91 cm long and 9 kg. The maximum age reported to date is 24 years. Females grow faster than males and achieve larger sizes. As males age, their growth slows. Most males are reported to mature at age two and lengths of about 30 cm. Some males do not mature until age four. Most females mature at age three and lengths of about 35 cm. All females are mature at age four. A simple age and growth study for starry flounder from the Baynes Sound area using modern determination methods would be most useful.

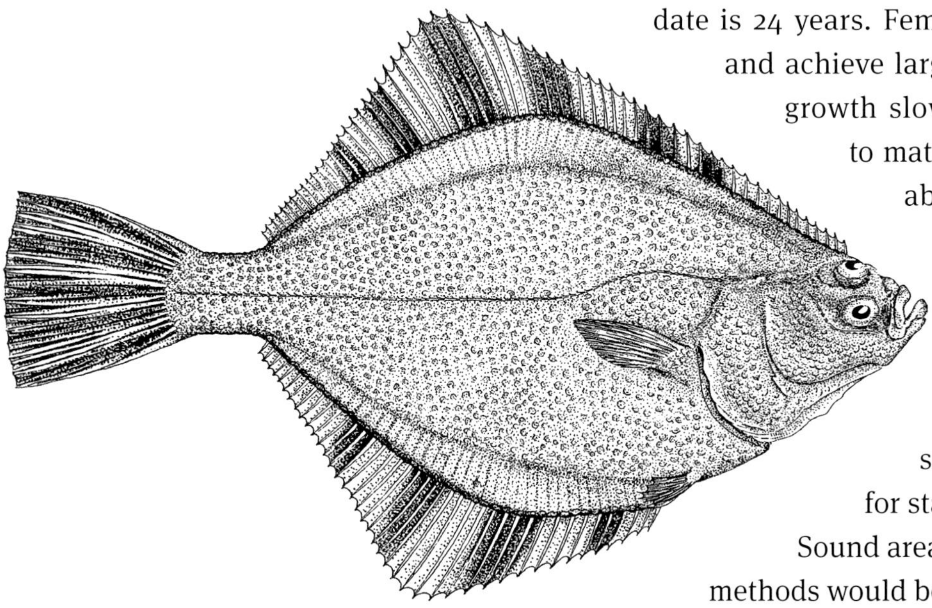

C-O SOLE

Pleuronichthys coenosus COMMON ●

IN THE STRAIT OF GEORGIA

The C-O sole is a commonly found flatfish that is often seen by scuba divers throughout the Strait of Georgia, but it is not abundant. It is mostly a shallow-water species and is not common in deeper areas that would be fished for shrimp. The Squamish People fished this species around False Creek, Capilano River and the head of Indian Arm.

DESCRIPTION

The C-O sole gets its name from a dark bar at the base of the caudal fin that looks like an inverted *C*, and a spot posterior to this, in the middle of the caudal fin, that looks like an *O*. The fish is generally dark brown on the eyed side, but there is a wide range of colours and blotches. There is a prominent black spot in the middle of the body. Juveniles can be mostly white. The body shape is oval, and the dorsal and anal fins are broad. The blind side is creamy white. The mouth is small, and the large and bulging eyes are separated by a prominent ridge. The dorsal fin originates on the blind side at about the level of the upper lip. The accessory dorsal branch of the lateral line is long, extending past the midpoint of the body.

Andy Murch

LIFE HISTORY

This species is reported at depths of up to 350 m throughout its range, but is most often found at less than 15 m on soft sediments near rocky areas. The maximum length is 36 cm. The C-O sole is distributed from northern Baja California to Sitka in southeast Alaska, with the largest abundances at the south-central part of its range. Most spawning is from March to October, but it may occur throughout the year. The young feed on gammarid amphipods and small crustaceans. As they grow, their diet adds crabs, shrimps, clams and polychaetes. The C-O sole is good to eat, but it is small and the skin is tough, making it difficult to fillet. This is another species on the long list of those that need to be studied in the Strait of Georgia.

CURLFIN SOLE

Pleuronichthys decurrens

UNCOMMON ◆

Gregory Jensen

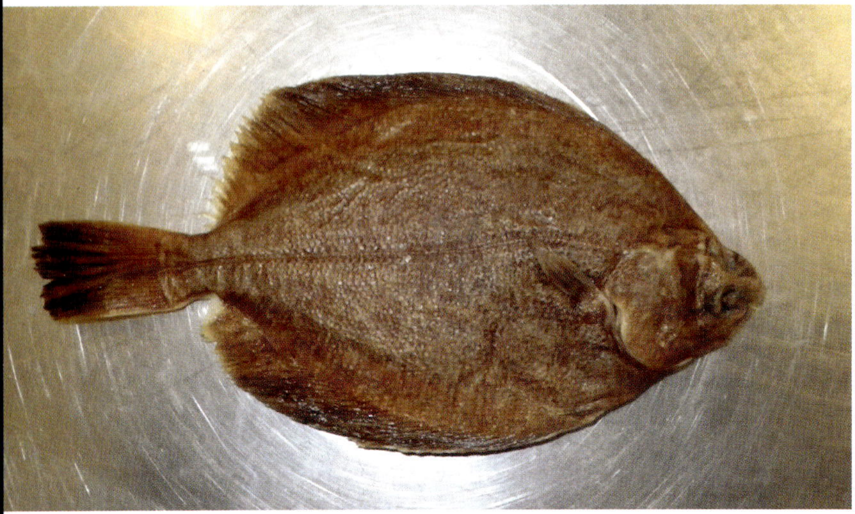

Aaron Baldwin

IN THE STRAIT OF GEORGIA

Small numbers of curlfin sole were captured in experimental bottom trawl surveys in 2003, 2005 and 2006 in the southern Strait of Georgia. There are no reports from any other sources, including shrimp trawl surveys throughout the strait and scuba observations.

DESCRIPTION

The blind side of the curlfin sole is creamy white. This is a deeply oval-shaped fish that is brown, black or grey and finely spotted on the eyed side. There is a high bony ridge between the eyes; the caudal fin is rounded. This is one of the two flatfish that has the anterior of the dorsal fin starting on the blind side with the first 9 to 12 dorsal rays (the C-O sole is the other). The eyes are very large and the mouth is small. The accessory dorsal branch of the lateral line is slightly longer than the mid-body.

LIFE HISTORY

The curlfin sole occurs on soft bottoms from the southeast Bering Sea and Gulf of Alaska to southern Baja California at depths from shallow water to about 535 m. However, it is usually close to shore at depths of less than 20 m. It grows to a length of 36 cm and a weight of about 770 g. The maximum age may be about 10 years. Spawning peaks in the spring, but is reported to occur throughout the year. This is not a species that is fished commercially.

SAND SOLE

Psettichthys melanostictus UNCOMMON ◆

IN THE STRAIT OF GEORGIA

Sand sole are found in shallow water from 5 to 30 m throughout the Strait of Georgia. They are most common in the southern strait in the transboundary area and off the mouth of the Fraser River, where they occur occasionally in bottom research trawl catches. Sand sole have also been reported from other estuaries, like at the Nanaimo and Cowichan Rivers and off Courtenay. It is possible that more sampling close to shore along sandy beaches may reveal a greater abundance. They have been reported in Burrard Inlet only by scuba divers. Sand sole appear to be more common in Puget Sound than in the Strait of Georgia.

Gregory Jensen

DESCRIPTION

Adult sand sole are distinguished by the large mouth with the maxilla extending below the middle of the lower eye. The first eight dorsal fin rays are elongated and free from the membrane. There is an accessory dorsal branch of the lateral line that is relatively short, extending below the first 18 to 22 dorsal fin rays. The teeth are large, with several anterior teeth appearing long. The fish is light green or brown or even grey on the eyed side. This side has widely spaced eyes and an extensive speckling of fine black spots. The blind side is a creamy white.

LIFE HISTORY

The sand sole ranges from the southeastern Bering Sea to Southern California. It is common on soft bottoms at depths of less than 70 m. It has been reported at depths up to 325 m, but most are found at less than 150 m. The maximum length is 63 cm, but most adult fish are between 31 and 35 cm. An age of 10 years has been estimated as the maximum. Males mature at age two and a

Vicky Okimura

length of 20 cm. Females are mature at two to three years old and 28 cm long. Three-year-old male and female fish average 25 cm and 28 cm, respectively. A 37 cm female can produce 1.4 million eggs. Spawning occurs in shallow water of 20 to 30 m from January to April, mostly in March. Some spawning has been reported as late as July. The eggs are pelagic and rise to the surface as they develop. Juveniles metamorphose at a length of about 3 cm, possibly after about 60 to 68 days in the larval stage. However, the juveniles may not settle on the bottom immediately, remaining for a time in a pelagic area.

Sand sole feed on mysid shrimps, amphipods, decapods, polychaetes and molluscs. The prominent teeth are an indication that sand sole are predators of other fish. Adults and larger fish will prey on juvenile sand sole, sanddabs, sculpins and herring, as well as crabs, squids, polychaetes and molluscs. This is another flatfish that needs to be studied in the Strait of Georgia, and particularly in the Parksville–Qualicum area. Spawning areas should be documented and the age of the spawning fish determined.

Gregory Jensen

Molas

Molas are in the order Tetraodontiformes and the family Molidae. Molas are very large pelagic fishes; there are about five species found in tropical and temperate open oceans around the world. Molas are amazing species that, once seen, will never be forgotten. They are the largest of all bony fishes, with a maximum weight of 2.3 t and length of 4 m. They are mostly found near the surface but can dive to depths of up to 1,100 m, and tagging shows they also can migrate long distances. They appear to feed by diving frequently during the day to the plankton layers at depths of about 90 to 150 m, where a major prey appears to be gelatinous plankton. Aquarium studies have shown that molas' growth rates are exceptionally fast. Their round shape is a consequence of a unique tail structure. There is no caudal peduncle; it is replaced by a posterior structure called a clavus, giving the fish a truncated appearance. The skeleton is mostly cartilage and the small number of vertebrae are fused, eliminating the normal swimming motion seen in other fishes. Instead, these fish flap their strong dorsal and ventral fins. They move very quickly when they want to. There is one species of mola that is very rarely reported in the Strait of Georgia.

OCEAN SUNFISH

Mola mola

VISITOR ▼

IN THE STRAIT OF GEORGIA

There is only one old report of an ocean sunfish in the Strait of Georgia. The species was found near Quadra Island across from Campbell River. There are a few more reports from Puget Sound, including some more recent reports. Any specimens that are found in the Strait of Georgia are visitors that were transported from elsewhere. As oceans warm off the coast of British Columbia, it is likely that we will see more ocean sunfish in the strait.

Andy Murch

DESCRIPTION

The ocean sunfish is oval in shape; it appears to be all head and no body. There are no pelvic or caudal fins, but the dorsal and anal fins are large. They are a grey to silvery colour with rough, leathery skin. They can be huge, growing to lengths of over 4 m and weights of up to 2.3 t.

LIFE HISTORY

Ocean sunfish occur worldwide. This is a fish that mostly remains a mystery to biologists, but taxonomists consider it to be a highly evolved species. Female ocean sunfish can produce more than 300 million eggs during spawning; this is more than any other vertebrate on the planet. The ocean sunfish has a unique appearance and is readily identified by its shape, large size and habit of lazily floating at the surface during the day. There are estimates of the fish living to over 20 years, but this is mostly an educated guess. They appear to be poor swimmers when seen casually floating on the ocean surface, but they are fast and powerful swimmers when disturbed. Recent information has shown that ocean sunfish will dive deeper than 800 m. They feed mostly on jellyfish using the beak-like teeth in their small mouths. We have seen them for sale in fish markets in Korea.

ACKNOWLEDGEMENTS

WE PRODUCED THIS BOOK WITH THE HELP AND SUPPORT OF MORE PEOPLE THAN we could possibly acknowledge. Information we use comes from sources from researchers to beach walkers, with all of it contributing to an understanding of the life of a species. Hopefully, our readers will find opportunities to contribute new information.

People and organizations providing photographs are acknowledged for each image. Alaska Seafood, Archipelago, Aaron Baldwin, Jon Bryan, Jon Buchanan, Gregory Jensen, Jeremy Jodoin, Danny Kent, Tom Laidig, Valerie Lyttle, Catherine Mecklenburg, Andy Murch, Lee Newman, Janna Nichols, Alexei Orlov, Bill Pennell, Scott Simpson and Scott Stevenson provided numerous photographs. The US National Oceanic and Atmospheric Administration (NOAA) and TINRO, the research centre in Vladivostok, Russia, also provided a number of photographs. We very much appreciate reviews of drafts of the book by Dana Haggerty, Danny Kent and Andy Lamb. Editors for Harbour Publishing also helped improve the reporting. Rusty Sweeting was the chief scientist who collected many years of information about the fishes of the Strait of Georgia. Lana Fitzpatrick helped to organize this information as well as to produce the book. We relied on material contributed to the groundfish and shellfish databases at the Pacific Biological Station by researchers such as Sandy McFarlane, Jackie King, Chrys Neville, Nev Venables, Malcolm Wyeth, Lynne Yamanaka, Dana Haggarty, Greg Workman, Jim Boutillier, Dennis Rutherford and old-timers such as Ed Barraclough and Keith Ketchen. Maria Cornthwaite and Lorri Granium helped find information. We used species accounts published in peer-reviewed literature, but we do not cite or list these references to make the book readable for a general public. We do provide a short list of key publications at the back of the book. The volumes of the recent *Fishes of the Salish Sea* by T.W. Pietsch and J.W. Orr have extensive bibliographies of studies relevant to the fishes of the Strait of Georgia. We appreciate the help and contributions of Bob Armstrong, Jim Auzins, Steve Axtell, Tara Bak, Aaron Baldwin, Ann Beamish, Dylan Belliveau, Jen Blaine, Jon Bryan, Scott Buchanan, Mike Byerly, Adam Dick, Jackson W.F. Chu, Keith Clements, Goff Connor, Justin Cox, Ernie Cooper, Lindsay Dealy, Jeremiah Degenhardt, William Dulac, Andrew Edwards, Ryan Fields, Heather Hall, John Holmes,

Andrew Howe, Rick Harbo, Gregory Jensen, Jeremy Jodoin, Eiko Jones, Ken Jones, Tiernan Kingzett, J. Kirkhart, Karen Kloempken, Michael Lister, Tara Logan, Camilla Lombard, Kirk Lombard, Joan Lopez, Gary Lundquist, John Dennis Lundquist, Heather McDermott, T. Mecklenburg, Theo Moddler, Jim Murphy, Lee Newman, Vicky Okimura, Michael Oliver, Brian Recalma, Mike Rohford, Andy Scheffler, Andrea Scheffler, Ben Shanafelt, Tommy Sheridan, Rick Starr, Jennifer Strachan, Richard Thomson, Andrew Trites, Mark Trudel, John Whitelaw, Ron Wolf, Jon Wong, David Young, Mark Yuasa and Patrik Zetterburg. We received financial support for the publication of the book from the Raincoast Conservation Foundation, including generous donations from Rudy North and Michael Borden. The Pacific Salmon Foundation and Isobel Pearsall, Brian Riddell and Mike Meneer were strong supporters of this book. We also received financial support from the Pacific Biological Station and the Department of Fisheries and Oceans.

GLOSSARY

adhesive disc. A disc composed of body parts that is on the bottom of some fishes and used to attach to objects

adipose fin. A small fleshy fin without spines or rays that occurs posterior to the dorsal fin

alevin. A young salmon or trout that still feeds off the yolk

ammocoete. The larva stage of lamprey before they metamorphose into an adult appearance

amphipods. Crustaceans in the order Amphipoda that are common in the zooplankton and in the diets of many fishes

anadromous. The behaviour of fishes that hatch in fresh water, spend some time in the ocean and return to fresh water to spawn

anal fin. The fin that is behind the pectoral and pelvic fins on the bottom of the fish in front of the tail

barbel. A fleshy appendage in the mouth area

batch spawning. Spawning by releasing eggs over a prolonged period rather than all at once

benthic. Occurring close to the bottom of the ocean

benthopelagic. Living near the bottom of the ocean, usually in deep areas

bifid. Two-pronged

blind side. A flatfish's side that has no eyes

brackish. Describes low-salinity water, such as water in the vicinity of rivers

branchiostegal rays. Parts of the lower outer margin of the gill cover or operculum that create a pleated membrane

brittle star. A relative of starfish with long slender arms

broodline. A group of fish that hatch at the same time

bycatch. Species caught unintentionally in a commercial catch that are not of value and that are usually discarded at sea

caridean. A type of shrimp that is common around the world and eaten by fishes

cartilaginous. A description of skeletal tissue that is elastic and flexible, and a reference to the group of fishes that are without bones

catadromous. Living in fresh water and spawning in the ocean

caudal fin. Tail fin

caudal peduncle. The part of a fish's body in front of the caudal fin

cephalopods. A small group of marine animals that includes squids, octopuses and cuttlefishes

circumpolar. Distributed throughout the Arctic and subarctic

cirri. Hair-like projections; whiskers

claspers. Modified portions of the pelvic fins of male sharks, rays and chimaeras that transfer semen into females' reproductive organs

clavus. Part of the posterior body of an ocean sunfish that acts like a rudder

cloaca. A posterior, ventral opening for reproductive and excretory tracts

copepods. Crustaceans in the order Copepoda that are common in the zooplankton and in the diets of many fishes

copulation. Internal fertilization of eggs with sperm

crustaceans. A large group of invertebrates that includes crabs and shrimps; a common diet item and seldom identified to the species level

cryptic species. Morphologically similar species that have been determined to be distinct species

cumaceans. Small crustaceans with a distinctive appearance of a large head-like area followed by a long narrow abdomen. They occur in the surface layers of mud or sand bottoms.

cusp. A point or projection on a tooth

cusplets. Small sharp projections on a tooth

decapods. Crustaceans in the order Decapoda, including crabs and shrimps, that are common prey items for fish

demersal. Living close to the bottom of the ocean

denticles. Scales of cartilaginous fishes that are tooth- or thorn-shaped

diatoms. Microscopic, unicellular planktonic plants in the phytoplankton

dichromatic. Describes species in which males and females are different colours

dorsal fin. A fin on the top of a fish that consists of hard (bony) and soft (fleshy) rays

dorsal spine. A spine at the front of a dorsal fin

echinoderms. A group of species that includes starfish, sea urchins, sea cucumbers, brittle stars and sand dollars

egg case. A case made of keratin, which is a type of protein, that contains the developing embryo of cartilaginous fishes

epibenthic. Living on the surface of the ocean bottom

escapement. Pacific salmon that escape being caught in a fishery and are able to spawn in fresh water

euphausiids. Shrimp-like species also known as "krill" that are in the zooplankton. They are common in the diets of fishes

finlets. Small fins that have only a few rays and occur in a series on the top or bottom of a fish's body, behind the dorsal or anal fins

fin rays. Hard (bony) or soft (fleshy) structures that support the tissue of a fin

flatfish. A term used for flounders and soles that describes their flattened shape

forage fish. Species of fish such as smelts or herring that are commonly consumed by other fishes

fry. A young anadromous salmon that has recently emerged and is about to migrate to the ocean

fusiform. Tapering toward each end

gamete. A reproductive cell; the ovum in females and sperm in males

genital papilla. A structure behind the anus that is used to release eggs or sperm

gestation period. The time for the full development of an embryo

gill arch. A bony arch that supports the gill filaments and gill rakers

gill filament. Fleshy projections that support blood vessels used in respiration

gill raker. A bony projection on the anterior edge of the gill arch opposite the gill filaments

gill slit/gill pore. The external opening that allows water from the gill chamber to exit

gravid. Full of eggs

gut area. The length or area of the body cavity that contains the intestine

homing. The ability of a salmon to find its way from the high seas to the exact area of its birth in fresh water

ichthyoplankton. Larval fish or fish eggs within plankton

infraoral lamina. A band of teeth below the mouth opening in lampreys

infra-orbital ridge. A ridge of bone below the eye

interorbital space. The space between the eyes

intertidal. The shore area that is above water at low tide and below water at high tide

invertebrate. An animal with no backbone

isopods. A large group of small crustaceans that are common in the ocean and in the diets of fishes

krill. Small shrimp-like plankton that are abundant in the ocean

larvacean. A tunicate and type of plankton that is usually small and transparent

lateral line. A series of pores along the side of a fish that are connected internally to a sensory system

leptocephalus. An eel larva that is clear, transparent and frequently flat

longline. A series of baited hooks attached to a long line used to fish on bottom or near the surface

luminous glands. Small glands, also called photophores, that produce chemical light

maxilla. The main bone in the upper jaw of fishes, also called the maxillary bone

megalops. A crab larva that is very common in the diets of young Pacific salmon

meristic. Body structures that are divided into segments that can be counted

mesopelagic. Living at water depths of about 200 to 1,000 m

metamorphosis. A change in structure as the body and organs transition from a larval to an adult appearance

midwater. The ocean depth zone generally considered to be from about 200 to 1,000 m

milt. Sperm

mysids. Shrimp-like crustaceans that are common in the diets of fishes

nauplii. The initial larval stage of many crustaceans

nekton. Organisms that occur in the plankton and can control their movement

nematodes. Roundworms

nictitating membrane. A lower movable eyelid in some sharks

ocelli. Dark spots surrounded by light pigment on the dorsal surface of skates

olfactory. Relating to the sense of smell and nasal organs

operculum. Cheek or gill cover that is a composite of a number of bones

oral disc. Tissue around the opening of the mouth that is used to attach to objects

otolith. A bone in the inner ear

overfishing. Removing enough fish from the fishery that the population is prevented from replenishing itself

oviparous. Producing eggs that develop and hatch outside the body

ovoviviparous. Producing eggs that are fertilized and hatch inside the female

palatine teeth. Teeth on a pair of bones on the roof of the mouth and outside of the vomer bases

pandalid shrimp. A family of shrimp that are common diet items of fishes and commercially important

papillae. Small, fleshy projections

parr marks. Dark areas of different shapes on the sides of young salmon that disappear before or shortly after the salmon enters the ocean

parturition. The act of giving birth

pectoral fins. The first paired fins behind the head and uppermost on the body

peduncle. The narrowing of a fish's body immediately in front of the caudal fin

pelagic. Resident within the water column and not on or near the bottom or close to the surface

pelvic fins. Paired fins behind and below the pectoral fins

peritoneum. The thin membrane that lines the flesh around the body cavity

photophores. Small light-producing organs

phytoplankton. Plant plankton that is a major food source for zooplankton

plankton. A variety of small floating plants and animals that mostly drift with the currents

polychaetes. A class of worms that are mostly marine and common in the diets of fishes

population. A general term to describe a large aggregate of a species

pre-opercular spine. A spine on the bone at the front of the cheek or operculum

prickles. Scales that are in the shape of sharp points

protogynous. Describes a hermaphroditic animal that first lives as a female and then becomes a male

pteropods. A free-swimming type of sea snail that lives near the surface of the ocean and is a major item in the zooplankton

pup. A shark prior to being released from the female or shortly after release

pyloric caeca. Blind sacs of various shapes and lengths at the posterior of a stomach and beginning of the intestine

remotely operated vehicle. An unoccupied underwater vehicle that can be fitted with a camera and manoeuvred underwater by a remote operator

rostral plate. A small plate at the tip of the snout

Salish Sea. A recently adopted name for the combined areas of the Strait of Georgia, Puget Sound and the Juan de Fuca Strait

scaphopod. A type of mollusc that resembles a tusk; sometimes called a tusk shell

school. A group of closely spaced fish that moves as a single entity

scuba diving. Swimming underwater with a self-contained underwater breathing apparatus

scute. A type of scale that has a thickened ridge or point

sea cucumber. A marine animal with a soft tubular body found on the ocean bottom

sea pen. A marine colonial animal that is attached to the ocean bottom and sometimes resembles a quill

sexual dimorphism. Relates to the difference in appearance between males and females of the same species

sibling species. Species that look similar but do not breed with each other

spinules. Small spines, or pointed bar-like structures, that may occur on larger spines

spiracle. A small opening behind the eye that allows oxygen and water flow into the mouth cavity of some fish

stellate scales. Scales that are star-like, with ridges radiating from the centre

stock. A term similar to "population" but referring more specifically to an aggregate within a population that is fished separately from the rest

subtidal. The area below the lowest intertidal tide and always covered in water

subyearling. A salmon that left fresh water in the same year it emerged from the gravel

supraoral lamina or teeth. Teeth in the back of the mouth of a lamprey

swim bladder. A membranous bladder within the body below the backbone that is usually filled with gases and regulates the fish's position in the water column

symphyseal knob. Enlarged flesh at the tip of the lower jaw or chin

t. An abbreviation for both ton (an imperial unit) and tonne (a metric unit)

trawl. A fishing net that is bag-shaped and towed behind a boat from the surface

tubercles. Small, rounded structures that appear on the body and, in fish, can be temporarily associated with spawning

urogenital pore. The opening where reproductive material and waste products leave the body

untrawlable. Refers to an ocean bottom that is too rough to allow a trawl to fish without damage

ventral fin. Another name for pelvic fin

viviparous. Giving birth to live young instead of releasing eggs

vomer. A bone at the top of the mouth

vomerine teeth. Teeth on the vomer

voucher specimen. A preserved specimen that becomes a verifiable record in an accessible location

year class. All fish produced in the same year

yearling. A salmon that spent one winter in fresh water after emerging from the gravel

young-of-the-year. All fish in their first year of life

zooplankton. Animal plankton and a major food source for fishes

Beamish, R., and G. McFarlane. 2014. *The Sea Among Us: The Amazing Strait of Georgia*. Madeira Park, BC: Harbour Publishing.

Hart, J.L. 1973. *Pacific Fishes of Canada*. Ottawa: Fisheries Research Board of Canada.

Lamb, A. and P. Edgell. 2010. *Coastal Fishes of the Pacific Northwest*. 2nd ed. Madeira Park, BC: Harbour Publishing.

Love, M.S., M. Yoklavich and L.K. Thorsteinson. 2002. *The Rockfishes of the Northeast Pacific*. Berkeley: University of California Press.

Love, M.S. 2011. *Certainly More than You Want to Know about the Fishes of the Pacific Coast: A Postmodern Experience*. Santa Barbara, CA: Really Big Press.

McFarlane, G. and J. King. 2020. *Sharks, Skates, Rays and Chimeras of British Columbia*. Victoria, BC: Royal BC Museum.

McPhail, J.D. 2007. *The Freshwater Fishes of British Columbia*. Edmonton: University of Alberta Press.

Mecklenburg, C.W., T.A. Mecklenburg and L.K. Thorsteinson. 2002. *Fishes of Alaska*. Bethesda, MD: American Fisheries Society.

Pietsch, T.W., and J.W. Orr. 2019. *Fishes of the Salish Sea*. Seattle: University of Washington Press.

INDEX

About this index: Formal species names (in italics) point to the main entry for a species and locations discussing species names. For further notes on a particular species, see the common name. Common names are found under the fish category (e.g., for Dover sole, *see under* sole). Natural geographic features have the feature part of the name second (e.g., for Point Atkinson, *see* Atkinson, Point). Pages in boldface include an illustration.